Resins for Aerospace

Clayton A. May, EDITOR
Lockheed Missiles and Space Company, Incorporated

Based on a symposium sponsored by the ACS Division of Organic Coatings and Plastics Chemistry at the ACS/CSJ Chemical Congress (177th ACS National Meeting), Honolulu, Hawaii, April 3–6, 1979.

ACS SYMPOSIUM SERIES 132

AMERICAN CHEMICAL SOCIETY
WASHINGTON, D. C. 1980

Library of Congress CIP Data

ACS/CSJ Chemical Congress, Honolulu, 1979.
Resins for aerospace.
(ACS symposium series; 132 ISSN 0097-6156)

Includes bibliographies and index.

1. Aerospace industries—Materials—Congresses. 2. Gums and resins, Synthetic—Congresses.
I. May, Clayton A. II. American Chemical Society. Division of Organic Coatings and Plastics Chemistry. III. American Chemical Society. IV. Nippon Kagakukai. V. Title. VI. Series: American Chemical Society. ACS symposium series; 132.

TL699.G85A18 1979 629.1'028 80-15342
ISBN 0-8412-0567-1 ACSMC8 132 1-501 1980

PRINTED IN THE UNITED STATES OF AMERICA

ACS Symposium Series

M. Joan Comstock, *Series Editor*

FOREWORD

The ACS SYMPOSIUM SERIES was founded in 1974 to provide a medium for publishing symposia quickly in book form. The format of the Series parallels that of the continuing ADVANCES IN CHEMISTRY SERIES except that in order to save time the papers are not typeset but are reproduced as they are submitted by the authors in camera-ready form. Papers are reviewed under the supervision of the Editors with the assistance of the Series Advisory Board and are selected to maintain the integrity of the symposia; however, verbatim reproductions of previously published papers are not accepted. Both reviews and reports of research are acceptable since symposia may embrace both types of presentation.

CONTENTS

REINFORCED PLASTICS

INSTRUMENTAL CHARACTERIZATION TECHNOLOGY

PREFACE

This book is divided into four sections since the subject matter encompasses a variety of technical expertise. Part I is devoted to Resin Chemistry; Part II, Adhesives, Coatings, and Sealants; Part III, Reinforced Plastics; and Part IV to Instrumental Characterization Technology. The aerospace industry has been, and is, contributing much to instrumental characterization methods as evidenced by Part IV and substantial portions of Part III. This activity is based on a need to know as much as possible about the materials being used to assure high quality hardware.

Aerospace materials and technology have had a pronounced impact on our everyday lives. Perhaps the classic example is the early work of N. A. DeBruyne and co-workers, who developed the Redux bonding system that played a heavy role in aircraft construction during World War II. Today we find adhesives everywhere—in the homes where we live, in the offices and factories where we work, in the automobiles, buses, trucks, and trains of our surface transportation systems, and in the planes in which we fly. It is thus fitting that the American Chemical Society and the Chemical Society of Japan have asked the workers in this industry to discuss a variety of aspects of their technology.

The editor is deeply grateful to Dr. Dave Kaelble of the Rockwell Science Center, Mr. George Schmitt of the Air Force Material Laboratory, and Mr. John Hoggatt of the Boeing Aerospace Company. They not only served as session chairpersons during the Symposium, but also helped procure the contributors who made the Symposium a success. Finally, a special note of thanks is due to Professor Kozo Kawata from the Institute of Space and Aeronautical Science at the University of Tokyo. Professor Kawata served as a co-chairperson representing CSJ, and his vigorous activity during the meeting is appreciated most gratefully.

Lockheed Missiles and
Space Company, Incorporated
Sunnyvale, CA 94086
February 26, 1980

CLAYTON A. MAY

RESIN CHEMISTRY

1

Status of High-Temperature Laminating Resins and Adhesives

P. M. HERGENROTHER and N. J. JOHNSTON

National Aeronautics and Space Administration, Langley Research Center, Hampton, VA 23665

The quest for high temperature polymers for use as functional resins (e.g., coatings and sealants) and structural resins (e.g., adhesives, composite matrices and foams) began in the mid-1950's. Early work on high temperature polymers concentrated primarily on thermal stability and paid little or no attention to cost, processability, performance or actual use temperature. As work progressed, it was quickly recognized that a successful polymer must exhibit a favorable combination of price, processability and performance and also that thermal stability differed from dimensional stability or heat resistance.

Work on most of the more exotic and costly polymers was soon discontinued as a result of this realism and the more current emphasis has generally been restricted to the development of materials for specific applications. In fact, this review will be confined to those high temperature polymers currently being developed as adhesives and composite matrices. The term "high temperature" in this context refers to the ability of a structural material to perform properly in air for thousands of hours at 232°C, hundreds of hours at 316°C and/or minutes at 538°C.

The ideal high temperature polymeric system for use as a production-line adhesive or laminating resin would exhibit a variety of desirable features such as the following: easy tape and prepreg preparation, good shelf-life, acceptable tack and drape, acceptable quality control procedures, low processing temperature and pressure, no volatile evolution, dense glueline or matrix (no voids), good mechanical performance over the desired temperature range, time, and environmental conditions, acceptable repairability and cost-effectiveness. Ironically, no commercially available system exists which actually offers any reasonable combination of these features. Instead, the few commercially available materials represent a compromise of these properties.

0-8412-0567-1/80/47-132-003$05.00/0

Historical Development of High Temperature Structural Resins

To place the current status of high temperature polymers in proper perspective, a brief chronological review on each of the more popular high temperature polymers is appropriate. In 1955, E. I. DuPont de Nemours and Company, Inc., was awarded the first of several patents (2) on aromatic polyimides (PI) considered to be the first family of high temperature polymers. Several companies such as Westinghouse, Monsanto, General Electric, Amoco and DuPont devoted a significant effort to PI development. In fact, more effort has gone into PI development than all of the other high temperature polymers combined and, now that the DuPont PI patents are expiring, it will be interesting to see if companies who are basic in certain PI monomers (e.g., Gulf Oil Chemicals Company who sells 3,3',4,4'-benzophenone tetracarboxylic acid dianhydride) will enter the PI marketplace. The major portion of the earlier PI work involved the preparation of soluble precursor polyamic acids from the reaction of aromatic tetracarboxylic acid dianhydrides and aromatic diamines and their subsequent cyclodehydration to PI (Eq. 1).

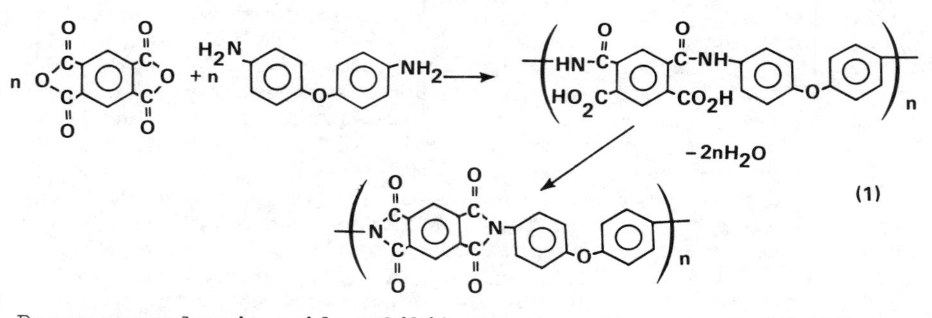

Precursor polyamic acids exhibit severe problems in shelf life and processing due to hydrolytic instability, water released during the cyclodehydration and general lack of melt flow. In spite of these problems, several PI made by this general synthetic route became commercially available about 1960, such as Pyre ML lacquers, Vespel molding materials and FM-34 adhesive.

Concurrent with the development of PI, the polybenzimidazoles (PBI) as initially reported by Marvel (3) were being evaluated. Polybenzimidazoles are generally prepared by the polycondensation of aromatic bis(o-diamines) with diphenyl esters of aromatic dicarboxylic acids (Eq. 2).

$-2n\ H_2O$

$-2n\ \Phi OH$

(2)

1

Status of High-Temperature Laminating Resins and Adhesives

P. M. HERGENROTHER and N. J. JOHNSTON

National Aeronautics and Space Administration, Langley Research Center, Hampton, VA 23665

The quest for high temperature polymers for use as functional resins (e.g., coatings and sealants) and structural resins (e.g., adhesives, composite matrices and foams) began in the mid-1950's. Early work on high temperature polymers concentrated primarily on thermal stability and paid little or no attention to cost, processability, performance or actual use temperature. As work progressed, it was quickly recognized that a successful polymer must exhibit a favorable combination of price, processability and performance and also that thermal stability differed from dimensional stability or heat resistance.

Work on most of the more exotic and costly polymers was soon discontinued as a result of this realism and the more current emphasis has generally been restricted to the development of materials for specific applications. In fact, this review will be confined to those high temperature polymers currently being developed as adhesives and composite matrices. The term "high temperature" in this context refers to the ability of a structural material to perform properly in air for thousands of hours at 232°C, hundreds of hours at 316°C and/or minutes at 538°C.

The ideal high temperature polymeric system for use as a production-line adhesive or laminating resin would exhibit a variety of desirable features such as the following: easy tape and prepreg preparation, good shelf-life, acceptable tack and drape, acceptable quality control procedures, low processing temperature and pressure, no volatile evolution, dense glueline or matrix (no voids), good mechanical performance over the desired temperature range, time, and environmental conditions, acceptable repairability and cost-effectiveness. Ironically, no commercially available system exists which actually offers any reasonable combination of these features. Instead, the few commercially available materials represent a compromise of these properties.

0-8412-0567-1/80/47-132-003$05.00/0

Historical Development of High Temperature Structural Resins

To place the current status of high temperature polymers in proper perspective, a brief chronological review on each of the more popular high temperature polymers is appropriate. In 1955, E. I. DuPont de Nemours and Company, Inc., was awarded the first of several patents (2) on aromatic polyimides (PI) considered to be the first family of high temperature polymers. Several companies such as Westinghouse, Monsanto, General Electric, Amoco and DuPont devoted a significant effort to PI development. In fact, more effort has gone into PI development than all of the other high temperature polymers combined and, now that the DuPont PI patents are expiring, it will be interesting to see if companies who are basic in certain PI monomers (e.g., Gulf Oil Chemicals Company who sells 3,3',4,4'-benzophenone tetracarboxylic acid dianhydride) will enter the PI marketplace. The major portion of the earlier PI work involved the preparation of soluble precursor polyamic acids from the reaction of aromatic tetracarboxylic acid dianhydrides and aromatic diamines and their subsequent cyclodehydration to PI (Eq. 1).

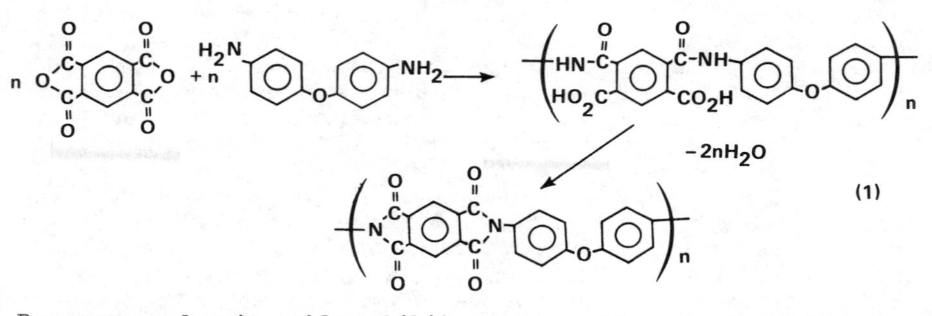

Precursor polyamic acids exhibit severe problems in shelf life and processing due to hydrolytic instability, water released during the cyclodehydration and general lack of melt flow. In spite of these problems, several PI made by this general synthetic route became commercially available about 1960, such as Pyre ML lacquers, Vespel molding materials and FM-34 adhesive.

Concurrent with the development of PI, the polybenzimidazoles (PBI) as initially reported by Marvel (3) were being evaluated. Polybenzimidazoles are generally prepared by the polycondensation of aromatic bis(o-diamines) with diphenyl esters of aromatic dicarboxylic acids (Eq. 2).

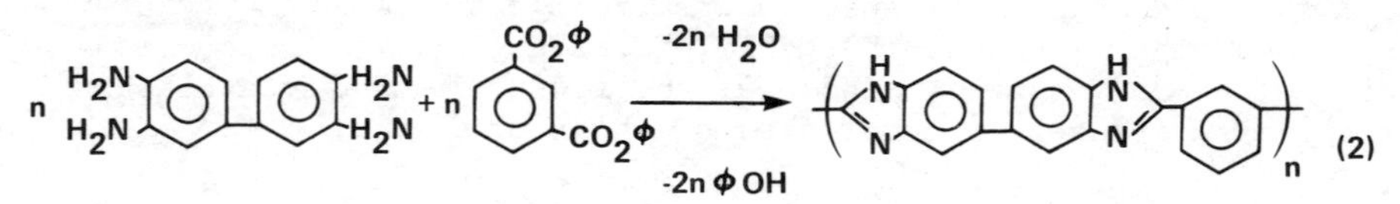

Considerable amounts of phenol and water are evolved which caused pronounced fabrication problems during the processing of composites and adhesives. PBI was commercially available in the mid 1960's (e.g., Imidite 850 from Narmco Materials) and is currently being evaluated for use as hollow fibers in water purification, flame resistant fibers and low density foams.

During the 1960's a variety of new polymers were reported as high temperature materials primarily on the basis of their weight loss performance as measured by thermogravimetric analysis (4,5). Few of these new materials were investigated as structural resins. One of several systems which did receive considerable attention, however, was the polyphenylquinoxalines (PPQ). These materials, as first reported in 1967 (6), are readily prepared from the polycyclocondensation of aromatic bis(o-diamines) and aromatic dibenzils (Eq. 3) as soluble, high molecular weight thermoplastics. Although they display excellent potential as high temperature structural resins, particularly as adhesives,

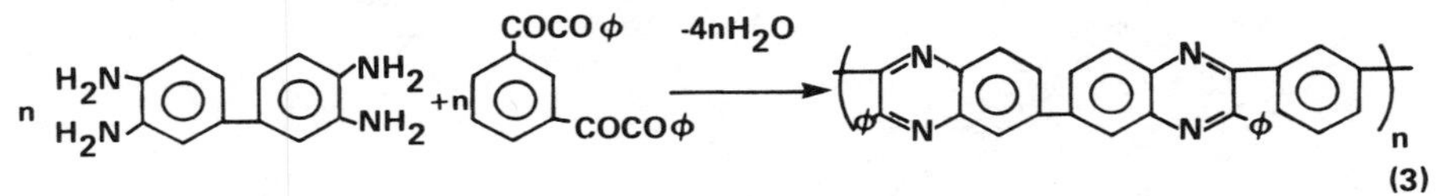

the PPQ are relatively expensive, require high temperature processing and at present are not commercially available.

As more potential applications developed for high temperature structural resins, alternate synthetic routes to PI were investigated to improve processability. These routes concentrated on alleviating the evolution of the volatiles responsible for fabrication problems. In the late 1960's TRW (7) developed a route to nadic end-capped imide oligomers which could be thermally polymerized through the unsaturation (Eq. 4). This technology became known as P13N with the P standing for polyimide, the 13 representing an average oligomeric molecular weight of 1300, and the N

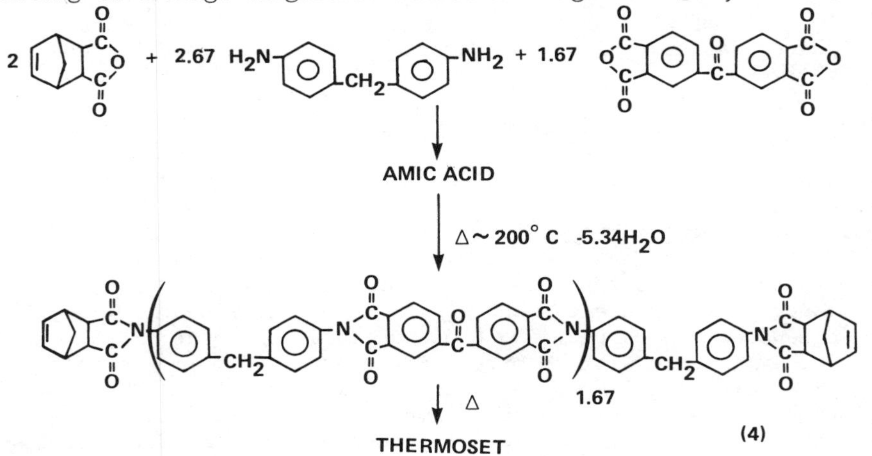

referring to nadic end-caps. The thermal chain extension, often referred to as pyrolytic polymerization, apparently occurs through a combination of addition reactions involving the vinyl groups of the nadimide moiety and maleimide/cyclopentadiene products formed from a reverse Diels-Alder reaction of the nadimide. This process is accomplished with the evolution of a small amount of cyclopentadiene. A significant improvement in processing over the polyamic acids was realized but the resultant resin was brittle. The P13N patents were purchased by the Ciba-Geigy Corporation who was unsuccessful in developing a P13N product line.

Work at Rhone-Poulenc on polyimides resulted in the development of maleimide end-capped oligomers which could be chain extended thermally and/or by reaction with an aromatic diamine (Eq. 5)(8). The Kerimid and Kinel products as marketed by Rhodia in the U. S. evolved from this technology. They exhibit excellent melt flow properties and easily form void-free parts. Although these materials are used in certain applications such

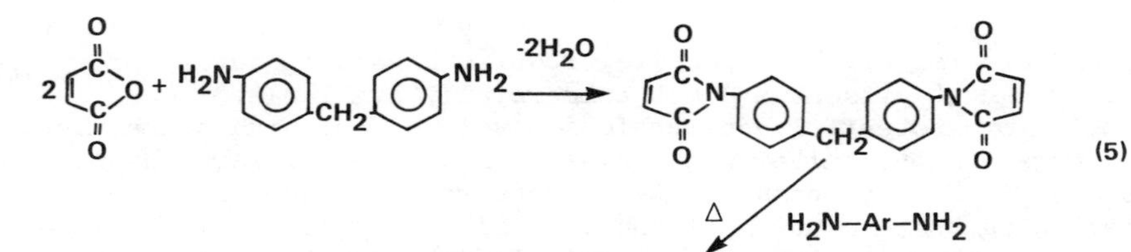

RESIN FROM ADDITION THROUGH DOUBLE BOND AND THROUGH MICHAEL REACTION OF AMINE WITH DOUBLE BOND

as circuit boards, they have not received acceptance as high temperature structural resins because of their limited thermal stability.

Current High Temperature Structural Resins

The P13N chemistry was extended at NASA-Lewis Research Center to the development of PMR-15, PMR standing for Polymerization of Monomeric Reactants (9). The chemistry is the same as in P13N except the average oligomeric molecular weight was increased to 1500 and the material initially used for prepreg formation is a mixture of monomers. The dianhydride of 3,3',4,4'-benzophenone tetracarboxylic acid and nadic anhydride are separately reacted with an excess of methanol to form the half-methyl esters. 4,4'-Methylenedianiline is added to a methanolic solution of the two half-esters. The resultant varnish is used to impregnate the reinforcement. Tack and drape are obtained in the prepreg through residual methanol which frequently evaporates and leaves the prepreg boardy. Esterification beyond the half-ester stage has been observed and has been the source of processing problems. Residual volatiles are removed during a B-stage cycle with additional cure occurring through the nadimide end-caps. PMR-15 has shown considerable potential for use in high temperature structural composites and is currently under evaluation in many laboratories. PMR-15 prepreg is commercially available.

In an attempt to improve processability and introduce tack and drape character into a PI prepreg without using solvent, work was initiated at NASA-Langley Research Center in the mid-1970's which climaxed in the development of LARC-160 (Eq. 6)(10). A mixture of nadic anhydride and 3,3'4,4'-benzophenone tetracarboxylic acid dianhydride is converted to the half-ethyl

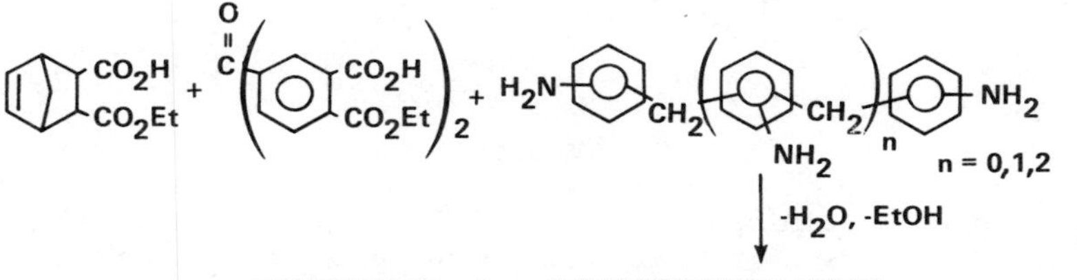

esters using essentially a stoichiometric quantity of ethanol. A liquid oligomeric amine, Jeffamine AP-22 or Ancamine-DL, is then blended with the two half-ethyl esters with the stoichiometry adjusted to provide an oligomer with an average molecular weight of 1600. The resultant mixture of monomers is a viscous liquid at room temperature and contains no residual solvent. The monomeric mixture is amenable to melt impregnation to provide prepreg with tack and drape without the use of residual solvent. In composite fabrication, the condensation volatiles are removed during a B-stage cycle to form imide oligomers which exhibit excellent flow. These are then thermally chain-extended and cross-linked through the nadimide end-groups. Composites of LARC-160 exhibit good performance at high temperatures. Both neat resin and prepreg can be obtained commercially. Because of its enhanced processability, LARC-160 is currently being evaluated as a matrix material in many laboratories. Fabrication development activities on both PMR-15 and LARC-160 are being sponsored by NASA-Langley at Boeing Aerospace Company and Rockwell International, respectively.

In 1969, Hughes Aircraft Company began work under Air Force funding on acetylene-terminated imide oligomers (Eq. 7) which could be thermally polymerized through acetylene end-groups

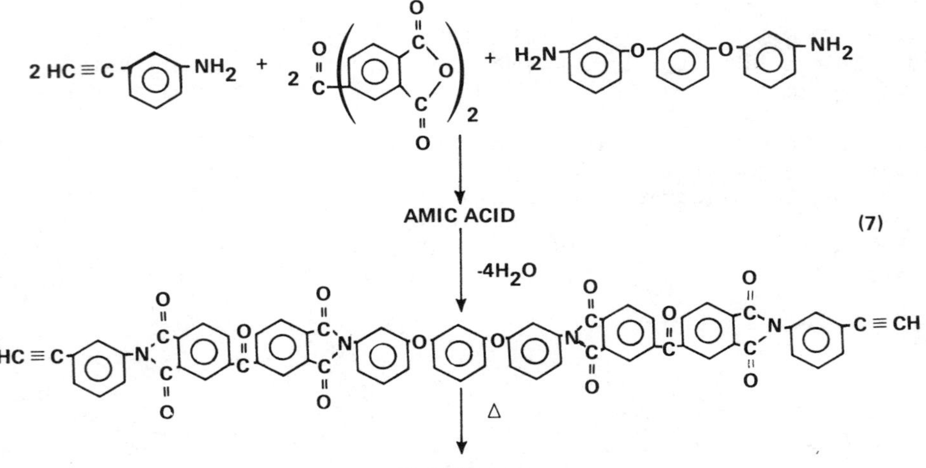

(11). This work culminated in the development of a material called HR-600. It was initially proposed that chain extension of HR-600 occurred through aromatization or acetylenic trimerization (12). Recent work has shown, however, that the thermal reaction of an ethynyl group provides a very complex mixture of products (13). This technology was licensed by Gulf Oil Chemicals Company and the HR-600 resin is now marketed as Thermid 600. Graphite reinforced composites from Thermid 600 have shown good performance at 316°C after aging. Further development of Thermid 600 as a laminating resin under AFML/NASA sponsorship was curtailed due to problems associated with inadequate flow and short gel time.

DuPont continued their work in the PI field and in 1972 announced the NR-150 series of PI (14). One of these materials, NR-150B2, is prepared by mixing 2,2-bis(3',4'-dicarboxyphenyl) hexafluoropropane with 95 mole % para- and 5 mole % meta-phenylenediamine in a suitable solvent such as N-methylpyrrolidone (Eq. 8). The monomeric varnish is used for the preparation of prepreg which retains tack and drape due to residual solvent. The fabrication of laminates [e.g., 30.5cm x 30.5cm x 0.32cm (12" x 12" x 0.125")] is difficult because of the removal of high boiling solvent and condensation volatiles; consequently, long cure and postcure cycles must be used. As with most composite materials, small amounts (<1%) of residual solvent in the laminate cause severe delamination during postcure and, if not removed, plasticize the resin and induce

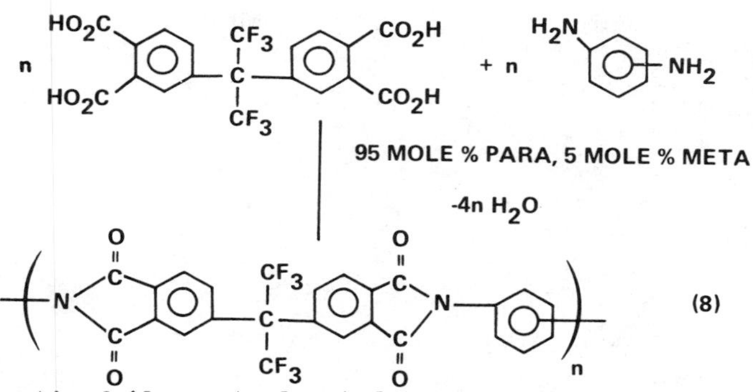

thermoplastic failure at elevated temperatures. Since small composites of NR-150B2 have shown good high temperature performance, NASA-Langley is sponsoring further work on fabrication development with Convair Division of General Dynamics Corporation. Prepreg is commercially available.

Composite Properties

Several laboratories have evaluated high temperature polymers as laminating resins and adhesives and have reported widely divergent results on the same material. This is to be expected since conditions which influence the performance of a structural resin vary from one laboratory to another. For example, composite properties are governed by factors such as prepreg quality, processing (*e.g.*, time, temperature and pressure), aging environment (*e.g.*, oven design, atmosphere and specimen configuration), reinforcement (*e.g.*, type, batch number and sizing), composite quality (*e.g.*, fiber volume and void content), and testing (*e.g.*, span-to-depth ratio in short beam shear test). Therefore, it is difficult to obtain and would be improper to expect a valid comparison on the performance of structural resins evaluated in different laboratories. A comparison is more reliable when the various resins are evaluated in the same laboratory under carefully controlled conditions.

In an attempt to obtain a meaningful comparison of the most promising high temperature matrix resins, laminates of the same dimensions and layup were fabricated from different resins and the same reinforcement by various companies with expertise in a particular system. These laminates were submitted to NASA-Langley then aged and tested under comparable conditions in order to provide the mechanical properties reported in Tables I and II (15). The quality of the laminates varied but the processing information supplied by each company and the test results indicate trends. Of the four resins evaluated, the order of decreasing thermal performance as indicated by short beam shear strengths is NR-150B2≃Thermid 600>PMR-15≃LARC-160. Ease of processing and

TABLE I- Effect of Elevated Temperature Exposure on Flexural Strength of Unidirectional HT-S Laminates

Flexural Strength,GPa[Kpsi], (% retention of RT properties)

Matrix	RT	260°C after 500 hr.@ 260°C in air	288°C after 500 hr. @288°C in air	316°C after 500 hr. @316°C in air
NR-150B2	1.43[207]	1.30[189] (91)	1.21[175] (85)	1.14[165] (80)
PMR-15	1.43[208]	1.84[267] (128)	1.24[180] (87)	0.97[140] (67)
Thermid-600	1.28[185]	1.17[169] (91)	0.99[144] (78)	1.04[151] (82)
LARC-160	2.13[309]	1.52[220] (71)	14.4[209] (68)	Not available

TABLE II- Effect of Elevated Temperature Exposure on Interlaminar Shear Strength of Unidirectional HT-S Laminates

Interlaminar Shear Strength,MPa[Kpsi](% retention of RT properties)

Matrix	RT	260°C after 500 hr. @260°C in air	288°C after 500 hr. @288°C in air	316°C after 500 hr @316°C in air
NR-150B2	70.3[10.2]	40.7[5.9] (58)	39.3[5.7] (56)	34.5[5.0] (49)
PMR-15	111.0[16.1]	53.8[7.8] (48)	47.6[6.9] (43)	17.9[2.6] (16)
Thermid-600	83.4[12.1]	60.0[8.7] (72)	51.0[7.4] (61)	41.4[6.0] (50)
LARC-160	95.8[13.9]	53.8[7.8] (56)	46.9[6.8] (49)	Not available

quality of prepreg followed the trend: LARC-160>PMR-15> Thermid-600>NR-150B2. With respect to cost the following trend exists: LARC-160<PMR-15<NR-150B2<Thermid 600. Since this initial screening, additional composite fabrication studies have been conducted which will be the subject of future reports.

Adhesive Properties

As with high temperature composites, the performance of high temperature adhesives has varied widely from one laboratory to another. This is due to the many factors which influence the performance of a thermally stable adhesive such as the adherend (e.g., type, surface treatment and thickness), adhesive tape quality, filler (e.g., type and amount), carrier, processing, aging environment, bond quality and testing. Four of the most promising high temperature adhesives were evaluated in an attempt to obtain a comparison of their initial performance and potential. These consisted of FM-34, a filled polyimide system from Bloomingdale Division of American Cyanamid Company; NR-056X, a modification of NR-150B2 from DuPont (16); LARC-13, a nadic endcapped imide oligomer similar to P13N except m,m'-methylenedianiline is used instead of p,p'-methylenedianiline (17); and a polyphenylquinoxaline (18). The latter two are under development at NASA-Langley. Titanium-to-titanium (Ti/Ti) and composite-to-composite (C/C) lap shear specimens and composite-to-PI glass core sandwich panels were fabricated without opitmization of resin or processing and tested. Preliminary results are presented in Table III. Many factors varied in this study such as filler content in the tape, surface treatment of the adherend, type and thickness of the composite adherends, failure mode, number of specimens tested, and scatter of test values. These variations prohibit a meaningful comparison of adhesive performance although the values in Table III are representative of what can be expected for each adhesive. FM-34 and NR-056X have comparable and acceptable strengths for all test specimen configurations. LARC-13 has lower Ti/Ti RT strength due to its brittle nature but strength at 316°C comparable to FM-34 and NR-056X. In addition, LARC-13 is more amenable to bonding unvented large areas than FM-34 or NR-056X. PPQ has outstanding performance in all areas except when Ti/Ti specimens were tested at 316°C. Thermoplastic failure occurred. However, excellent 316°C strengths were obtained on C/C specimens using the same adhesive. This information is preliminary and further work is in progress to develop high temperature adhesives.

The use of trade names or names of manufacturers in this paper does not constitute an official endorsement of such products or manufacturers, either expressed or implied, by the National Aeronautics and Space Administration.

TABLE III- PRELIMINARY HIGH TEMPERATURE ADHESIVE PROPERTIES

ADHESIVE	FM-34*	NR-056X*	LARC-13*	PPQ
Commercially Available Tape	Yes	No	No	No
Tape Properties				
Filler	Yes	Yes	Yes	No
Tack	Yes	No	No	No
Drape	Yes	No	No	No
Flow	Good	Good	Good	Fair
Shelflife	Poor	Poor	Good	Good
Volatiles,%	~14	~11	<1	<1
Final Processing Temp. °C	343	343	343	399
Pressure for Fabrication of LSS, MPa(psi)	1.4(200)	2.1(300)	0.34(50)	1.4(200)
Tg of Cured Adhesive, °C	322	328	306	318
Ti/Ti Tensile Shear Strength, MPa(psi)				
RT	22.7(3300)	24.8(3600)	13.8(2000)	30.3(4400)
288	NA	NA	NA	17.2(2500)
316 °C	11.7(1700)	11.9(1720)	11.5(1670)	2.1(300)(1)
316 °C(125hr)	10.3(1500)	9.0(1300)	11.3(1640)	3.4(500)(1)
C/C Tensile Shear Strength, MPa(psi)				
RT	17.5(2540)(2)	15.9(2300)(2)	17.0(2470)(2)	41.4(6000)(3)
316 °C	11.4(1660)	14.3(2080)	13.7(1980)	19.3(2800)
316 °C(125hr)	10.1(1460)	9.2(1340)	10.4(1510)	17.2(2500)
C/PI Glass Core FWT, MPa(psi)				
RT(unaged)	3.6(518)(2)	3.3(474)(2)	3.6(515)(2)	3.4(500)(3)
RT(after 125 hrs 316°C)	2.5(357)	2.5(363)	2.3(338)	NA
Capability for bonding unvented large areas	Poor	Poor	Fair	Good

(1) Thermoplastic Failure
(2) Predominantly Shear Failure in the Celion6000/PMR-15 Adherend
(3) Predominantly Shear Failure in the HTS/NR150-B2 adherend

*Reference 19

Literature Cited

1. NASA Grant NSG-1124 with Virginia Polytechnic Institute and State University, Blacksburg, VA 24061.
2. Edwards, W. M.; Robinson, I. M., U. S. Pat. 2,710,853 (1955) to E. I. DuPont de Nemours and Company.
3. Vogel, H. A.; Marvel, C. S., J. Polym. Sci., 1961,50,511.
4. Pezdirtz, G. F.; Johnston, N. J., in "Chemistry in Space Research", R. Landel and A. Rembaum, Ed., Am. Elsevier Pub. Co., Inc., New York 1972, Chapter 10.
5. Jones, J. I., Chem. Brit., 1970,6,251.
6. Hergenrother, P. M.; Levine, H. H., J. Polym. Sci., 1967,A-1,5,1453.
7. Lubowitz, H. R., U. S. Pat. 3,528,950 (1970) to TRW, Inc.
8. Mallet, M. A. J., Modern Plastics, June 1973, p. 78; Mallet, M. A. J.; Darmory, F. P., Am. Chem. Soc., Div., Coatings and Plastic Chem. Preprints, 1974,34,173.
9. Serafini, T. T.; Delvigs, P.; Lightsey, G. R., J. Appl. Polymer Sci., 1972,16,905.
10. St. Clair, T. L.; Jewell, R. A., National SAMPE Tech. Conf. Series, 1976,8,82; Sci. Adv. Mat'l & Proc. Eng. Ser., 1978, 23,520.
11. Bilow, N.; Landis, A. L.; Miller, L. J.; U. S. Pat. 3,845,018 (1974) to Hughes Aircraft Company.
12. Landis, A. L.; Bilow, N.; Boschan, R. H.; Lawrence, R. E.; Aponyi, T. J., Am. Chem. Soc., Div. Polymer Chem. Preprints, 1974,15(2),537.
13. Hergenrother, P. M.; Sykes, G. F.; Young, P. R., Am. Chem. Soc., Div. Petroleum Chem. Preprints, 1979,24(1),243.
14. Gibbs, H. H., 17th National SAMPE Symposium, 1972,17,III-B-6.
15. Davis, J. G., Jr., Composites for Advanced Space Transportation Systems-(CASTS), Technical Report for Period July 1, 1975-April 1, 1978. NASA TM 80038.
16. Blatz, P. S., Adhesives Age, 1978, 21,39.
17. St. Clair, T. L.; Progar, D. J., Sci. Adv. Mat'l & Proc. Eng. Ser., 1979,24(1),1081.
18. Hergenrother, P. M.; Progar, D. J., Adhesives Age, 1977, 20,38.
19. Progar, D. J., NASA-Langley Research Center, unpublished data.

RECEIVED January 30, 1980.

2

Status Review of PMR Polyimides

TITO T. SERAFINI

National Aeronautics and Space Administration, Lewis Research Center, Cleveland, OH 44135

Until recently the application of polymer matrix composite materials has been limited to structural components having use-temperature requirements which could be met by epoxy resins. Although high temperature resistant polymer matrix composites provided an opportunity to design structures having nearly a two-fold increase in use-temperature, the chemistry and severe processing requirements of early technology (condensation-type) high temperature polymers made it impractical and difficult to fabricate high quality structural components. In contrast, fiber reinforced epoxy resins can easily be processed using a variety of techniques at relatively low temperatures and pressures.

Studies conducted at the NASA Lewis Research Center led to the development of a class of polyimides known as PMR (for in situ polymerization of monomer reactants) polyimides (1,2,3,4). In the PMR approach, the reinforcing fibers are impregnated with a solution containing a mixture of monomers dissolved in a low boiling point alkyl alcohol solvent. The monomers are essentially unreactive at room temperature, but react in situ at elevated temperatures to form a thermo-oxidatively stable polyimide matrix. These highly processable addition-type polyimides can be processed by either compression (5) or autoclave (6) molding techniques and are now making it possible to realize much of the potential of high temperature polymer matrix composites.

Our research has identified monomer reactant combinations for two PMR polyimides differing in chemical composition. The earliest or "first generation" PMR material is designated PMR-15 and a more recently developed "second generation" material is designated PMR II (7). Prepreg materials employing PMR-15 are commercially available from the major suppliers of prepreg materials. The development of a modified PMR-15 has been reported (8).

The purpose of this paper is to review the current status of first and second generation PMR polyimides. The following topics are reviewed: (1) synthesis and properties, (2) processing, and (3) applications.

Discussion

Synthesis and Properties. Condensation type aryl polyimides are generally synthesized by reacting aryl diamines with aromatic dianhydrides, aromatic tetracarboxylic acids or dialkyl esters of aromatic tetracarboxylic acids. The diamine/dianhydride reaction is preferred for preparing polyimide films whereas the latter two reactions are generally preferred for preparing polyimide matrix resins. The solution used to impregnate fiber reinforcement materials is prepared by dissolving the reactants in aprotic high boiling point solvents such as N,N-dimethylformamide (DMF) or N-methyl-2-pyrrolidone (NMP). During composites fabrication, volatilization of the solvent and condensation reaction by-products results in high void content composites having inferior mechanical properties and thermo-oxidative stability.

Investigators at the Systems Group of TRW, Inc., working under NASA sponsorship, developed an approach to prepare polyimides by means of an addition reaction (9). Their approach consisted of synthesizing low molecular weight amide-acid prepolymers whose chain ends were terminated, or end-capped, with norbornenyl groups. Addition polymerization of the norbornenyl groups occurred at elevated temperatures (275°-350° C)(527°-662° F) without the evolution of volatile materials making it possible to synthesize low void composites. The prepolymer approach, however, did have several shortcomings. These included: (1) the use of DMF, (2) variable solution stability, and (3) less than desirable thermo-oxidative stability at 316° C (600° F).

Another approach was developed in our laboratories for preparing fiber reinforced addition-type polyimides. Our approach eliminated the need for prepolymer synthesis and circumvented most of the shortcomings associated with the use of prepolymers. In our approach a dialkyl ester of an aromatic tetracarboxylic acid, an aromatic diamine and a monoalkyl ester of 5-norbornene-2,3-dicarboxylic acid (NE), are dissolved in a low boiling point alkyl alcohol, such as methanol or ethanol, and the solution is used to impregnate the reinforcing fibers. The number of moles of each monomer reactant is governed by the following ratio:

$$n: (n + 1): 2$$

Where n, (n + 1) and 2 are the number of moles of the dialkyl ester of the aromatic tetracarboxylic acid, the aromatic diamine and NE, respectively. In situ polymerization of the monomer reactants (PMR) occurs upon heating the impregnated fibers.

In the initial study (1) which established the feasbility of the PMR approach, it was noted that composites made from monomer solutions containing the dimethyl ester of 3,3',4,4'-benzophenonetetracarboxylic acid (BTDE), 4,4'-methylenedianiline (MDA) and NE exhibited a higher level of thermo-oxidative stability than did composites prepared from a monomer solution consisting of the dimethyl ester of pyromellitic acid, MDA and NE. The unexpected observation was confirmed in a subsequent study (10) and the optimum number of moles of BTDE (n) which provided the best overall balance of processing characteristics and thermo-oxidative stability was found to be 2.087, corresponding to a PMR polyimide having a formulated molecular weight (FMW) of 1500. The FMW is considered to be the average molecular weight of imidized prepolymer that could have been formed if amide-acid prepolymer had been synthesized. The equation for the FMW of a PMR polyimide prepared from n moles of BTDE (n + 1) moles of MDA and 2 moles of NE is:

$$FMW = n\ MW_{BTDE} + (n + 1)MW_{MDA} + 2\ MW_{NE} - 2(n + 1)\ MW_{H_2O} + MW_{CH_3OH}$$

Where MW_{BTDE}, MW_{MDA}, etc., are the molecular weights of the mono-er reactants and by-products. It is now common practice to denote the stoichiometry of a PMR resin by dividing the FMW by 100. PMR matrices employing BTDE are referred to as "first generation" materials. The first generation PMR matrix prepared from BTDE, MDA and NE having an FMW of 1500 is widely known as PMR-15. Prepreg materials based on PMR-15 are commercially available from the major prepreg suppliers. The structures of the monomers used in PMR-15 are shown in Table I.

These early studies (1,10) also clearly demonstrated the efficacy and versatility of the PMR approach. By varying the chemical nature of either the dialkyl ester acid or aromatic diamine, or both, and the monomer reactant stoichiometry, PMR matrices having a broad range of processing characteristics and properties could easily be synthesized. A modified PMR-15, called LARC-160, has been developed by substituting an aromatic polyamine for MDA (8). Other studies (11,12) have shown that the PMR approach has excellent potential for "tailor making" matrix resins with specific properties. Figure 1 shows the effect of FMW on resin flow for PMR/HTS graphite fiber composites. It can be seen that significantly higher resin flow can be achieved by reducing the FMW. However, as shown in Figure 2, the PMR compositions which exhibit increased resin flow are less thermo-oxidatively stable at 288° C (550° F). The lower resin flow and increased thermo-oxidative stability in going from PMR-10 to PMR-15 clearly show the sensitivity of these properties to imide ring or alicyclic contents. The reduction in resin flow with increased FMW also serves to quantitatively account for the intractable nature of linear high molecular weight condensation polyimides.

Replacement of BTDE with the dimethyl ester of 4,4'-(hexafluoroisopropylidene)-bis (phthalic acid) (HFDE) significantly improved the thermo-oxidative stability of "first generation" PMR resins (12). However, the initial 316° C (600° F) mechanical properties of HFDE/MDA/NE PMR polyimide composites were considerably lower than the corresponding properties of BTDE/MDA/NE PMR polyimides. Graphite fiber reinforced PMR polyimide composites

TABLE I

MONOMERS USED FOR PMR 15 POLYIMIDE

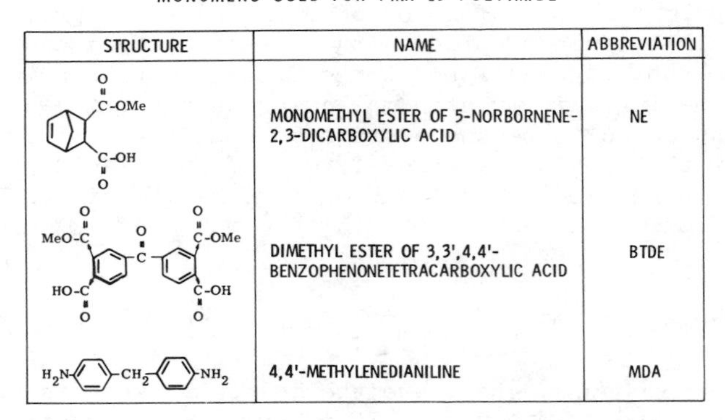

STRUCTURE	NAME	ABBREVIATION
(norbornene monoester structure)	MONOMETHYL ESTER OF 5-NORBORNENE-2,3-DICARBOXYLIC ACID	NE
(benzophenone diester structure)	DIMETHYL ESTER OF 3,3',4,4'-BENZOPHENONETETRACARBOXYLIC ACID	BTDE
$H_2N-C_6H_4-CH_2-C_6H_4-NH_2$	4,4'-METHYLENEDIANILINE	MDA

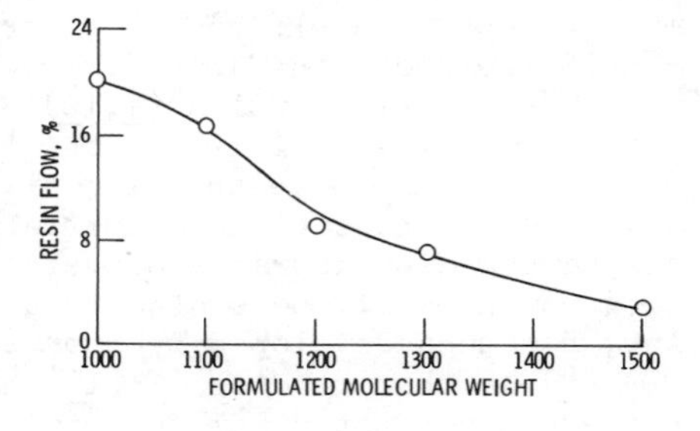

Figure 1. Percent resin flow for PMR PI/HTS graphite fiber composites

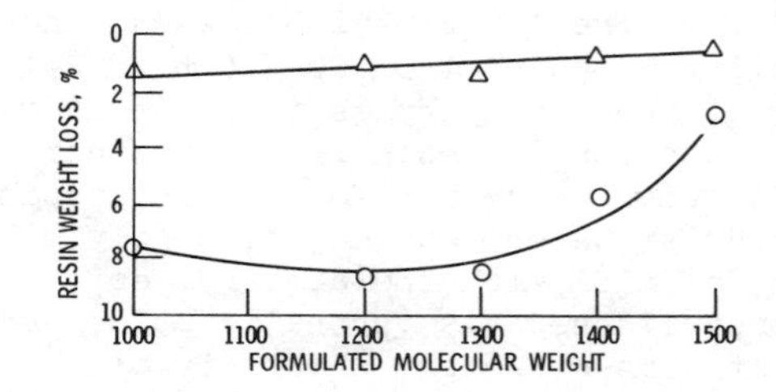

Figure 2. Percent resin weight loss for PMR PI/HTS graphite fiber composites after 600-hr exposure in air at (△) 232°C (450°F) and (○) 288°C (550°F)

prepared from a monomer solution consisting of HFDE, p-phenylenediamine and NE at an FMW of 1267 (n = 1.67) were found to exhibit significantly improved thermo-oxidative stability and retention of mechanical properties at 316° C (600° F) compared to PMR-15 composites (7). The HFDE-PMR compositions are referred to as "second generation" materials to differentiate them from the "first generation" BTDE-PMR materials. The "second generation" resin consisting of HFDE, PPDA and NE with n = 1.67, is known as PMR II. The structure of the monomers used in PMR II are shown in Table II. The interlaminar shear strength of PMR II (n = 1.67) and PMR-13 (a first generation composition with n = 1.67) HTS graphite fiber composites are compared in Figure 3. It can be seen that the PMR II composites have at least twice the useful 316° C (600° F) life of the earlier PMR composites. PMR II was compared to PMR-13 because they each contain an identical number of imide rings.

Further improvement in the performance of PMR polyimides at elevated temperatures has been made possible by the recent development of graphite fibers with improved thermo-oxidative stability. Figure 4 compares the weight loss characteristics of PMR-15 composites made with HTS-1, HTS-2, and Celion 6000 graphite fibers after isothermal exposure in air at 316° C (600° F). The HTS-1 composite data are from Reference 10 and the HTS-2 and Celion 6000 composite are from Reference 13. The data presented in the figure clearly show the significantly improved elevated temperature stability of the HTS-2 and Celion 6000 composites compared to the HTS-1 composites.

Figure 5 compares the interlaminar shear strength retention after exposure in air at 316° C (600° F) of PMR-15/HTS-1, PMR-15/HTS-2 and PMR-15/Celion 6000 composites. It can be seen that both the HTS-2 and Celion 6000 composites exhibited 100 percent retention of their initial 316° C (600° F) interlaminar shear strength during the first 1000 hours of exposure. The shear strength then slowly decreased with further exposure. These mechanical property data and the composite weight loss data shown in the previous figure clearly show that the useful life of PMR-15 composites at 316° C (600° F) made with high strength, intermediate modulus graphite fibers such as HTS-2 and Celion 6000 is at least 1000 hours.

Composites Processing. High pressure (compression) and low pressure (autoclave) molding cycles have been developed for fabrication of PMR composites. Although the thermally induced cross-linking addition cure reaction of the norbornenyl group occurs at temperatures in the range of 275° to 350° C (527° to 662° F), nearly all of the processes developed use a maximum cure temperature of 316° C (600° F). Cure times of 1 to 2 hours followed by a free standing post-cure in air at 316° C (600° F) for 4 to 16 hours, are also normally employed. Compression molding cycles generally employ high rates of heating (5° to 10° C/min) and pres-

TABLE II

MONOMERS USED FOR SECOND GENERATION PMR POLYIMIDES

STRUCTURE	NAME	ABBREVIATION
C(=O)-OMe, C(=O)-OH	MONOMETHYL ESTER OF 5-NORBORNENE-2, 3-DICARBOXYLIC ACID	NE
MeO-C(=O), CF_3, C, CF_3, C(=O)-OMe, HO-C(=O), C(=O)-OH	DIMETHYL ESTER OF 4, 4'-(HEXAFLUOROISOPROPYLIDENE)-BIS(PHTHALIC ACID)	HFDE
H_2N-, -NH_2	p-PHENYLENEDIAMINE	PPDA

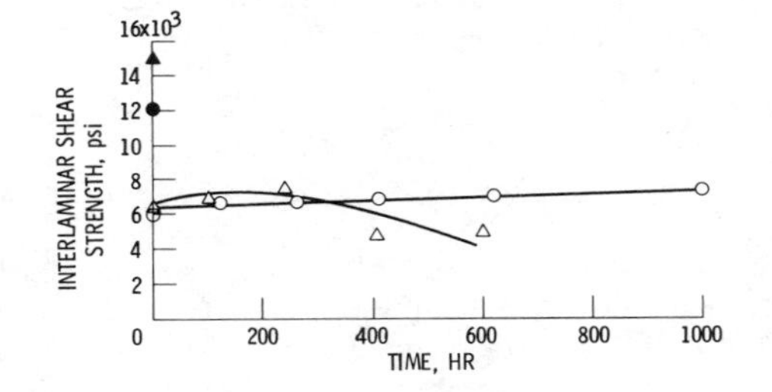

Figure 3. Interlaminar shear strength of PMR PI(n = 1.67)/HTS graphite fiber composites exposed and tested at 316°C (600°F): (△) BTME/MDA/NE, (○) HFDE/PPDA/NE, solid symbols denote room-temperature tests

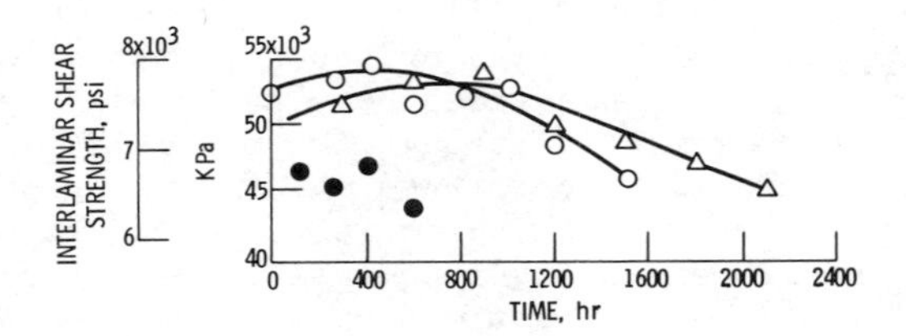

Figure 4. Weight loss of PMR 15/graphite fiber composites exposed in air at 316°C (600°F); fiber: (●) HTS-1, (○) HTS-2, (△) Celion 6000

sures in the range of $3.45x10^6$ to $6.9x10^6$ N/m^2 (500 to 1000 psi). Vacuum bag autoclave processes at low heating rates (2° to 4° C/min) and pressures of $1.38x10^6$ N/m^2 (200 psi) or less have been successfully used to fabricate void-free composites. The successful application of autoclave processing methodology to PMR polyimides results from the presence of a thermal transition, termed "melt-flow," which occurs over a fairly broad temperature range (6). The lower limit of the melt-flow temperature range depends on a number of factors including the chemical nature and stoichiometry of the monomer reactant mixture, and the prior thermal history of the PMR prepreg. Differential scanning calorimetry studies have shown the presence of four thermal transitions which occur during the overall cure of a PMR polyimide (14). The first, second and third transitions are endothermic and are related to the following: (1) melting of the monomer reactant mixture below 100° C (212° F), (2) in situ reaction of the monomers at 140° C (284° F), and (3) melting of the norbornenyl terminated prepolymers in the range of 175° to 250° C (347° to 482° F) referred to as the melt-flow temperature range. The fourth transition, centered near 340° C (644° F) is exothermic and is related to the addition crosslinking reaction. To a large extent the excellent processing characteristics of PMR polyimides can be attributed to the presence of these widely separated and chemically distinct thermal transitions.

Applications. Because of their excellent processing characteristics and commercial availability, PMR-15 polyimide materials have been or are being used in a number of diverse structural components. Some of these components are listed in Table III and a brief description of several is presented. The QCSEE (for Quiet Clean Short Haul Experimental Engine) inner cowl (Fig. 6) is for an experimental turbofan engine developed for NASA-Lewis by General Electric (15). The cowl has a maximum diameter of about 90 cm and has successfully undergone more than 100 hours of ground-engine tests. The compressor blade skins are for sparshell blades having a 30 cm chord and a 150 cm span. The oil tank bracket was fabricated by TRW, Inc., using a chopped graphite fiber molding compound. The shuttle orbiter aft body flap currently being developed by Boeing is approximately 2 m wide by 6.5 m long.

Concluding Remarks

The in situ polymerization of monomer reactants (PMR) approach is a powerful method for fabricating high performance polymer matrix composites. The PMR approach offers a number of significant advantages to fabricators and users of polyimide/fiber composites. Foremost among these are superior high temperature properties and processing versatility. Because of their excellent processability, high performance and commercial availability, PMR

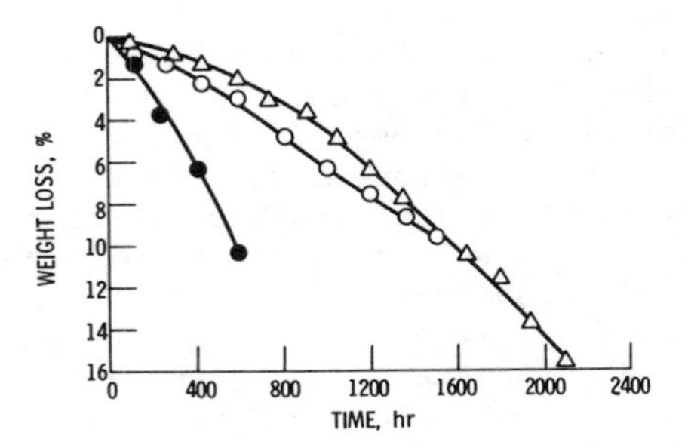

Figure 5. Interlaminar shear strength of PMR 15/graphite fiber composites exposed and tested in air at 316°C (600°F); fiber: (●) HTS-1, (○) HTS-2, (△) Celion 6000

TABLE III

APPLICATIONS OF PMR-15 POLYIMIDES

COMPONENT	AGENCY	CONTRACTOR
QCSEE INNER COWL	NASA-LEWIS	GENERAL ELECTRIC
SUPERSONIC WIND TUNNEL COMPRESSOR BLADE SKINS	AIR-FORCE	HAMILTON-STANDARD
OIL TANK BRACKET FOR F100 ENGINE	NAVY	PRATT & WHITNEY/TRW
SHUTTLE ORIBTER AFT BODY FLAP	NASA-LANGLEY	BOEING
AUGMENTOR DUCT OF F100 ENGINE	AIR FORCE	COMPOSITES HORIZONS
FAN BLADES FOR AN ULTRA-HIGH SPEED AXIAL FLOW FAN STAGE	NASA-LEWIS	PRATT & WHITNEY/TRW

Figure 6. PMR 15 QCSEE inner cowl

polyimides are becoming accepted as engineering materials for fabrication of high temperature resin/fiber structural components.

Literature Cited

1. Serafini, T. T.; Delvigs, P.; Lightsey, G. R. J. Appl. Polym. Sci., 1972, 16, 905.
2. Serafini, T. T.; Delvigs, P.; Lightsey, G. R. July 10, 1973, U.S. Pat. 3,745,149.
3. Serafini, T. T.; Delvigs, P. 1973, J. Appl. Polym. Sci., Appl. Polym. Symp., No. 22, 89.
4. Serafini. T. T. "Proc. of the 1975 International Conference on Composite Materials," Scala, E., Ed.; American Institute of Mining, Metallurgical and Petroleum Engineers: New York, 1976; Vol. 1, p. 202.
5. Cavano, P. J.; Winters, W. E. 1976, NASA CR-135113.
6. Vannucci, R. D. 1977, NASA TM-73701.
7. Serafini, T. T.; Vannucci, R. D.; Alston, W. B. 1976, NASA TM-71894.
8. St. Clair, T. L.; Jewell, R. A. "Proc. of Twenty-Third National SAMPE Symposium and Exhibition," Society for the Advancement of Material and Process Engineering: Azusa, Calif., 1978; p. 520.
9. Burns, E. A.; Lubowitz, H. R.; Jones, J. F. 1968 NASA CR-72460.
10. Delvigs, P.; Serafini, T. T.; Lightsey, G. R. 1972, NASA TN D-6877.
11. Serafini, T. T.; Vannucci, R. D. 1975, NASA TM X-71616.
12. Vannucci, R. D.; Alston, W. B. 1976, NASA TM X-71816.
13. Delvigs, P.; Alston, W. B.; and Vannucci, R. D. 1979, NASA TM-79062.
14. Lauver, R. W. 1977, NASA TM-78844.
15. Ruggles, C. L. 1978, NASA CR-135279.

RECEIVED April 8, 1980.

3

The Synthesis of a New Class of Polyphthalocyanine Resins

T. M. KELLER and J. R. GRIFFITH

Polymeric Materials Branch, Chemistry Division, Naval Research Laboratory, Washington, DC 20375

The principal driving force behind composite development is the requirement for lighter weight structural materials. New design concepts using easily fabricated, fiber-reinforced composites that combine superior stiffness with a high strength-to-weight ratio are needed to replace metal structure. Presently, epoxies and polyimides are being used but each has disadvantages. Conventional epoxy-based composites and adhesives are limited to 120°C maximum service. Other problems associated with these polymers include their brittleness, water absorptivity and engineering reliability.

In recent years considerable advances have been achieved in the synthesis of thermally stable polymers. Frequently thermal and oxidative stability in such materials has been realized by the combination of aromatic and heteroaromatic units; for example, the polybenzimidazoles (1), polybenzoxazoles (2), polybenzthiazoles (3), polyquinazolinediones (4), polyquinazolones (5), polyimides (6,7), etc. However, polar and rigid structures of high symmetry, which is inherent to the aromatic and heteroaromatic rings, are responsible for the general lack of processability of these polymers. In general more tractable polymeric precursors such as polyamide acids (in the case of polyimides) are formed into a desired shape and then converted *in situ* to the final thermally stable polymers. Because the final conversion generally involves a chemical reaction which releases a volatile product such as water, the use of these polymers is often limited to applications such as coatings, films and adhesives.

In our continuing investigation of phthalocyanines (8-13), a new class of phthalonitrile monomers in which either an alkoxy or a phenoxy linkage connects the terminal phthalonitrile units has been synthesized. These monomers are prepared by the simple nucleophilic displacement of a nitro substituent, which is activated by cyano groups, from an aromatic ring by either an alkoxide or a phenoxide unit. The reaction has made it possible to synthesize a large number of structurally different polyphthalocyanines containing ether linkages and alkylene and aromatic spacing units. The products from the reaction have also been hydrolyzed to the corresponding tetraacids which are readily converted to the corresponding bis (anhydrides) (14).

Nucleophilic aromatic displacement of activated nitro groups has been the subject of many investigations (15,16,17). In general a nitro group can be readily displaced when it is positioned ortho or para to another substituent capable of stabilizing a negative charge. Numerous examples of nitro displacement involving activation by sulfone, carboxamide, ketone, phenyl, ester, aldehyde and cyano substituents have appeared in the literature (18-23).

Results and Discussion

The fundamental approach to generation of new polymeric materials involves a determination of the complex of properties desired in the product and selecting those molecular constituents for these structures which can reasonably be expected to yield the desired results. In the case of thermally stable polymers, it is necessary to avoid the inclusion of chemical structures which are known to decompose or oxidize at temperatures below the projected levels. Water-resistant polymers should be composed of a maximum of hydrophobic structure and a minimum of hydrophilic groups consistant with other desired properties.

Our interest in polymers with high thermal and oxidative stabilities, low flammability with high char formation, chemical resistance and low water absorptivity prompted our study of ether-containing polyphthalocyanines. The linking structure between the two terminal phthalonitrile units must play a crucial role in

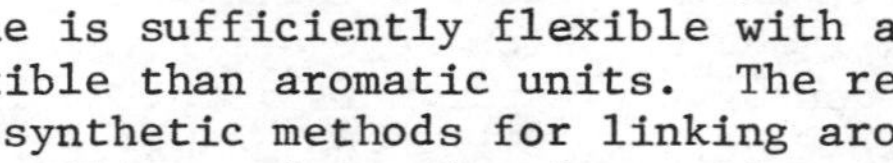

NC, NC — Linking Structure — CN, CN

determining properties for they control the flexibility of the polymer chain and determine the overall polarity of the macromolecule. Thus, to obtain the properties required, these links should be selected such that good stability is retained and the polymer backbone is sufficiently flexible with aliphatic moieties being more flexible than aromatic units. The real problem has been to devise synthetic methods for linking aromatic nuclei with sufficient versatility and freedom from side reactions to effect the synthesis of a wide range of high molecular weight polymers in the hope that some of these would show the properties required. The flexible ether linkage was a natural choice for consideration due to the thermal stability particularly of diaryl ethers and to the chemical resistance of the ether bond in general.

Highly aromatized monomers in which a phenoxy linkage connects the terminal phthalonitrile moieties are easily synthesized in high yield. A variety of monomers depending on the properties desired can be prepared by this method. The reaction involves

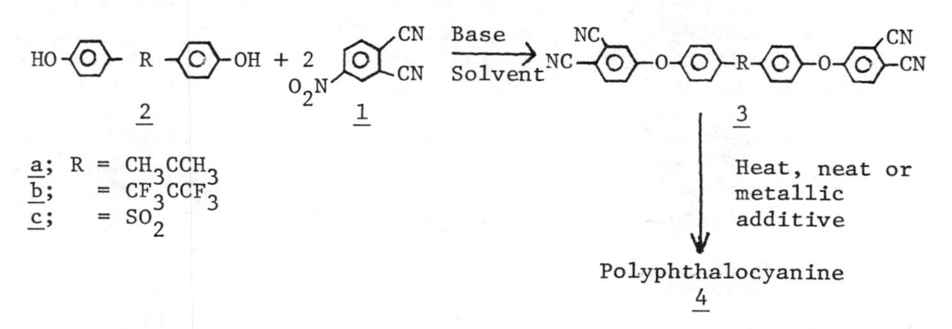

the nucleophilic displacement of the labile nitro substituent from 4-nitrophthalonitrile 1 by a bisphenol 2 in the presence of a base and in a dry dipolar aprotic solvent under an inert atmosphere. Higher yields of 3 are obtained when the reaction is carried out at room temperature.

Phthalonitrile monomers with alkoxy linking units can be prepared by the same method, but the reaction is more sluggish and the yields are lower. If a weak base such as anhydrous potassium carbonate is used, an amount in excess of stoichiometry is preferred and it must be added in increments to ensure a complete reaction. It is theorized that the cessation is due to the

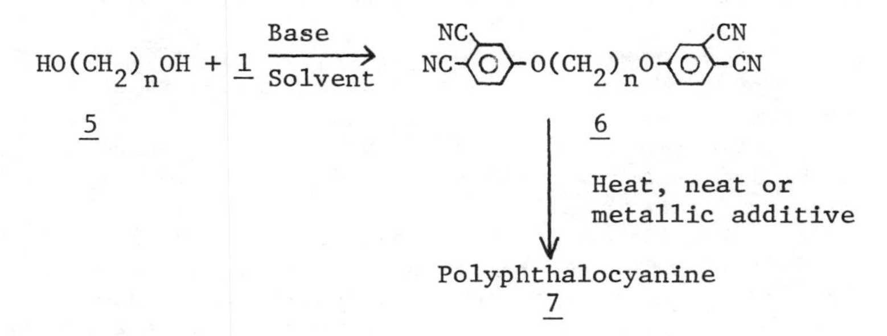

surface of the carbonate becoming coated during the course of the reaction. When a strong base such as sodium hydroxide is used, the disodium salt of 5 is prepared and the by-product (water) is removed before 1 is added. To ensure a high conversion to 6, the reaction must be carried out at elevated temperatures.

From our observation it appears that the nitro displacement reactions described above occur by means of the classical addition-elimination mechanism of nucleophilic aromatic substitution (24). The high yields and absence of side reactions, other than those leading to nitro group displacement, support this mechanism, although an alternate route involving radical anion intermediates cannot be ruled out.

Polymerization of 3 and 6 occurs by a cyclic addition reaction without formation of volatile by-products to produce solid, void-free products, 4 and 7, respectively. Phthalocyanine formation is believed to be the principal reaction due mainly to the terminal phthalonitrile units of the monomers and to the development of the green color during the polymeric progression, but other cyano-addition reactions may also occur. When heated neat, gelation occurs more readily for the aliphatic monomer 6 than for the more aromatized monomer 4. The cure time of both types of monomers can be greatly reduced by the addition of metallic additives.

The ease of polymerization of the alkoxy-linked phthalonitriles 6 depends on the length of the linkage between the terminal phthalonitrile moieties. At a given temperature, the more flexible or longer spacing units require less time to cure. Regardless of the temperature, the heating is continued until the melt solidifies to an extremely hard material. The polymerization can be carried out in an oxygen-containing, inert or vacuum atmosphere. A postcure at a temperature from 210°C to 240°C is used to improve the strength of the resin.

In the case of the highly aromatized linking units, polymerization is more difficult and requires much higher temperatures for gelation. The monomers 3 are cured at 260-290°C and require several days of continuous heating before a viscosity increase is detected. The slow rate of polymerization could be attributable to the rigidity of the linking structure which reduces the mobility of the reaction sites.

The phthalonitrile monomers can be polymerized stepwise to distinct stages. The method comprises heating the monomers at a specified temperature until the viscosity starts to increase due to the onset of phthalocyanine formation (B-stage). The prepolymer can then be cooled to a frangible solid and can be stored indefinitely without further reaction. The prepolymer can either be remelted and heated until solidification occurs (C-stage) or can be pulverized and then processed in any desired form. The optimum cure for any resin at a particular temperature is determined empirically by testing the structural strength over a range of cure times.

The polyphthalocyanines 4 and 7 show high thermal and oxidative stabilities (see Figure 1). Polymer 7 withstood temperatures greater than 200°C without degradation, and 230°C with slight degradation, but decomposed when heated above 250°C for an extended period. Polymer 4 was stable for an extended period at 280°C before any weight loss occurred. Surprisingly, 4 initially showed a small weight increase at 250°C, which is probably due to oxygen absorption, and then slowly lost weight (1% after approximately 4 months) when held at this temperature for prolonged periods. The greater thermal stability of 4 relative to 7 must be attributed to the nature of the linkage with diaryl ethers being more stable than aryl-alkyl ethers.

The water absorptivity of a polyphthalocyanine will also depend on the linking structure between the phthalocyanine nuclei. Polar groups located on the linking structure would be expected to show a stronger attraction for water than nonpolar groups. Polymers 4a, 4b and 7, which contain no polar units, show a similar and low affinity for water (see Figure 2). On the other hand, 4c has a much stronger attraction for water which must be related to the polar sulfone group being present in the molecule.

Conclusions

We have shown the synthetic utility of nucleophilic displacements of a nitro group activated by cyano functions. In our studies, highly aromatized low-cost bisphenol systems are being emphasized in an effort to maximize the char density of the resins. Exposure and removal of samples of these polymers from a high temperature flame has demonstrated that these resins are self-extinguishing. The synthesis of these resins is short and simple and takes advantage of relatively inexpensive starting materials. In addition to the structural variations, the cost to produce these resins should be competitive with that of other high temperature polymeric systems. The diether-linked polyphthalocyanines provide a new matrix resin system with long-term operational capability in excess of 250°C with insensitivity to high humidity and with the ability to retain reinforcing fibers during or following exposure to a fire environment.

Experimental

Synthesis of Bis (3,4-Dicyanophenyl) Ether of Bisphenol A 3a. A mixture of 125 g (0.55 mol) of bisphenol A, 275 g (2.0 mol) of anhydrous potassium carbonate, 190 g (1.1 mol) of 4-nitrophthalonitrile and 900 ml of dry dimethyl sulfoxide was stirred and heated at 55-60°C for 4 hours under a nitrogen atmosphere. After cooling the product mixture was slowly poured into 2000 ml of cold dilute hydrochloric acid. The precipitate was isolated by suction filtration and washed with water until neutral. The crude, dried product was pulverized and washed thoroughly with hot absolute ethanol which removed the impurities. The pure product (230 g, 87%), m.p. 196-199°C, being insoluble in ethanol was collected and analyzed; ir (KBr) 3080-3020 (=CH), 2238 (CN), 1580 (C=C), 1500-1470 (aromatic), 1310-1170 cm^{-1} (CO); nmr ($CDCl_3$) δ 1.73 (singlet, 6H), 7.63 (multiplet, 14H); Anal. Calcd for $C_{31}H_{20}N_4O_2$: C, 77.48; H, 4.20; N, 11.66; O, 6.66. Found: C, 77.21; H, 4.26; N, 11.54; O, 6.93.

A second mixture of 10 g (0.04 mol) of bisphenol A, 3.6 g (0.09 mol) of 50% sodium hydroxide, 70 ml of dimethyl sulfoxide and 30 ml of benzene was stirred at reflux for 3 hours under a nitrogen atmosphere and the water was azeotroped from the mixture

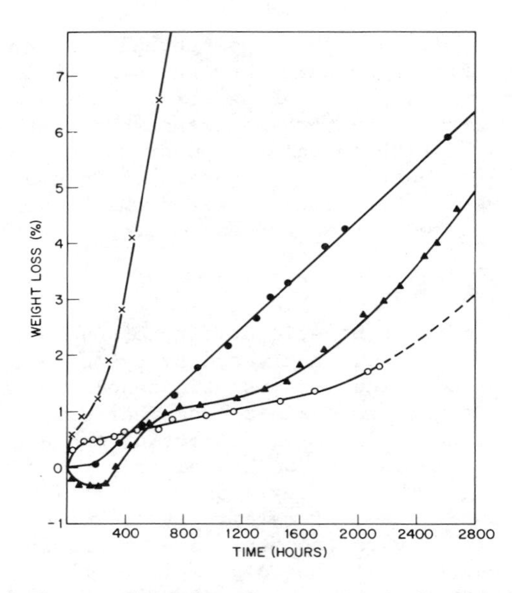

Figure 1. Weight loss at elevated temperature of diether-linked polyphthalocyanines: (×) 1,12-dodecanediol-linked polymer at 250°C; (●) bisphenol A-linked polymer at 280°C; (▲) bisphenol S-linked polymer at 280°C; (○) bisphenol A6F-linked polymer at 280°C

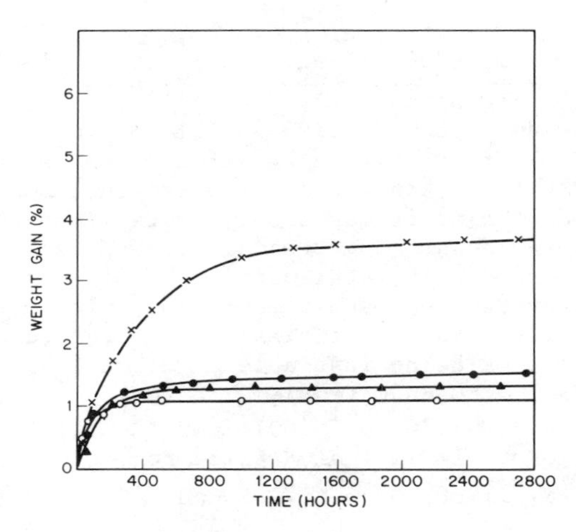

Figure 2. Water absorption of diether-linked polyphthalocyanines on immersion in water at room temperature: (×) bisphenol S-linked polymer; (●) 1,12-dodecanediol-linked polymer; (▲) bisphenol A-linked polymer; (○) bisphenol A6F-linked polymer

with a Dean-Stark trap. The benzene was removed by distillation and 15.7 g (0.09 mol) of 4-nitrophthalonitrile was added to the reaction mixture at room temperature. The resulting dark mixture was stirred at room temperature overnight. The mixture was poured into 300 ml of cold water and the white solid which separated was collected by suction filtration, washed with water, dried and washed with hot absolute ethanol to yield 20.7 g (98%) of the desired product, m.p. 196-199°C.

Synthesis of Bis (3,4-Dicyanophenyl) Ester of Hexafluoroacetone Bisphenol A 3b. A mixture of 10.1 g (0.03 mol) of hexafluoroacetone bisphenol A, 10.4 g (0.06 mol) of 4-nitrophthalonitrile, 12.4 g (0.09 mol) of anhydrous potassium carbonate and 60 ml of dry dimethyl sulfoxide was stirred under a nitrogen atmosphere at 70-80°C for 6 hours. The cooled product mixture was poured into 300 ml of cold dilute hydrochloric acid. The pale brown product which separated was collected by suction filtration and washed with water until neutral. The crude material was recrystallized from acetonitrile to give 13.8 g (78%), m.p. 230-233°C, of the desired product; ir (KBr) 3105-3020 (=CH); 2238 (CN), 1590 (C=C), 1520-1485 (aromatic), 1320-1140 cm^{-1} (CF_3CO); nmr (acetone-d_6) δ 7.73 (multiplet, 14H); ^{19}F nmr ($CFCl_3$ external ref.) - 63.42 ppm (singlet, 6F); Anal. Calcd for $C_{31}H_{14}F_6N_4O_2$: C, 63.29; H, 2.38; F, 19.37; N, 9.52; O, 5.44. Found: C, 63.38; H, 2.62; F, 19.37; N, 9.61; O, 5.02.

A second mixture containing 67.2 g (0.21 mol) of hexafluoroacetone bisphenol A, 16.5 g (0.4 mol) of 50% aqueous sodium hydroxide, 300 ml of dimethyl sulfoxide and 75 ml of benzene was stirred at reflux for 15 hours under a nitrogen atmosphere and the water was removed with a Dean-Stark trap. The benzene was removed by distillation and 69.4 g (0.4 mol) of 4-nitrophthalonitrile was added to the reaction mixture at room temperature. The resulting dark mixture was stirred at room temperature for 12 hours under a nitrogen atmosphere. The cooled mixture was then poured into 800 ml of cold water and the pale brown product was collected by suction filtration. Recrystallization from acetonitrile yielded 107 g (91%) of product.

Synthesis of Bis (3,4-Dicyanophenyl) Ether of Bisphenol S 3c. A mixture of 51 g (0.2 mol) of bisphenol S, 16.4 g (0.4 mol) of 50% aqueous sodium hydroxide, 450 ml of dimethyl sulfoxide and 100 ml of benzene was stirred at reflux for 6 hours. The water and benzene was removed with a Dean-Stark trap. The reaction content was cooled to room temperature and 69.4 g (0.4 mol) of 4-nitrophthalonitrile was added in one sum. The resulting mixture was stirred for 12 hours at room temperature under a nitrogen atmosphere and then poured into 1500 ml of cold water. The slightly colored solid which separated was collected by suction filtration, washed with water and dried. The product was then washed with 400 ml of hot ethanol to afford 99.2 g (98%) of product, m.p. 231-233°C; ir (KBr) 3100-3080 (=CH), 2238 (CN), 1600-1560 (C=C), 1470 (aromatic), 1310-1100 cm^{-1} (SO_2, CO); nmr ($DMSO-d_6$) δ 7.81 (multiplet, 14H); Anal. Calcd for $C_{28}H_{14}N_4O_4S$: C, 66.92; H, 2.81; N, 11.15; O, 12.74; S, 6.38. Found: C, 66.65; H, 2.85; N, 11.29; O, 12.67; S, 6.29.

Synthesis of 1,12-Bis (3,4-Dicyanophenoxy) Dodecane 6. A mixture of 1,12-dodecanediol (2.2 g, 0.01 mol), anhydrous potassium carbonate (4.1 g, 0.03 mol), 35 ml of dimethylformamide and 25 ml of benzene was heated at reflux under a nitrogen atmosphere for 12 hours. A small quantity of water was collected in a Dean-Stark trap. The benzene was then removed by distillation and the reaction mixture cooled to room temperature. 4-Nitrophthalonitrile (4.0 g, 0.02 mol) was added in one sum which resulted in an immediate color change to a deep blue. After 10 minutes at room temperature, the reaction medium had turned to a yellowish-orange and remained this color throughout the entire reaction. The mixture was stirred and heated at 100°C for 16 hours under a nitrogen atmosphere. The cooled reaction mixture was poured into 200 ml of cold dilute hydrochloric acid and extracted with three, 75 ml portions of chloroform. The combined extract was washed with water, dried over anhydrous sodium sulfate, charcoaled and concentrated at reduced pressure. Recrystallization of the crude product from ethanol-water yielded 3.1 g (70%) of 6, m.p. 104-107°C; ir (KBr) 3110-3040 (=CH), 2930-2850 (CH), 2238 (CN), 1600 (C=C), 1490-1470 (aromatic), 1350-1250 (aromatic CO), 1100-1010 cm^{-1} (aliphatic CO); nmr ($CDCl_3$) δ 1.52 (multiplet, 20H), 4.01 (triplet, 4H), 7.55 (multiplet, 6H); Anal. Calcd for $C_{28}H_{30}N_4O_2$: C, 74.00; H, 6.65; N, 12.33; O, 7.04. Found: C, 73.71; H, 6.69; N, 12.39; O, 7.21.

Polymerization of 3. Samples (1-2 g) of 3 were placed in planchets and heated at 280°C for 7 days. A viscosity increase, which indicated that polymerization was progressing, occurred very slowly. After gelation (3-4 days) the samples were postcured for 3 additional days to ensure complete polymerization and to toughen the polymers.

Samples of 3 and stoichiometric amounts of stannous chloride dihydrate were heated at 220-250°C for 24 hours. After the monomers melted, the samples quickly turned green along with an immediate dissolution of the salt. The viscosity increased rapidly with gelation occurring in 5-15 minutes.

Polymerization of 6. A sample (1-2 g) of 6 was melted and heated at 220°C for 48 hours. Gelation was extremely slow (30 hours) at this temperature. The polymeric material was postcured at 240°C for 24 hours which enhanced the toughness of the material.

Another sample of 6 was heated at 240°C for 24 hours. Gelation had occurred after 6 hours at this temperature.

A mixture of 6 (1.5 g, 3.3 mmol) and a stoichiometric amount of stannous chloride dihydrate (0.36 g, 1.6 mmol) was placed in a test tube. The monomer 6 melted at 105-110°C. At 170-175°C the salt dissolved and the reaction medium became green immediately. The temperature was increased to 215°C and the sample was heated at this temperature for 24 hours. Gelation had occurred after 15 minutes at 215°C.

Acknowledgment

We wish to thank Dr. C. F. Poranski, Jr., Naval Research Laboratory, for recording and analyzing the nmr spectra.

Literature Cited

1. Vogel, H.; Marvel, C. S.; J. Polym. Sci., 1961, 50, 511.
2. Kubota, T.; Nakanishi, R.; J. Polym. Sci., 1964, 2, 655.
3. Hergenrother, P. M.; Levine, H. M.; J. Polym. Sci., A3, 1965, 1665.
4. Kurihara, M.; Hagiwara, Y.; Polym. J., 1970, 1, 425.
5. Kurihara, M.; Yoda, N.; J. Polym. Sci. (B), 1966, 4, 11.
6. Jones, J. I.; Ochynski, F. W.; Rackley, F. A.; Chem. and Ind., 1962, 1686.
7. Varma, I. K.; Goel, R. N.; Varma, D. S.; J. Polym. Sci., 1979, 17, 703.
8. Griffith, J. R.; O'Rear, J. G.; Walton, T. R.; "Phthalonitrile Resin in Copolymers, Polyblends, and Composites", Advanc. Chem. Ser., 1975, 142, 458.
9. Walton, T. R.; Griffith, J. R.; Applied Polymer Symposium, 1975, 26, 429.
10. Walton, T. R.; Griffith, J. R.; O'Rear, J. G.; Adhesion Science and Technology, 1975, 9b, 665.
11. Walton, T. R.; Griffith, J. R.; O'Rear, J. G.; 168th National American Chemical Society Meeting, Organic Coatings and Plastics Preprints, Sept. 1974, 34, 446.
12. Walton, T. R.; Griffith, J. R.; O'Rear, J. G.; 174th National American Chemical Society Meeting, Organic Coatings and Plastics Preprints, Sept. 1977, 37 (2), 180.
13. Keller, T. M.; Griffith, J. R.; 176th National American Chemical Society Meeting, Organic Coatings and Plastics Preprints, Sept. 1978, 39, 546.
14. Takekoshi, T.; Wirth, J. G.; Heath, D. R.; Kochanowski, J. E.; Manello, J. S.; Webber, M. J.; 177th National American Chemical Society Meeting, Polymer Preprints, April 1979, 20 (1), 179.
15. Williams, F. J.; Donahue, P. E.; J. Org. Chem., 1977, 42, 3414.
16. Beck, J. R.; Sobizak, R. L.; Suhr, R. G.; Yahner, J. A.; J. Org. Chem., 1974, 39, 1839.
17. Relles, H. M.; Orlando, C. M.; Heath, D. R.; Schluenz, R. W.; Manello, J. S.; Hoff, S.; J. Polym. Sci., 1977, 15, 2441.

18. Radlmann, E.; Schmidt, W.; Nischk, G. E.; Makromol. Chem., 1969, 130, 45.
19. Govin, J. H.; Chem. Ind. (London), 1967, 36, 1525.
20. Spence, T. W. M.; Tennant, G.; J. Chem. Soc., Perkin Trans., 1972, 1, 835.
21. Beck, J. R.; J. Org. Chem., 1972, 37, 3224.
22. Beck, J. R.; Yahner, J. A.; J. Org. Chem., 1974, 39, 3440.
23. Knudsen, R. D.; Snyder, H. R.; J. Org. Chem., 1974, 39, 3343.
24. Bunnett, J. F.; Q. Rev., Chem. Soc., 1958, 12, 1.

RECEIVED February 15, 1980.

Silicone Amine Cured Fluoroepoxy Resins

J. R. GRIFFITH and J. G. O'REAR

Polymeric Materials Branch, Chemistry Division, Naval Research Laboratory, Washington, DC 20375

Compounds which contain large quantities of fluorine are frequently incompatible with those of a hydrocarbon nature, and liquid materials of the two types form separate phases when mixed. This incompatibility presents a problem regarding the cure of the heavily fluorinated liquid epoxies previously synthesized at NRL (1) because heavily fluorinated aliphatic amines are not generally either stable nor reactive. For higher temperature cures, some fluoroanhydrides have been synthesized and work well (2), but for cures near room temperature, compatible curing agents are not so plentiful.

In many respects, silicone compounds are similar in their properties to fluorocarbons and advantage can be taken of this since polyamino silicones can be induced to become compatible with fluoroepoxy resins quite readily, and the polymers produced when these two types of materials react are exceptional in several important respects.

Discussion

The series of fluorinated diglycidyl ethers represented by the following general formula are all clear, colorless liquids at ambient temperatures:

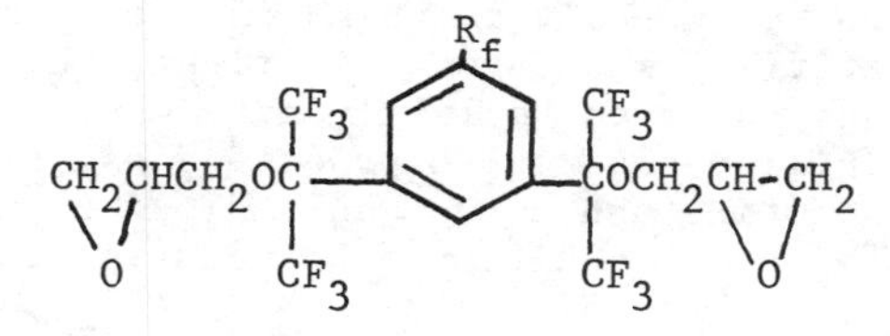

(I)

Amine-bearing siloxanes of the following type are also liquids:

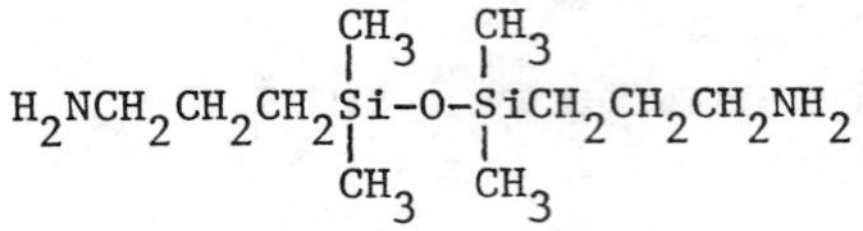

$$H_2NCH_2CH_2CH_2\overset{CH_3}{\underset{CH_3}{Si}}\text{-O-}\overset{CH_3}{\underset{CH_3}{Si}}CH_2CH_2CH_2NH_2 \quad \text{(II)}$$

In the conventional reaction between glycidyl ethers and amines, these compounds are difunctional and tetrafunctional respectively, which requires two moles of the fluoroepoxy for each of the silicone amine in a stoichiometric blend. This composition produces a tightly crosslinked network, but the following modification of the silicone amine produces an elastomeric composition because of the reduced functionality of the amine.

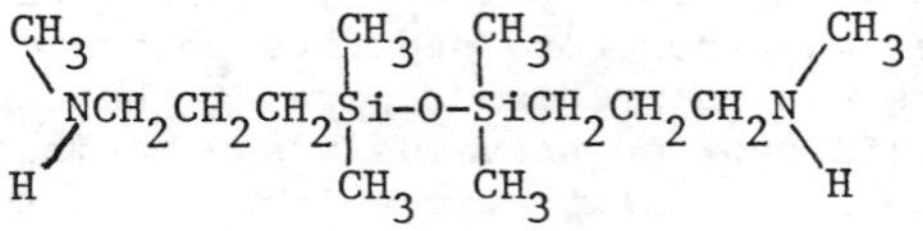

$$(CH_3)(H)NCH_2CH_2CH_2\overset{CH_3}{\underset{CH_3}{Si}}\text{-O-}\overset{CH_3}{\underset{CH_3}{Si}}CH_2CH_2CH_2N(CH_3)(H) \quad \text{(III)}$$

Blends of these two types of amines give curing agents capable of producing plastics upon reaction with the fluoroepoxy which are intermediate in properties between the extremes. The material produced from (I) and (III) alone is a nearly linear polymer in which fluorocarbon and silicone alternate along the polymer chains.

Experimental

A 25 X 150 mm test tube was charged with 14.0 g of (I) ($R_f = C_6F_{13}$) and 2.0 g of (II). Care was taken to expose the amine to the atmosphere for a minimum time in order to avoid carbon dioxide absorption. A small Teflon-coated magnetic stirring bar was dropped into the test tube and it was sealed tightly with a rubber stopper. The test tube was then clamped in a vertical position with the lower one-half immersed in a silicone oil bath at 50°C, and rapid stirring was begun. At first, the composition was incompatible and appeared "milky". After stirring for 20 minutes, the incompatibility cleared to produce a colorless transparent syrup. At this time the syrup would cloud if cooled to room temperature, but after an additional 20 minutes stirring at 50°C, it would remain clear at 25°C.

This prepolymer syrup was divided into two parts. One portion was dissolved in trifluorotrichloroethane as a solvent and used to produce protective coatings of high hydrophobicity. The other portion was allowed to cure for 24 hours at 25°C during which gelation occurred to produce a clear, nearly colorless

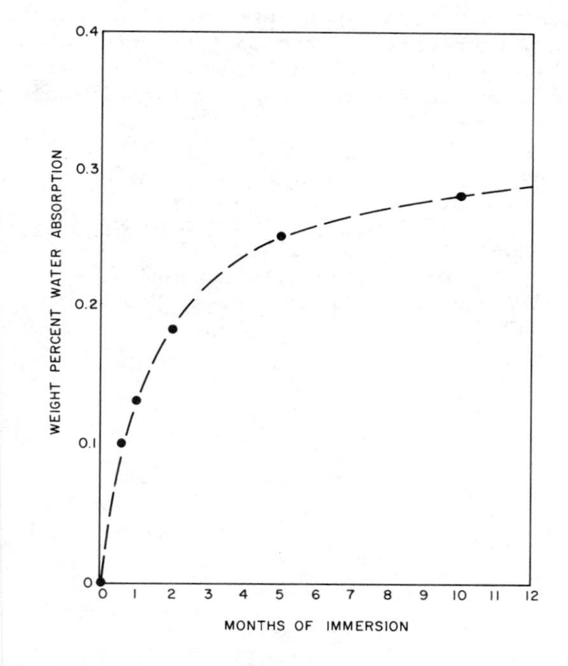

Figure 1. Water absorption of a silicone amine-cured fluoroepoxy during 1 yr of constant immersion

plastic. The strength of this plastic was substantially enhanced by a postcure at 60°C for 5 hours.

Conclusions

The materials presented here are easily processed resins of the epoxy class which are composed of two of the most water-resistant polymer types known. Consequently, it is not surprising that the water absorption of such resins in the cured form would be minimal, and, indeed, Figure 1 illustrates the very low water absorption of such a resin during a year's constant immersion. In the precured state the resin components are liquids of very low surface tension and, as such, are excellent wetting fluids. It is suggested, therefore, that composite structures which entail difficult soak-in problems or which encounter a serious problem of water degradation in service could profitably use these systems.

Literature Cited

1. O'Rear, J. G.; Griffith, J. R.; Organic Coatings and Plastics Preprints, April 1973, 33, No. 1, 657.
2. Griffith, J. R.; O'Rear, J. G.; Reardon, J. P.; Adhesion Science and Technology, Plenum Press, New York, 1975, p. 429.

RECEIVED February 15, 1980.

5

Phenylated Polyimidazopyrrolones

Polymerization of a Phenylated Bis(phthalic anhydride) with Aromatic Tetraamines

FRANK W. HARRIS, RODNEY M. HARRIS, MICHAEL KELLER, and WILLIAM A. FELD

Department of Chemistry, Wright State University, Dayton, OH 45435

Polyimidazopyrrolones have been shown to retain useful mechanical properties at elevated temperatures, after severe chemical treatment, and after unusually high exposure to ionizing radiation (1,2). Although this combination of properties suggests numerous potential aerospace applications, no practical uses for these polymers have emerged. This is primarily due to difficulties associated with their fabrication that result from their extremely high Tg's and their very limited solubilities.

One approach to obtaining solubility in linear aromatic polymers has been to incorporate pendent phenyl groups along the polymer backbone. For example, phenylated polyphenylenes (3) and phenylated polyquinoxalines (4,5) are soluble in common organic solvents, which is in marked contrast to the insolubility displayed by their parent polymers. Soluble phenylated polyimides have been prepared in this laboratory by the polymerization of phenylated dianhydrides with aromatic diamines (6,7). The objective of this research was the synthesis of phenylated polyimidazopyrrolones. The approach involved the polymerization of a phenylated dianhydride with aromatic tetraamines.

RESULTS AND DISCUSSION

Monomers. 4,4'-(Oxydi-1,4-phenylene)bis(3,5,6-triphenylphthalic anhydride) (I) was prepared from 3,3'-(oxydi-1,4-phenylene)bis (2,4,5-triphenylcyclopentadienone) and maleic anhydride by the known procedure (6,7). 3,3',4,4'-Tetraaminobiphenyl (IIa) and 3,3',4,4'-tetraaminodiphenyl ether (IIb) were obtained from Burdick and Jackson Laboratories. These monomers were recrystallized from water and then sublimed immediately prior to use (8).

0-8412-0567-1/80/47-132-039$05.00/0

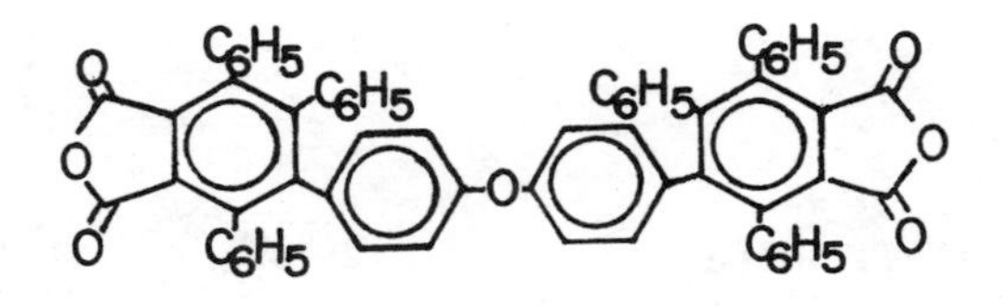

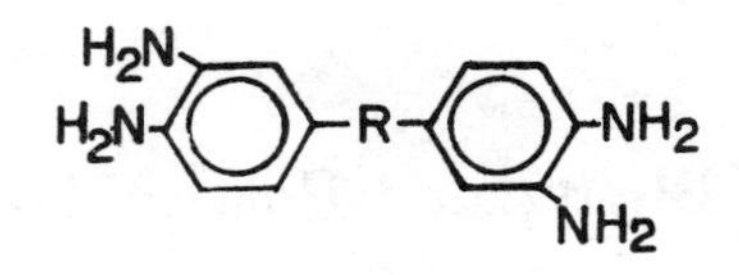

Model Compounds. In order to determine suitable experimental conditions for the polymerization of I and II, a series of model reactions was carried out. The model compounds obtained were thoroughly characterized so that their structures could be correlated with those of the desired polymers. The initial conditions used were those previously employed in the preparation of phenylated polyimides (6,7). Thus, the reaction of 3,4,5,6-tetraphenylphthalic anhydride (III) with 1,2-diaminobenzene (IV) in refluxing *m*-cresol containing isoquinoline afforded a 91% yield of the bright-yellow model compound (V). Similarly, the reaction of IIb with III under these conditions provided a 95% yield of a bright-yellow product. High pressure liquid

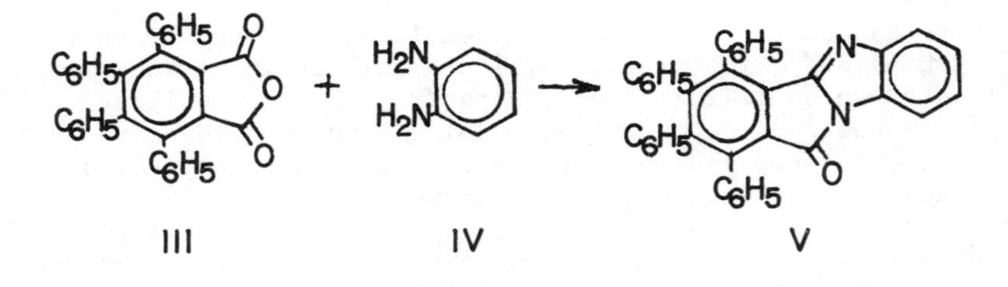

chromatography (HPLC), however, showed that the product was a complex mixture consisting of at least six components. Since such a mixture could arise from incomplete ring closures, the product was heated under high vacuum at 275-300°C for 4 hr. The infrared spectrum of the heat-treated sample was nearly identical to that of V indicating that complete cyclization had occurred. HPLC, however, still showed the presence of three components. Since different orientations of the reactants in the condensation reaction are possible, it is likely that the product consists of a mixture of the isomers VIa-c.

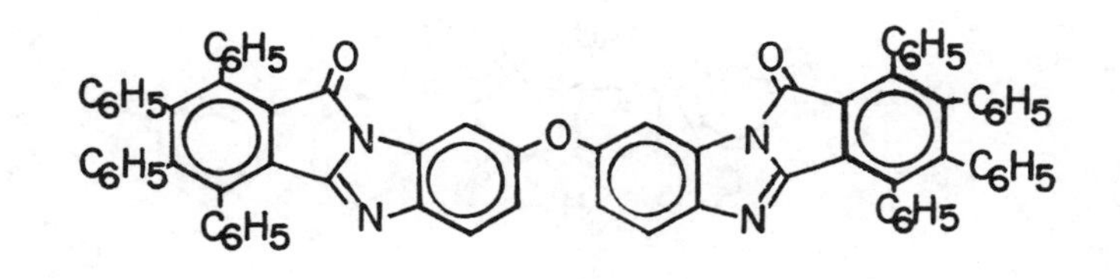

VI a

IIb + III →

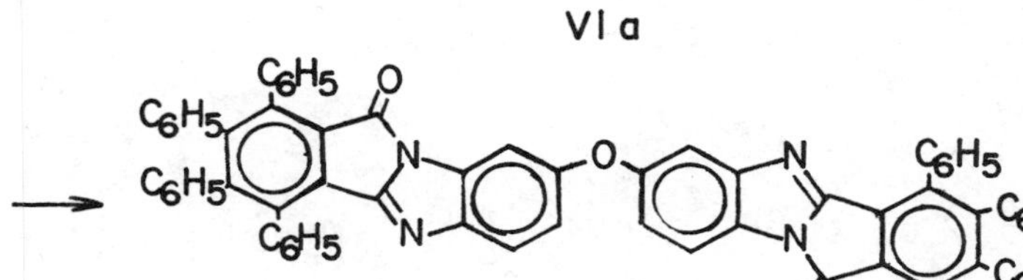

VI b

VI c

Polymerizations. Since the model reactions gave high yields in *m*-cresol containing isoquinoline, the polymerization of I and IIa was carried out in this solution. After a short heating period, a dark-yellow polymer precipitated from the reaction medium. This product was completely insoluble in organic solvents. Several subsequent attempts to polymerize I and IIa also resulted in insoluble products. Polymerizations of I and IIb in the *m*-cresol/isoquinoline mixture, however, did afford soluble products as long as the combined monomer concentration was no higher than approximately 3% (Table I). In these cases the polymerization mixtures were stirred at ambient temperature for 1 hr and then heated at 202-203°C for 4 hr. The water that evolved from the ring closures was continuously removed by distillation. The reaction mixtures were added to absolute ethanol to precipitate the dark-yellow polymer VII in nearly quantitative yields. Since the polymer's infrared spectra indicated that approximately 5% of the imidazopyrrolone rings were not closed, the samples were heated under high vacuum at 275-300°C for 4 hr to affect complete cyclization. The material with the highest inherent viscosity (VIIc) was obtained from the polymerization

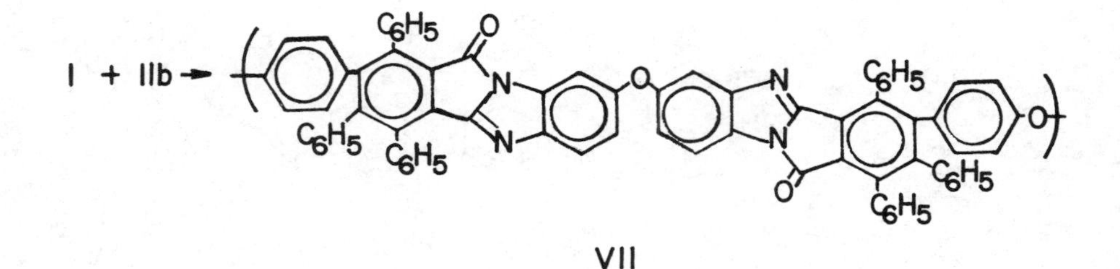

TABLE I

Preparation of Phenylated Polyimidazopyrrolones

Run No.	% Solids (w/v)	Polymer η_{inh}[a]
VIIa	1.79	0.27
VIIb	1.91	0.31
VIIc	3.00	0.73
VIId	6.25	gel
VIIe	12.50	gel

a. Determined in sym-tetrachloroethane at 30°C with a concentration of 0.250 g/dl.

mixture containing 3% solids.

The phenylated-polyimidazopyrrolone samples are soluble in chlorinated-hydrocarbon solvents. Thin films of sample VIIc cast from chloroform are yellow, tough, and flexible. These films also adhere tenaciously to glass. Films prepared with lower molecular weight material, however, are brittle and exhibit poor adhesive properties. The infrared spectra of the films show carbonyl absorptions at 1735 cm^{-1}, which is characteristic of polyimidazopyrrolones (1), and can be compared directly to the spectra of the model compounds. Thermogravimetric analysis (TGA) thermograms of VII obtained in nitrogen and air show 10% weight losses at 590 and 520°C, respectively (Figure 1). Hence, the polymer appears to be slightly more thermally stable than the previously prepared polyimidazopyrrolones (1).

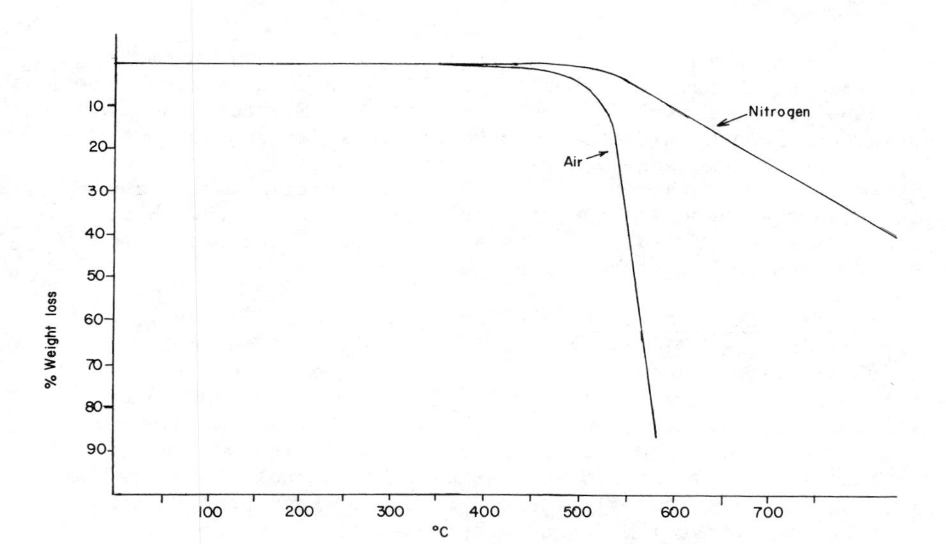

Figure 1. Thermogravimetric analysis of Polymer VII ($\Delta T = 5°C/min$)

CONCLUSIONS

Polyimidazopyrrolones that are soluble in chlorinated-hydrocarbon solvents can be prepared by the polymerization of a phenylated bis(phthalic anhydride) with aromatic diamines. The introduction of pendent phenyl groups along the polyimidazopyrrolone backbone, however, does not in itself result in solubility. In order to attain solubility, several flexible linkages, such as aryl-ether linkages, must also be simultaneously incorporated in the backbone. The polymerization of the dianhydride and the diamines must also be carried out in very dilute solutions in order to avoid crosslinking.

EXPERIMENTAL

Melting points were determined on a DuPont 900 Thermal Analyzer equipped with a DSC cell. Infrared spectra were recorded on a Perkin-Elmer 457 spectrophotometer. Elemental analyses were performed by Midwest Microlab, Inc., Indianapolis, Indiana. Thermogravimetric analyses were carried out on a DuPont 950 Thermogravimetric Analyzer. Inherent viscosities were measured in a Cannon-Ubbeholde microdilution viscometer No. 75. HPLC analyses were carried out on an Altex Model 332 Liquid Chromatograph using an ODS reverse-phase column.

Model Compound V: A solution of 1.500 g (3.315 mmol) of III, 0.3581 g (3.318 mmol) of IV, and 3.0 g of isoquinoline in 60 ml of deoxygenated *m*-cresol was heated at reflux under nitrogen for 18 hr. During this time the solution turned bright yellow. Approximately 50 ml of the *m*-cresol was removed under reduced pressure, and the precipitate that formed was collected by filtration. The solid was washed with ethanol and dried to give 1.5757 g (90.4%) of crude product. Recrystallization from benzene-ethanol afforded a bright-yellow powder: mp 347-349°C; ir (KBr) 1735 cm^{-1} (s,C=O).

Anal. Calcd for $C_{38}H_{24}N_2O$: C, 86.99; H, 4.62

Found: C, 86.72; H, 4.73.

Model Compound VI: A solution of 1.000 g (2.210 mmol) of III, 0.2687 g (0.1167 mmol) of II, and 1.0 g of isoquinoline in 21 ml of deoxygenated *m*-cresol was heated at reflux under nitrogen for 17 hr. After approximately 18 ml of the *m*-cresol was removed under reduced pressure, the residue was dissolved in 5 ml of chloroform and then added to ethanol to precipitate 1.1707 g (95%) of a dark-yellow solid. The solid was placed in a 250-ml, round-bottom flask that was immersed in a Woods-metal bath. The flask was evacuated and then heated at 275-300°C for 4 hr. The product was recrystallized 4 times from benezene-ethanol to afford a bright-yellow powder: mp 265-270°C; ir (KBr) 1735 cm^{-1} (s,c=o).

Anal. Calcd for $C_{76}H_{46}N_4O_3$: C, 85.85; H, 4.36

Found: C, 86.02; H, 4.14.

General Polymerization Procedure: The tetraamine and deoxygenated m-cresol containing 0.05 g/ml of isoquinoline were placed in a 100-ml, 3-necked flask equipped with a magnetic stirring bar, a nitrogen inlet, a short-path distillation apparatus and a stopper. After the tetraamine had dissolved, the dianhydride was added in several portions over 1 hr. The final portion of the dianhydride was dissolved in 1 ml of m-cresol, and the solution added to the reaction mixture. The stopper was replaced with an addition funnel, and the temperature of the flask was slowly increased until distillation commenced. The volume of the mixture was kept essentially constant by continually replacing the distillate with m-cresol containing 0.05 g/ml of isoquinoline. The distillation-addition cycle was carried out for 4 hr. After the yellow, viscous solution was allowed to cool, it was slowly added to vigorously-stirred absolute ethanol, collected, and air dried to afford a nearly quantitative yield of product. The dark-yellow polymer was then heat treated to insure complete cyclization. Thus, a film of the polymer was cast from a 5% chloroform solution on the inside of a 250-ml, round-bottom flask by slowly removing the solvent under reduced pressure. The flask was immersed in a Woods-metal bath, evacuated, and heated at 275-300°C for 4 hr. The heat-treated polymer was precipitated from chloroform with absolute ethanol, collected, and dried under vacuum at 153°C for 48 hr.

LITERATURE CITED

1. Bell, V.L. in "Encyclopedia of Polymer Science and Technology," Vol. II, Mark, H. F., Gaylord, N. G.and Bikales, N. M. ,Eds., Wiley, New York, 1969, p. 240.
2. Bell, V. L. and Jewell, R. A., J. Polym. Sci., Part A-1, 1967, 5, 3043.
3. Mukamal, H., Harris, F. W. and Stille, J. K., J. Polym. Sci., Part A-1, 1967, 5, 2721.
4. Wrasidlo, W., and Augl, J. M., J. Polym. Sci., Part A-1, 1969, 7, 3393.
5. Hergenrother, P. M., and Levine, H. H., J. Polym. Sci., Part A-1, 1967, 5, 1453.
6. Harris, F. W., Feld, W. A., and Lanier, L. H., J. Polym. Sci., Polym. Letters Ed., 1975, 13, 283.
7. Harris, F. W., Feld, W. A., and Lanier, L. H., in "Applied Polymer Sumposium No. 26," Platzer, N., Ed., Wiley, New York, 1975, p. 421.
8. Foster, R. T., and Marvel, C. S., J. Polym. Sci., Part A, 3, 417.

RECEIVED February 8, 1980.

6

Synthesis and Properties of Fluoroalkylarylene-siloxanylene (Fasil) High-Temperature Polymer

HAROLD ROSENBERG and EUI-WON CHOE[1]

Nonmetallic Materials Division, Air Force Materials Laboratory, Wright–Patterson Air Force Base, OH 45433

The discovery of new polymeric materials for sealant and seal applications with inherently wider use temperature range, greater chemical stability and longer operational life than state-of-the-art elastomers has proven to be an unyielding challenge to the Air Force for over a decade. Of particular concern has been the need for new viscoelastic polymers with the requisite chemical/fuel resistance, high-temperature stability, low-temperature flexibility, adhesion to metal substrates and ready processability for improved aircraft integral fuel tank sealants with broad temperature capability. The background, requirements and status of research programs aimed at this specific materials objective, with an emphasis on filleting sealants, was reviewed recently[1].

This quest has now led to the synthesis and evaluation of a new fluorine-containing organosilicon polymer, FASIL, which offers considerable promise as a candidate base material for a broad use-temperature, long-life integral fuel sealant of both the channel and filleting types. Of particular interest is the potential shown by this elastomeric polymer for the formulation of a non-curing, reversion-resistant $-54^{o}C$ ($-65^{o}F$) to $232^{o}C$ ($450^{o}F$) fuel tank channel sealant with long-term utility. This development was the planned outgrowth of an extensive investigation into the synthesis and characterization of new thermally-and chemical resistant viscoelastic alkarylene- and arylenepolysiloxanylene polymers. Two classes of candidate polymers, methyl- and

[1] Present address: Celanese Research Company, Summit, NJ 07901

3,3,3-trifluoropropyl-substituted poly(m-xylylenesiloxanylenes) and the corresponding poly(m-phenylenesiloxanylenes), were selected for study as the most promising representatives of such macromolecular systems. Emphasis was placed in both types on di-, tri- and tetrasiloxanylene subclasses since these would be expected to yield polymers with the balance of physical, chemical and mechanical properties required for fuel tank sealant applications. In order to achieve the latter, it was necessary to develop appropriate structure/property correlations as a prerequisite to the specific tailoring of molecular structures. Representative members, appropriately substituted, of both polymer classes were synthesized and characterized with respect to their glass-transition temperature, thermal stability and solubility or fuel resistance. From the data obtained, required structure/property relationships were established to permit the prediction of structures with the optimum combination of properties.

The synthesis and characterization[2], thermal behavior[3], and solubility or fuel resistance[4] of the first class of polymers, i.e., the poly(m-xylylenesiloxanylenes), has been reported previously. In this paper some results of the investigation of the second class, i.e., the poly(m-phenylenesiloxanylenes), I, with

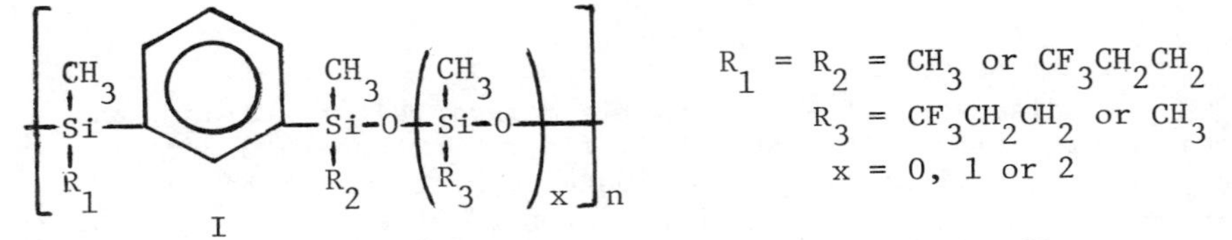

emphasis on the synthesis and characterization of the fluoroalkyl-substituted members of the subclasses which led to the selection of FASIL as the candidate polymer of choice, are summarized.

Result and Discussion

Synthesis. As in the case of the poly(m-xylylenesiloxanylenes)[2], in order to obtain appropriately-structured representative members for characterization and evaluation, three subclasses or families of m-phenylenesiloxanylene polymers were considered for synthesis. These were the poly(m-phenylenedisiloxanylenes, II; -trisiloxanylenes, III; and -tetrasiloxanylenes, IV. The specific polymers in these subclasses selected for preparation were those considered likely to provide the most useful information in connection with structure/property correlations and the tailoring of the macromolecules. In order to synthesize these polymers, various methyl- and 3,3,3-trifluoropropyl-substituted 1,3-bis(silyl)benzene intermediates and monomers, including bis-chlorosilanes, bis-ethoxysilanes and bis-silanols, together with bis-amino-silanes and -disiloxanes, were initially prepared and characterized[5]. The bis-silanols were homopolymerized in refluxing benzene, using a 1,1,3,3-tetramethylguanidine salt as catalyst, to yield the poly(m-phenylenedisiloxanylenes), II. When the two bis-silanols, 1,3-bis(hydroxydimethylsilyl)- and 1,3-bis[hydroxymethyl(3,3,3-trifluoropropyl)-silyl] benzene were copolymerized with appropriately substituted bis(dimethylamino)silanes (Figure 1) or -disiloxanes (Figure 2), two families of methyl- and 3,3,3-trifluoropropyl substituted poly(m-phenylenetrisiloxanylenes), III, and -tetrasiloxanylenes, IV, were obtained. Recently, in an extension of the newly developed silanol-acetoxysilane polycondensation reaction for the synthesis of dimethyl-substituted poly(arylenesiloxanylenes),[6] it was found possible to synthesize the corresponding methyl (3,3,3-trifluoropropyl) analogs. For example, by the reaction of 1,3-bis-[hydroxymethyl(3,3,3-trifluoropropyl)silyl] benzene with the appropriately-substituted diacetoxysilanes or diacetoxydisiloxanes, III($R_1 = R_2 = R_3 = R_4 = CF_3CH_2CH_2$) and IV($R_1 = R_2 = R_3 = R_4 = CF_3CH_2CH_2$) were obtained. Since a number of the newly synthesized polymers [e.g.,III($R_1 = R_2 = CF_3CH_2CH_2$, $R_3 = CH_3$)] could not be cured by conventional methods to provide vulcanizates for fuel swell evaluation, vinylsilyl groups were introduced into such polymers in order to provide sites for crosslinking. This was accomplished by the addition of methylvinylbis(dimethylamino) silane as 3 mole percent of the total silylamine monomer composition in the reactions shown in Figure 1 and 2. The vinyl groups incorporated in the resulting polymers would be expected to be distributed randomly along the chains.

Physical Properties. The synthesized polymers, as with their m-xylylene analogs, were fully characterized and evaluated with respect to glass-transition temperature, thermal stability and fuel resistance or polymer insolubility. Viscosities, together with number average molecular weights obtained by either vapor phase or membrane osmometry, were recorded for the relatively low molecular weight polymers sought. Glass-transition temperatures were determined by means of differential scanning calorimetry (DSC), with $\Delta T = 20^\circ C/min$, and have been reported in part.[7] Thermal characterization of the polymers involved thermogravimetric analysis (TGA) under vacuum at $\Delta T = 5^\circ C/min$. Data obtained from the TGA curves, including values for T_{25} (temperature at which 25% weight loss is recorded), were discussed for certain of the polymers, together with that for their m-xylylene analogs, in an earlier report[3]. In order to determine fuel resistance, polymers were first cured with di-t-butylcumyl peroxide at $170^\circ C$ and 2000 psi. In the case of those polymers with a high content of fluoroalkyl groups, samples of polymers containing the aforementioned 3 mole percent of vinyl groups were used for preparation of the vulcanizates. Volume swell ratios of the volcanizates were determined after immersion of the samples for 72 hours at room temperatures in hydrocarbons, such as isooctane and JP-4 jet fuel. The results obtained with regard to

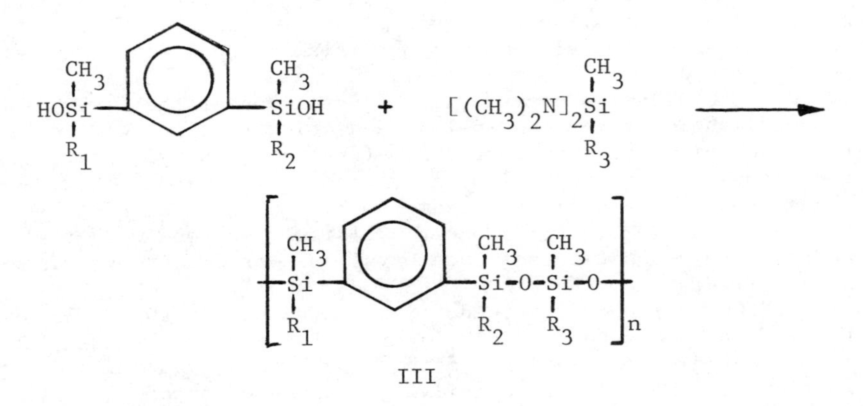

$R_1 = R_2 = CF_3CH_2CH_2$ or CH_3; $R_3 = R_4$ = Same as or different R_1.

Figure 1. Copolymerization of bis*silanols with* bis*(dimethylamino)silanes*

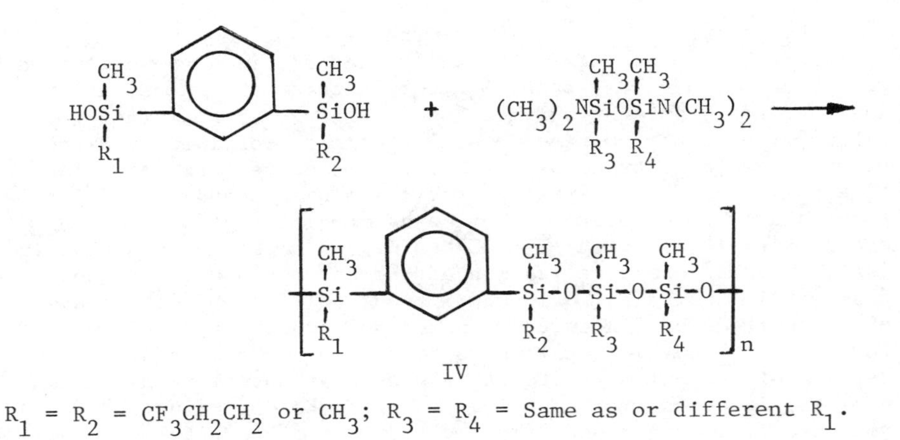

Figure 2. Copolymerization of bis*silanols with* bis*(dimethylamino)disiloxanes*

the solvent resistance of the polymers through the use of vulcanizates has been corroborated, in part, in studies on solvent interactions with both the m-xylylene- and, to a lesser extent, the m-phenylenesiloxanylene polymers. Using gas-liquid chromatography (GLC) to determine infinite dilution activity coefficients, structure/solubility relationships were derived which correlated well with volume swell data on volcanizates[4].

From the results obtained and structure/property correlations[2,3,4] previously established in our investigation of their m-xylylene analogs, it was possible to reduce the total number of poly(m-phenylenesiloxanylenes) required for synthesis and property evaluation. More specifically, from the earlier fuel resistance/polymer solubility studies, a fluorine content of 30% or greater by weight appeared necessary to provide required fuel resistance to the polymer systems under investigation. Therefore, in the poly(m-phenylenesiloxanylenes) only those polymers with a high ratio of 3,3,3-trifluoro-propyl to methyl substituents could be expected to possess desired fuel resistance. Thus, the only members of this class with potential for fuel tank sealants were further restricted to five in number, consisting only of those with two or more fluoroalkyl groups per mer unit. Further detailed evaluation of this smaller group of polymers in both uncured and cured states and with emphasis on isothermal aging and solvent resistance at elevated temperatures was then carried out. This resulted in the final selection of poly[m-phenylene-1,3,5,7-tetramethyl-1,3,5,7-tetrakis(3,3,3-trifluoropropyl)tetrasiloxanylene] as the most promising member not only of the group but of both classes of polymers investigated.

FASIL Evaluation. The aforementioned fluoroalkylarylenesiloxanylene polymer, IV($R_1 = R_2 = R_3 = R_4 = CF_3CH_2CH_2$) has been given the name FASIL, an acronym derived from the general polymer class name. In addition to being readily synthesized from commercially-available starting materials (cf. Experimental), the polymer was found to exhibit a better overall combination of requisite physical, chemical and mechanical properties for fuel-resistant sealant applications than any of the newly synthesized siloxanylene or other candidate sealant polymers. Additional quantities of FASIL were synthesized for use in formulation and evaluation studies as a fuel tank channel sealant. Preliminary results of FASIL evaluation as a formulated sealant, including accelerated broad-temperature range cycling tests from $-54^{\circ}C$ ($-65^{\circ}F$) up to $232^{\circ}C$ ($450^{\circ}F$) and titanium stress corrosion testing, indicate that FASIL possesses superior properties (Figure 3) and life expectancy to other elastomeric polymers evaluated as a candidate for a wide-temperature range fuel tank channel sealant material. In addition to holding promise as a channel sealant polymer, structural modification of FASIL as a two-part system is expected to permit its use for the formulation of filleting sealants. Under a contract with Midwest Research Institute, FASIL-

MAJOR	MINOR
• Long-Term Thermal Stability [Up to 260°C (500°F)]	• Reversion Resistance [204°C (400°F), 7 days total
• Low-Temperature Flexibility [-54°C (-65°F)]	• Extrudable
• Excellent Adhesion to Al & Ti	• No Stress Corrosion on Ti Substrates
• Chemical Stability High Resistance to JP-4 Fuel Nonhydrolyzable	• Reasonable Cost Synthesis Based on Commerc.-Available Starting Materials

Figure 3. Properties of FASIL required for broad-temperature fuel tank sealants

type vinyl-terminated oligomers, chain extenders and crosslinking agents have been synthesized to enable the early development of a FASIL-based filleting sealant. Lastly, from the properties obtained in the evaluation of its vulcanizates, FASIL offers further potential as a base polymer for chemical-resistant seal applications.

Experimental

The following examples of FASIL monomer and polymer synthesis are given to illustrate the general procedures used in the preparation of the m-phenylenesiloxanylene polymers.

1,3-Bis[hydroxymethyl(3,3,3-trifluoropropyl)silyl]benzene. To a stirred mixture of 2.6 g of sodium hydroxide, 11 ml of methanol and 1.2 ml of water was added 7 g(0.016 mole) of 1,3-bis[ethoxymethyl(3,3,3-trifluoroprophyl)silyl]benzene [obtained from the reaction of the corresponding bis-chlorosilane with sodium ethoxide or from the in-situ Grignard reaction of m-dibromobenzene, magnesium and diethoxymethyl-(3,3,3-trifluoropropyl)-silane], followed by the addition of a solution of 2.69 g of sodium hydroxide and 12 ml of water. The resulting mixture was stirred for 30 minutes and then poured into a solution of 17 g of potassium dihydrogen phosphate in 30 g of water and ice. After the oily product was extracted with diethyl ether, the ether extract was washed with water, dried over anhydrous magnesium sulfate and evaporated on a rotary evaporator. Fractional distillation of the clearly oily residue through a short-path distillation column gave 5.6 g (91.5%) of 1,3-bis[hydroxymethyl-(3,3,3-trifluoropropyl)silyl]benzene: bp 140-142 /0.1 Torr. Found for $C_{14}H_{20}F_6O_2Si_2$: C, 43.06; H, 5.16.

1,3-Bis(dimethylamino)-1,3-dimethyl-1,3-bis(3,3,3-trifluoropropyl)disiloxane. To a 2-l. four-necked flask, cooled to -27°, equipped with a mechanical stirrer, Dry Ice-acetone condenser, thermometer and addition funnel, and containing 500 ml of petroleum ether (b.p. 30-60°), was added 18 g (0.4 mole) of anhydrous dimethylamine under a nitrogen atmosphere. A solution of 36.7 g (0.1 mole) of 1,3-dichloro-1,3-dimethyl-1,3-bis(3,3,3-trifluoropropyl)disiloxane was added to the dimethylamine solution over a 1-hr period while the reaction temperature was maintained at -27°. The mixture was stirred for an additional 30 minutes and then was allowed to warm to room temperature. Excess dimethylamine was passed through a hydrochloric acid solution and the amine salt removed by filtration. Fractional distillation of the filtrate gave 26 g (68%) of 1,3-bis(dimethylamino)-1,3-dimethyl-1,3-bis-(3,3,3-trifluoropropyl)disiloxane: bp 237-240°. Found for $C_{12}H_{26}F_6N_2OSi_2$: C, 37.49; H, 6.82.

Poly [m-phenylene-1,3,5,7-tetramethyl-1,3,5,7-tetrakis (3,3,3-trifluoropropyl) tetrasiloxanylene] with vinyl groups. Into a 100-ml. three-necked flask, equipped with thermometer, stirrer, nitrogen inlet, and condenser which was connected to a solution of hydrochloric acid, was placed 5.0 grams (12.8 mmoles) of 1,3-bis[hydroxymethyl(3,3,3-trifluoropropyl)silyl]benezene, 4.275 grams (11.16 mmmoles) of 1,3-dimethyl-1,3-bis(3,3,3-trifluoropropyl)-1,3-bis(dimethylamino)disiloxane, 54.9 mg (0.36 mmoles) of methylvinylbis(dimethylamino)silane, and 5 ml. of toluene. The mixture was warmed to 110° during which period dimethylamine began to evolve. After 15 minutes of warming, the remainder of the diaminosilanes, 0.475 grams (0.12 mmoles) of 1,3-dimethyl-1,3-bis-(3,3,3-trifluoropropyl)-1,3-bis(dimethylamino)disiloxane and 6.1 mg (0.04 mmoles) of methylvinylbis(dimethylamino)silane, was added dropwise over a period of 5 minutes. The mixture was heated for 1.5 hours (or for a total of two-hours reaction time) and then hydrolyzed with 1 ml. of water. Water was removed as an azeotrope. The polymer was dissolved in 3 ml of toluene, filtered, precipitated with 15 ml. of methanol, washed twice with 30 ml. of methanol, and dried at 260° and 0.1 Torr for 16 hours to yield 4.2 grams of poly[m-phenylene-1,3,5,7-tetramethyl-1,3,5,7-tetrakis(3,3,3-trifluoropropyl)-tetrasiloxanylene](with 3 mole percent of vinylmethylsiloxy groups incorporated randomly onto the polymer backbone) as a very viscous oil, molecular weight (VPO in THF) 3,700; T_g (DSC, $\Delta T=20^{o}$/min.) = -49°C; transparent; $\eta_{inh} \doteq 0.02$ dl/g.

Acknowledgment

The support of one of us (E.W.C) by the Air Force Systems Command through an NRC Postdoctoral Research Associateship is gratefully acknowledged. The authors thank Lt Col Russell M. Luck for assistance in the polymer synthesis and are indebted to Warren Griffin, Elastomers and Coatings Branch, AFML, for the sealant formulation and evaluation studies.

Literature Cited

1. Anspach, W. F. Chemtech, 1975, 752.
2. Rosenberg, H. and Choe, E. W. Coatings and Plastics Preprints, 1977, 37(1), 166; ibid., Carraher, C. E., Sheats, J. E., and Pittman, C. U., Jr., Eds. "Organometallic Polymers", Academic Press: New York, N. Y., 1978.
3. Goldfarb, I. J., Choe, E. W. and Rosenberg, H., Coatings and Plastics Preprints, 37(1), 172 (1977); ibid., Carraher, C. E., Sheats, J. E., and Pittman, C. U., Eds. "Organometallic Polymers", Academic Press: New York, N. Y., 1978.
4. Bonner, D. C., Chen, K. C. and Rosenberg, H., Polymer Preprints, 1976, 17(2), 372; ibid., Harris, F. W., and Seymour, R. B., Eds. "Structure-Solubility Relationships in Polymers",

Academic Press: New York, N. Y., 1977.
5. Rosenberg, H. and Choe, E. W., Abstracts of Papers, 172nd Nat. Meeting, Amer. Chem. Soc., San Francisco, CA, August 1976, FLUO 41.
6. Rosenberg, H. and Nahlovsky, B. D. Polymer Preprints, 1978, 19(2), 625.
7. Ehlers, G. F. and Fisch, K. R. "Correlations Between Polymer Structure and Glass-Transition Temperature," AFML-TR-75-202, Part 1, October 1975.

RECEIVED January 8, 1980.

ADHESIVES, COATINGS, AND SEALANTS

7

Analysis of Adhesive Fracture Testing Methods for Aerospace Use

K. KAWATA, H. FUKUDA, N. TAKEDA, and A. HONDO

Institute of Space and Aeronautical Science, University of Tokyo, 4-6-1 Komaba, Meguro-ku, Tokyo 153, Japan

The determination of adhesive bond strength and testing methods associated with the bond strength are the important subjects in composite materials and structures, especially their application in aerospace use. In aerospace materials, peeling tests are important as well as usual bond shearing test. The apparent strength obtained from the bond shearing test is $\tau_{cr} = P/S$ where P is the breaking load (kg) and S is the bonded area (mm^2) while the peeling strength obtained from the peeling test is P/b where P is the breaking load (g) and b is the width (mm). These two so called "strengths" are different in dimensions and have not been related to each other. As a result, for practical problems these two tests had to be conducted individually to qualify these two strengths.

In the present paper, these two "strengths" are related by the analyses from the unified standpoint. That is, the critical strength based upon the energy balance concept (1) is used to establish a relation between these two strengths of adhesive bonded joint and to demonstrate the interchangeability of these data (2, 3). There are many types of adhesive joints or bond strength testing methods as shown in Fig. 1. T-test is treated similarly with θ degree peeling test (2). Single-lap (3), tapered-lap and scarf-joint (4) are treated similarly. Analysis of this type may be extended to pull-out test (4).

When we survey historically briefly, we cite the pioneering works on energy in adhesive peeling by T. Hata (5). Adhesive fracture energy was used by H. Dannenberg (6), B. M. Malyshev and R. L. Salganik (7), and J. D. Burton, W. B. Jones and M. L. Williams (8) to study blister test.

In the followings, analytical results and comparisons with experimental results are mentioned for the tests shown in Fig. 1. Next, a new method to determine dynamic fracture toughness, that is, adhesive fractured surface energy (11) is proposed.

0-8412-0567-1/80/47-132-059$05.00/0

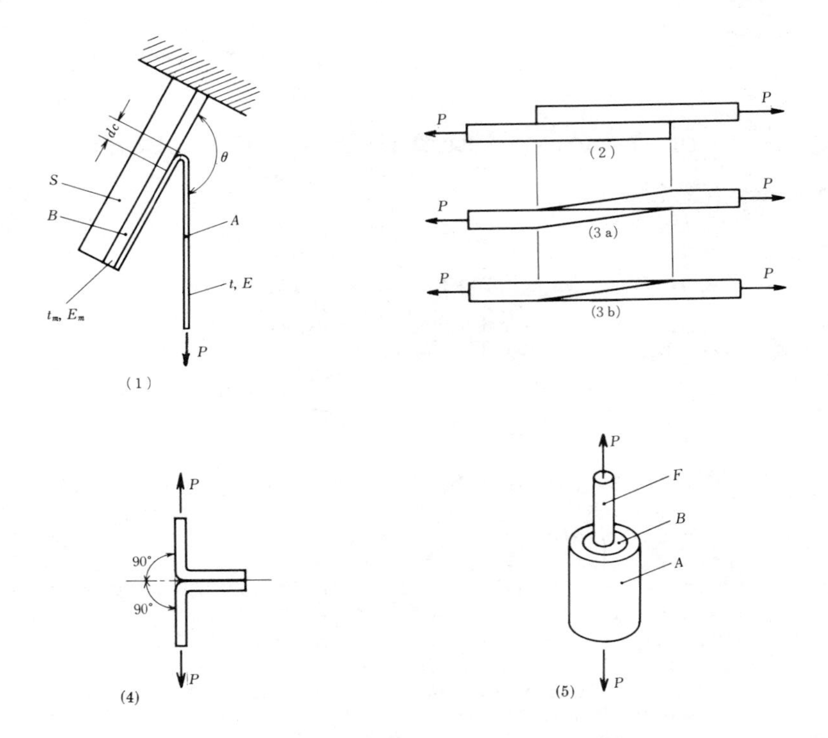

Figure 1. Adhesive bond strength testing methods.

(1) Θ degree peeling test: P, *load;* A, *bonding strip;* B, *metal plate;* S, *supporting plate for restricting bending of* B; t, t_m, *thickness;* E, E_m, *Young's modulus; suffix* m *indicates metal; width is* b. dc: *length of incremental peeling.* Θ: *peeling angle. (2) Single-lap adhesive joint shearing test. (3a) Tapered-lap adhesive joint shearing test. (3b) Scarf-joint shearing test. (4) T-test. (5) Pull-out test:* F, *fiber;* A, *adherend;* B, *adhesive.*

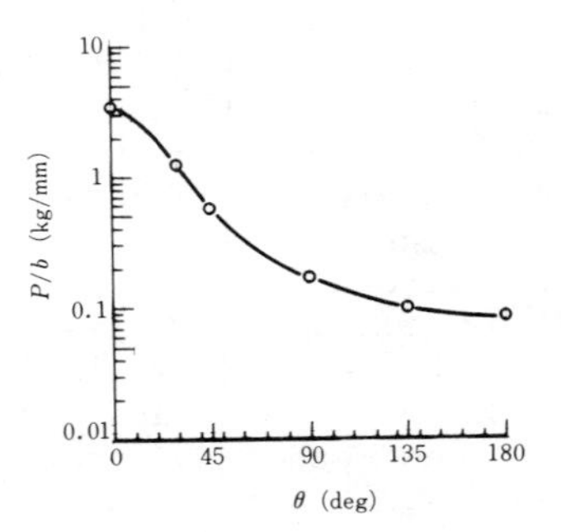

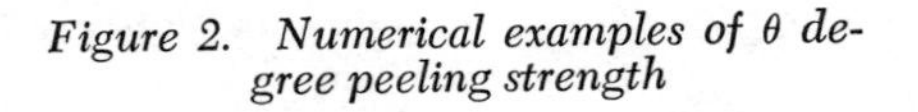
Figure 2. Numerical examples of θ degree peeling strength

θ Degree Peeling Test

Analysis of θ degree peeling of a soft material bonded to a plate with bending restrain is stated. The following assumptions are made: 1) A is soft and its flexural rigidity is negligible. 2) Stress is concentrated locally in the neighbourhood of dc and regions far from dc have no influence. The critical condition for peeling fracture is

$$dW + dU + dL = 0 \qquad (1)$$

where, the variations of surface energy, elastic strain energy, and potential energy of load due to peeling of length dc are denoted dW, dU, and dL, respectively. From (1), introducing adhesive fractured surface energy γ,

$$P^2\left(\frac{1}{Et} + \frac{\cos^2\theta}{E_m t_m}\right) + 2bt(1 - \cos\theta)P - 4\gamma b^2 = 0 \qquad (2)$$

is obtained. Solving (2), θ degree peeling strength is determined as follows:

$$P_{cr}/b = \begin{cases} \sqrt{4\gamma Et/(1 + \frac{Et}{E_m t_m})} & (\theta = 0 \text{ degree}) \\ 2\gamma/(1 - \cos\theta) & (\text{except in the neighbourhood of 0 degree}) \end{cases} \qquad (3)$$

The peeling strength for θ = 0, that is, the shearing strength varies with the rigidity of the bonded material, in contrast to the case except in the neighbourhood of 0 degree. This suggests an essential difference between peeling with a force component perpendicular to the plane of the adhesive and peeling with only a force component parallel to the plane of the adhesive. Non-zero degree peeling strengths are smaller than zero degree peeling strength that is pure shear strength by the order of 10 or 100, as seen in the numerical examples (Fig. 2).

T-test

Under the assumptions and the critical condition, same with θ degree peeling test, the following solution is obtained.

$$P_{cr}/b = \gamma \qquad (4)$$

Single-lap Adhesive Joint Shearing Test

A tensile load N produces axial forces N_1 and N_2 of the adherents (1) and (2) respectively, and a shearing stress τ in the adhesive layer. Assuming Hooke's law and disregarding bending effects due to the eccentric load path, shear fracture load is

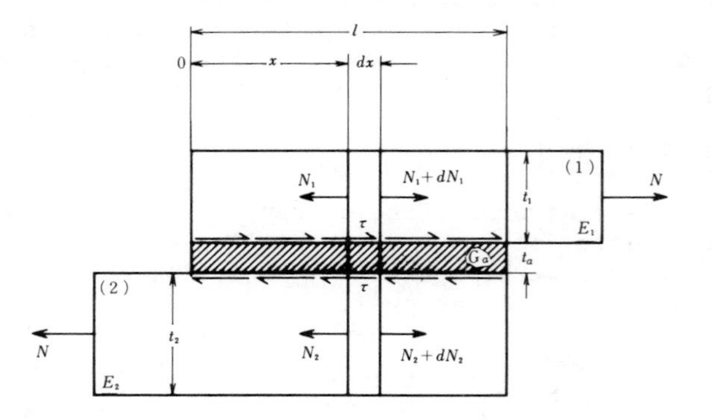

Figure 3. Stresses in a single-lap adhesive joint (unit width perpendicular to the paper): E, *Young's modulus;* t, *thickness of an adherend;* G_a, *shear modulus of adhesive;* t_a, *thickness of adhesive;* l, *overlap length;* N, *axial force,* τ, *shearing stress.*

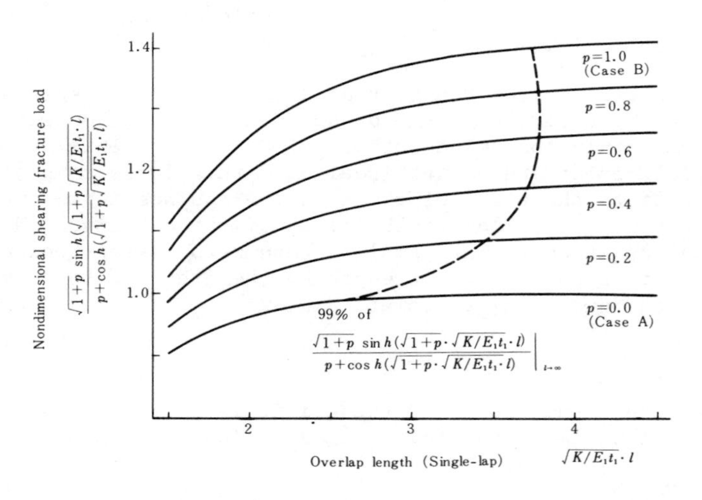

Figure 4. Nondimensional shearing fracture load vs. overlap length for single-lap joint

derived from the same critical condition.

$$N_{cr} = \sqrt{4\gamma E_1 t_1}\; f\left(\sqrt{\frac{K}{E_1 t_1}}\,\ell,\ p\right) \tag{5}$$

where, $$K = G_a/t_a \quad , \quad p = E_1 t_1/E_2 t_2 \leqq 1 \tag{6}$$

$$f\left(\sqrt{\frac{K}{E_1 t_1}}\,\ell,\ p\right) = \frac{\sqrt{1+p}\ \mathrm{sh}\left(\sqrt{1+p}\sqrt{\frac{K}{E_1 t_1}}\,\ell\right)}{p + \mathrm{ch}\left(\sqrt{1+p}\sqrt{\frac{K}{E_1 t_1}}\,\ell\right)} \tag{7}$$

(7) is designated as specimen characteristic factor. Case A, B, and C are shown in Table 1. Case A is an improved analysis of the 0 degree peeling test. The distinctive feature of single-lap adhesive joint is that as ℓ increases f converges to the limit of $(1+p)^{1/2}$. The present solution coincides fairly well with de Bruyne's experimental tendencies (9)(Fig. 5).

Table 1 Specimen Characteristic Factor

Case	$f\left(p,\ \sqrt{\frac{K}{E_1 t_1}}\,\ell\right)$
A. $E_1 t_1 \ll E_2 t_2$	$\tanh\left(\sqrt{\frac{K}{E_1 t_1}}\,\ell\right)$
B. $E_1 t_1 = E_2 t_2$	$\sqrt{2}\tanh\left(\sqrt{\frac{K}{E_1 t_1}}\,\ell/\sqrt{2}\right)$
C. General	$\dfrac{\sqrt{1+p}\,\sinh\left(\sqrt{1+p}\sqrt{\frac{K}{E_1 t_1}}\,\ell\right)}{p + \cosh\left(\sqrt{1+p}\sqrt{\frac{K}{E_1 t_1}}\,\ell\right)}$

Tapered-lap Adhesive Joint Shearing Test

This case (Fig. 6) is solved by the same way.

$$N_{cr} = \sqrt{4\gamma E t}\sqrt{\frac{K}{Et}}\,\ell \qquad (\text{for } E_1 = E_2 = E,\ t_1 = t_2 = t) \tag{8}$$

N_{cr} is proportional to ℓ, differing from single-lap joint (Fig. 7). The experimental data by de Bruyne (9) support this solution (Fig. 8).

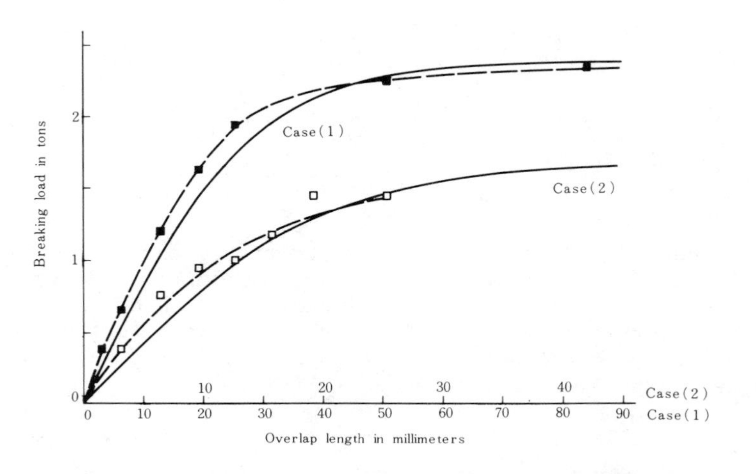

Figure 5. Comparison between (——) present solution and (■ □ — —) de Bruyne's experimental results

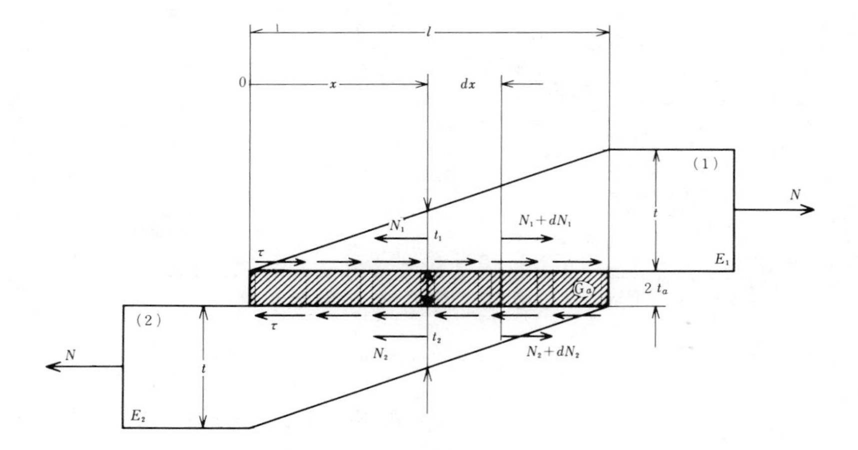

Figure 6. Stresses in a tapered-lap joint (unit width perpendicular to the paper)

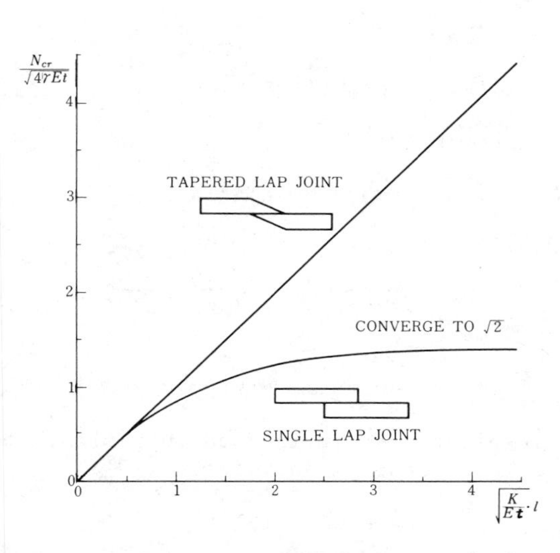

Figure 7. Nondimensional shearing fracture load vs. overlap length for single-lap and tapered-lap joints

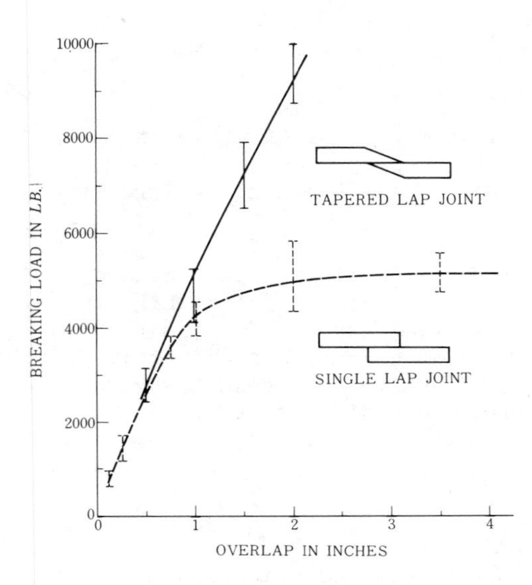

Figure 8. Experimental shearing fracture load of adhesive joints by de Bruyne

Pull-out Test

This axisymmetric case is also solved similarly (Fig. 9):

$$\sigma_{cr} = P_{cr}/\pi r^2 = \sqrt{4\gamma E_2/r}\ f(A\ell) \tag{9}$$

where,

$$A = \sqrt{2G/(E_2 br)} \cdot \sqrt{1+p}$$

$$f(A\ell) = (\sqrt{1+p}\ \mathrm{sh}A\ell)/(p + \mathrm{ch}A\ell) \tag{10}$$

$$p = E_2 r/\{2E_1 a(1 + (a/2r))\} > 0 \tag{11}$$

So, the relation of $\sigma_{cr}/\sqrt{4\gamma E_2/r}$ vs. $\sqrt{2G/(E_2 br)}\ \ell$ is the same with Fig. 4.

The Effect of Elastic-plastic Strip in Peeling Strength

When elastic-plastic strip is used in peeling test, the following correction factor is introduced (10):

$$P_{cr}/b = \begin{cases} \text{The same with (3)} & * \\ \{2\gamma/(1-\cos\theta)\}(1+\dfrac{t^2 Y}{4\rho\gamma}) & ** \end{cases} \tag{12}$$

* (θ = 0 degree)
** (except in the neighbourhood of 0 degree)

where, Y : yield stress of strip,
ρ : radius of curvature of strip at peeling point.

A New Method to Determine Dynamic Adhesive Fractured Surface Energy

As seen in the above mentioned, the value of γ is important. Using the specimens such as Fig. 1 (2) or Fig. 10, dynamic adhesive fractured surface energy γ_{ds} for shear or γ_{dt} for tear is measured by the method of comparing the energies of the impacting mass before and after the impact loading. By the energy method newly proposed, dynamic adhesive fractured surface energy is determined from the equation (13)(11),

$$E_0/S = 2\gamma_d \tag{13}$$

where, E_0 : absorbed energy,
S : narrowest cross sectional area.

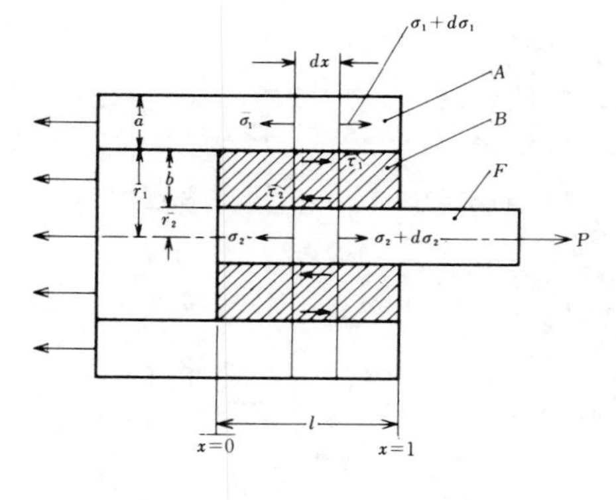

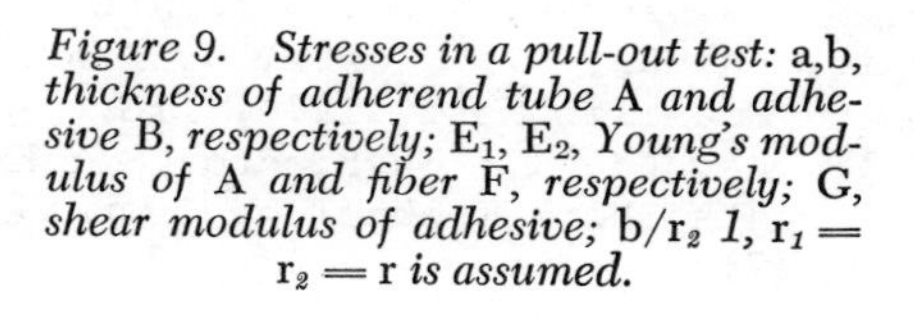

Figure 9. Stresses in a pull-out test: a,b, *thickness of adherend tube* A *and adhesive* B, *respectively;* E_1, E_2, *Young's modulus of* A *and fiber* F, *respectively;* G, *shear modulus of adhesive;* b/r_2 *1,* $r_1 = r_2 = r$ *is assumed.*

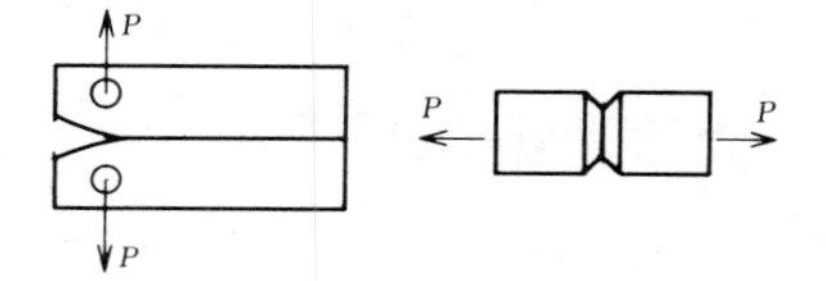

Figure 10. Dynamic tear test specimens for a new method to determine dynamic adhesive fractured surface energy

Conclusion

The above mentioned analyses based upon the unified standpoint give the relations among various testing methods. The relations of γ_s vs. γ_t and γ_{ds} vs. γ_{dt}, effect of viscoelasticity etc. would be next problems to be clarified more thoroughly.

Literature Cited

1. Griffith, A. A. Proc. 1st Int. Cong. Appl. Mech., 1924, 55.
2. Kawata, K.; Fukuda, H. Trans. Japan Soc. Compo. Mat., 1975, 1, 35.
3. Kawata, K.; Takeda, N. ibid., 1976, 2, 26.
4. Kawata, K.; Takeda, N. ibid., 1978, 4, 23.
5. Hata, T. Kobunshi Kagaku, 1947, 4, 67.
6. Dannenberg, H. J. Appl. Polym. Sci., 1961, 5, 125.
7. Malyshev, B. M.; Salganik, R. L. Int. J. Fract. Mech., 1965, 114.
8. Burton, J. D.; Jones, W. B.; Williams, M. L. Trans. Soc. Rheol., 1971, 15, 39.
9. de Bruyne, N. A. Aircraft Engng., 1944, 16, 115.
10. Fukuda, H.; Kawata, K. Proc. 20th Japan Congr. Mat. Res.-Metallic Mat., 1977, 118.
11. Kawata, K.; Hashimoto, S.; Kanayama, N. Collected Papers, 5th Shock Technology Symposium (University of Tokyo, 1978), 71.

RECEIVED January 8, 1980.

8

On the Mechanics of Peeling

HIROSHI FUKUDA and KOZO KAWATA

Institute of Space Aeronautical Science, University of Tokyo,
4-6-1 Komaba, Meguro-ku, Tokyo 153, Japan

In recent years adhesive bonding has become one of the primary joining methods(1). One reason is that high performance adhesives have been developed. Another reason is that composite materials have become to be used widely. The composite materials, especially FRP, themselves are the bonded structures. Therefore, it is important to make clear the mechanics of adhesion.

Tensile strength, shear strength, peel strength and so on are the primary factors which evaluate the characteristics of adhesive bonded structures from the mechanical point of view. Among them, we have been treating the mechanics of peeling for these several years. In the present paper, we will discuss two subjects of peeling, that is, i) the effect of peel speed on peel strength and ii) the effect of peel angle on peel strength.

Effect of Peel Speed on Peel Strength

Experimental Procedure. Concerning testing methods of peeling, the following three types are currently in use: the 180° peel test, the climbing drum test and the T-peel test(2). These methods may be sufficient to evaluate the quality of adhesive bonded structures in practical engineering situations. However, these testing methods are not sufficient for our experimental purpose because of the restriction of peel angle. Then all angle peel tester was made on an experimental basis.

Figure 1 shows the peel tester by the present authors by which arbitrary angle peel test is possible. A test plate for lengths up to 500mm and widths up to 120mm can be attached to the tester. Although this tester is fundamentally similar to that of Hata *et al.*(3), some refinements have been carried out. Kaelble *et al.*(4) made a device for peeling as an accessory to an Instron type testing machine. This device seems to be very convenient because experimental data under a constant peel speed

0-8412-0567-1/80/47-132-069$05.00/0

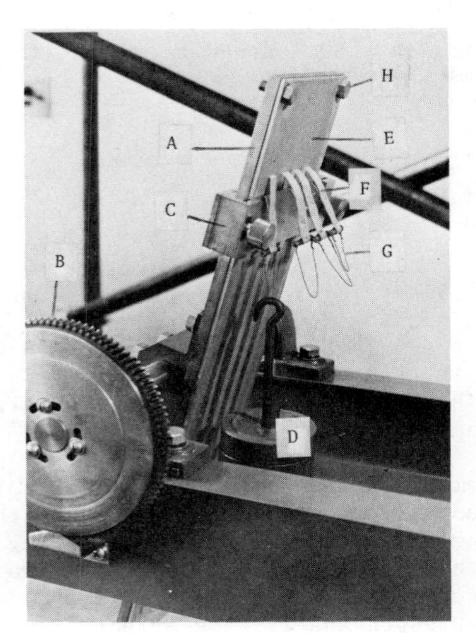

Figure 1. Peel tester and test specimen: A, *flat plate to which a plate* (E) *is bolted;* B, *knob by which the flat plate is rotated and fixed to an arbitrary angle;* C, *stopper;* D, *weight;* E, *base plate;* F, *aluminum foil;* G, *hook;* H, *bolts to attach the specimen to the flat plate.*

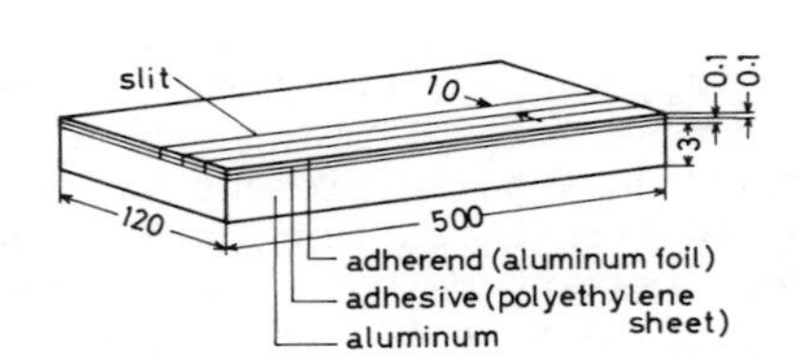

Figure 2. Construction and dimension of test plate

can be obtained easily. When a peel tester becomes large, however, it becomes difficult to attach it to an Instron testing machine and to control the peel angle constant. Then the dead load type tester was adopted as a first step. This type tester keeps the peel angle constant although the peel speed cannot be controlled.

The construction and the dimension of a test plate is shown in Fig.2. From a wide adhesive bonded plate, test specimens with the width of 10mm were made by cutting the parts of the foil and the adhesive with a knife. The peel angle was fixed to 90° in order to make the problem simple. The room temperature was 23±3°C.

After attaching the test plate to the peel tester, a certain amount of load was applied to one end of the adhesive foil. We have measured the time (t) during which a certain length (l) of foil is peeled. The peel speed (v) becomes l/t.

Results of Experiment. Figure 3 shows some raw data concerning the peel speed. As is shown in this figure, the peel speed shows fairly large variations. Since these large varia-thons are often observed(5), careful treatment is necessary to evaluate the strength of adhesive bonded structures.

Figure 4 shows the relationship between applied load and peel speed. The average value and the standard deviation were calculated in logarithm scale of peel speed. According to this figure, the peel speed doesn't increase gradually with increasing the applied load but has a gap at the load of about 4 Kg. This gap is shown by a fine dashed line in Fig.4.

An example of peeled surface at a small load (P=2500g) and a large load (P=4000g) are shown in Fig.5. In the case of small load, the peeled surface is very smooth. In the case of large load, on the contrary, the shell shape stripes with the interval of approximately 0.5mm can be observed. This phenomenon occured in almost all cases at the load greater than 4Kg. The peeling at the large load seems to occur unsteadily with the interval of stripes. Let the small load peeling and the large load one be named as steady peeling and unsteady peeling, respectively. The unsteady peeling occured even when we tested as carefully as possible. The results from Figs. 4 and 5 seems to show that the unsteady peeling becomes easier to occur than the steady one when the applied load exceeds an critical value. Figure 4 also indicates that the upper limit of the peel load and the lower one exists. This phenomenon has also been reported previously(3).

Viscoelastic Discussion of Steady Peeling. As is described in the former section, the peel strength depends greatly on the peel speed. This is based on the viscoelastic character of the adhesive. The rate dependency of peeling is discussed below taking into consideration the viscoelasticity of the adhesive.

The deformation in the neighborhood of peeled point is

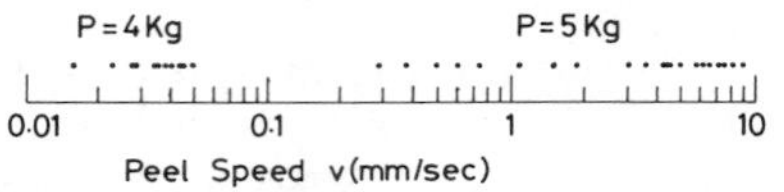

Figure 3. Example of data

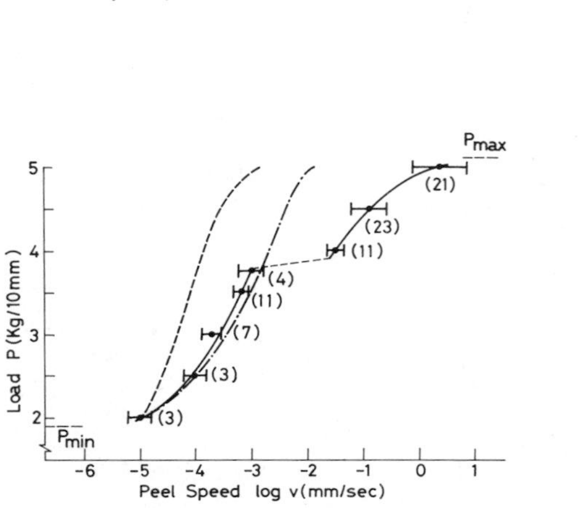

Figure 4. Applied load (P) vs. peel speed (v): (——) experimental; (– – –) 3-element model; (– · – ·) generalized Voigt model; dots, average peel speed; horizontal range, standard deviation; number in parenthesis, number of data tested.

Figure 5. Example of peeled surface

schematically shown in Fig.6, although the actual deformation is somewhat complicated. First of all, let the hatched part of the adhesive shown in Fig.6 be replaced by the three element model shown in Fig.7(a). If the hatched part sustains a constant load, the creep curve at that point becomes

$$E_0\varepsilon(t)/\sigma = 1 + K[1-\exp(-t/\tau_1)] \quad (1)$$

where, $K=E_0/E_1$, $\tau_1=\eta_1/E_1$. The curve named "3 element model" in Fig.8 shows eq.(1). Let the peeling be assumed to occur when the strain of the adhesive becomes some constant value, ε_0. Figure 8 can be interpreted that the ordinate indicates the applied stress instead of strain and the abscissa, the time when the strain becomes ε_0. The applied load and the stress of the adhesive is proportional as a first approximation. Therefore, the upper limit (P_{max}) and the lower one (P_{min}) in Fig.4 correspond to 1 and 1+K of the ordinate of Fig.8, respectively. The value of the ordinate corresponding to the intermediate load and the value of the abscissa corresponding to it can be obtained easily. Thus the relative peel speed at each load can be obtained by taking the inverse of the value of abscissa. In order to get the absolute peel speed, sufficient knowledge about the viscoelastic character of the adhesive is necessary. The dashed line in Fig.4 is calculated by taking P_{max}=5.1 Kg, P_{min}=1.9 Kg and using the experimental data at P=2 Kg. As shown in Fig.4, three element model is not appropriate to explain the experimental results.

Next, let's apply the generalized Voigt model shown in Fig. 7(b) as the viscoelastic character of the adhesive. In this case, the creep curve at the limit of $n\to\infty$ becomes(6)

$$\varepsilon(t)/\sigma = Y_0 + \int_0^{\infty} Y(\tau)[1-\exp(-t/\tau)]d\tau \quad (2)$$

where, Y_0= initial creep compliance, τ= reatrdation time, $Y(\tau)$= function which indicates retardation spectrum. After some calculation introducing the box idealization shown in Fig.8, eq.(2) becomes

$$\varepsilon(t)/\sigma = Y_0 + Y_b[\ln(\tau_m/\tau_3) - E_i(-t/\tau_3) + E_i(-t/\tau_m)] \quad (3)$$

where E_i is exponential integral function. The curve named as box distribution in Fig.8 is obtained by substituting $\tau_m/\tau_3=10^3$ in eq.(3). The dot-dashed curve in Fig.4 is obtained by a similar calculation to the case of three element model. The steady peeling can be explained by introducing the generalized Voigt model for the viscoelasticity of adhesive.

Transition From Steady Peeling to Unsteady Peeling. In the previous section we have considered the viscoelasticity only at the peeling point. In the actual peeling process, however, the deformation of the adhesive spreads to somewhat wide region. Therefore, it is necessary to analyze the problem shown in Fig.9.

Applying the generalized Voigt model, a fundamental equation

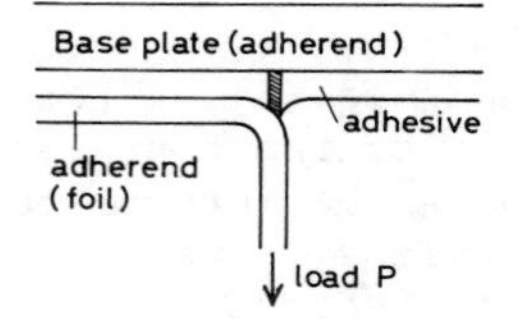

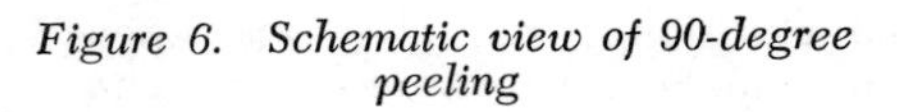

Figure 6. Schematic view of 90-degree peeling

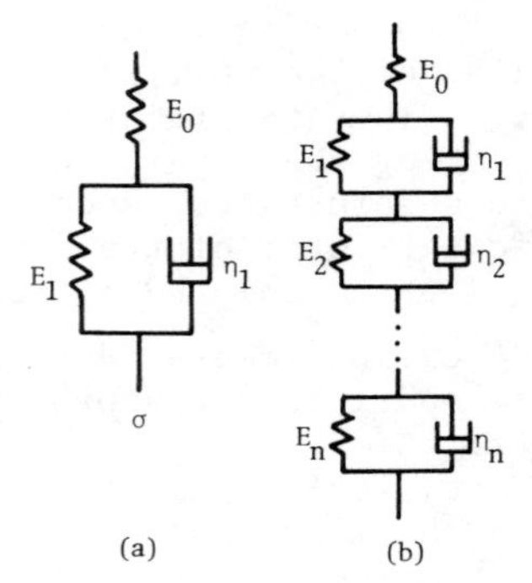

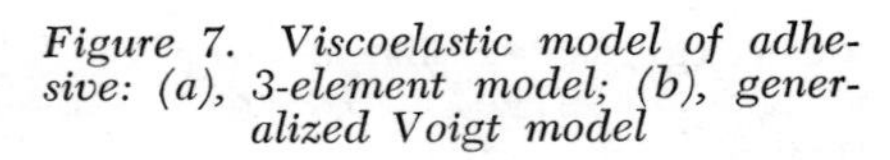

Figure 7. Viscoelastic model of adhesive: (a), 3-element model; (b), generalized Voigt model

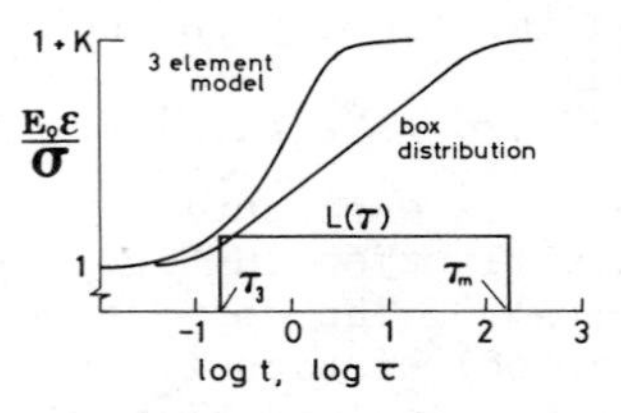

Figure 8. Time–strain curve of two viscoelastic models

becomes as follows although the detail of derivation is not shown:

$$A\frac{\partial^4 f(x,t)}{\partial x^4} + f(x,t) = -AE_0\int_0^\infty Y(\tau)\exp(-t/\tau)\left[\int_0^t \exp(t/\tau)\frac{\partial^4 f(x,t)}{\partial x^4}dt\right]d\tau \quad (4)$$

and

$$\varepsilon(x,t) = f(x,t)/E_0 + \int_0^\infty Y(\tau)\exp(-t/\tau)\left[\int_0^t \exp(t/\tau)f(x,t)dt\right]d\tau \quad (5)$$

where, $A=EIh/b$, EI= flexural rigidity of adherend, h= thickness of adhesive, b= width of specimen, f= force per unit length, ε= strain of adhesive. The right side of eq.(4) indicates the viscoelastic effect.

If $Y(\tau)$ is equal to zero, eq.(4) becomes

$$A\frac{\partial^4 f(x,t)}{\partial x^4} + f(x,t) = 0 \quad (6)$$

Therefore, eq.(4) is the fundamental equation of the bar supported by visco-elastic springs. If the time (t) is equal to zero, eq.(4) again reduces to eq.(6). Therefore, eq.(6) can be interpreted to show the initial state of the bar supported by visco-elastic springs. Considering the boundary conditions, eq.(6) becomes

$$f(x,0) = 2sP\exp(-sx)\cos(sx) \quad (7)$$

where, $s = \sqrt[4]{1/4A}$. This result coincides with another researcher's one(7).

Equation (5) is not only the function of x and t, but also the function of applied load, P. Steady peeling occurs when the reduction of potential energy of the dead weight coincides with the energy consumed to make the fracture surface, to bend the foil and so on. When the applied load is high, the peeling becomes unsteady. But it is possible to consider that the peeling is arrested by some reason, perhaps by the viscoelastic character of the adhesive. Therefore, it may be possible to explain the transition phenomena by discussing eqs. (4) and (5) in detail and by introducing an appropriate failure criterion. But it is the subject in future.

Effect of Peel Angle on Peel Strength

Experimental. In this section we will describe the effect of peel angle on peel strength under a constant peel speed. A test plate used here is composed of an aluminum base plate with a 3mm thickness, an aluminum foil with a 0.1mm thickness and an adhesive (Araldite standard) which attaches the foil to the base plate. The experimental procedure is similar to that described in the previous section although the peel angle is changed variously in the present section. The room temperature was about 25°C.

Some raw data are shown in Fig.10, where, the abscissa indicates the time required to peel the foil by 10mm. Dots and circles are the experimental values. Two symbols are used only to distinguish the values of the different peeling angle.

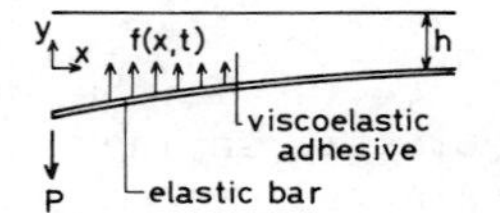

Figure 9. Bending of elastic bar supported by continuously distributed viscoelastic springs

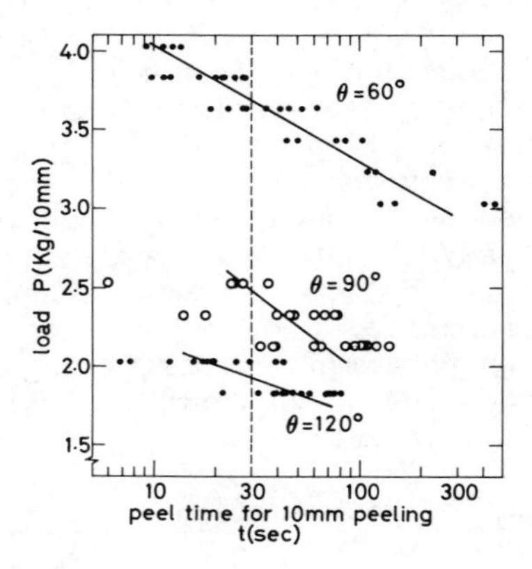

Figure 10. Examples of the relationship between applied load and peel time

As shown in Fig.10, the peel strength shows a large time dependency. Therefore, it is necessary to assign a peel time to a constant value because we don't discuss the effect of peel speed but the effect of peel angle. According to our experience, the time of 30 sec for 10mm peeling was the most preferable value for the experiment. Therefore, the average load at t=30sec is adopted for the "peel strength" hereafter. It corresponds to a peel speed of v=0.33 mm/sec. In the case of Kendall's experiment(5), the peel speed was in the region of 10^{-5}mm/sec to 10 mm/sec; and the ASTM Standard for 180° peel tests describes a peel rate of 2.54 mm/sec.

Figure 11 shows the relationship between the peel strength and the peel angle. Experiments were carried out for three specimens. Although some discrepancy was noted among the data of the three test specimens, all specimens showed the characteristic of the peel strength becoming minimum at a peel angle of about 120° and of the peel strength increasing again when the peel angle becomes larger. This phenomenon differs from the experimental results by Hata *et al.* (3),(7) and theoretical calculations by the present authors(8). The increasing of the peel strength near 180° seems to be based on the fact that some amount of applied energy (reduction of potential energy of dead load) is consumed as the plastic deformation energy of the alumunum foil.

Effect of Plastic Deformation. In the previous paper(8), the adherend was assumed to be elastic and therefore, it was assumed that the work done by the applied load was consumed as the elastic strain energy and the surface energy for peeling. In that case, the critical condition for peeling fracture becomes

$$dW+dU+dL=0 \tag{8}$$

where, dW, dU and dL indicate the increments of the surface energy, the elastic strain energy and the potential energy of load due to a peeling of length dl (*cf.* Fig.12), respectively. These values can be expressed as follows:

$$dW = 2\gamma b dl \tag{9}$$

$$dU = (P^2/2b)K_\theta dl \tag{10}$$

$$dL = -[P(1-\cos\theta) + (P^2/b)K_\theta]dl \tag{11}$$

$$K_\theta = 1/Et + \cos^2\theta/E_m t_m \tag{12}$$

where, γ and b are the surface energy of the fractured surface and the width of the foil, respectively. Other notations are shown in Fig.12.

In the actual peeling process, however, the radius of curvature of the adherend is very small at the root of the peeling. It indicates that the plastic deformation occurs. Therefore, the plastic deformation energy of the adherend (alumunum foil) must be taken into consideration. The critical condition of eq.(8) is thus modified to:

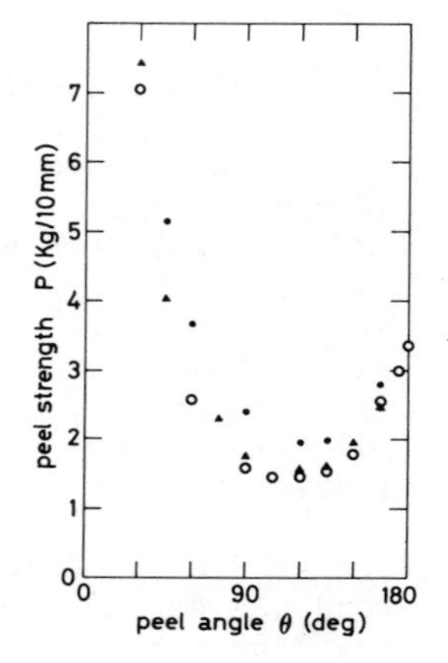

Figure 11. Relationship between the peel strength and the peel angle; three kinds of symbols indicate three test specimens

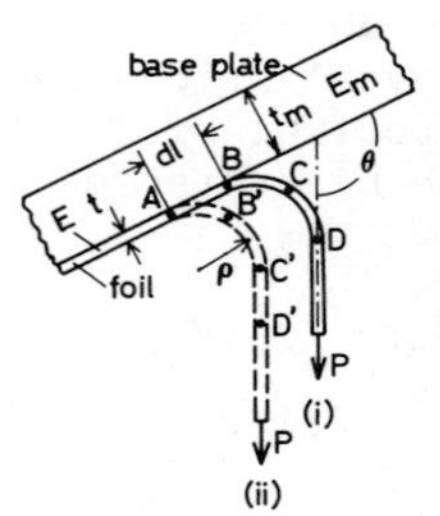

Figure 12. Schematic view of peeling and notations of θ-degree peeling

$$dW+dU+dL+dS=0 \tag{13}$$

where, dS is the increment of plastic strain energy. When a flat plate is bent beyond the elastic limit, it needs some amount of plastic strain energy. When the bent plate is rebent to its flat plate shape, it also needs some amount of plastic strain energy. dS is the total value of the above amount of energy. In order to make a rough estimation of dS, the following assumptions are adopted here: (i) the shape of the curved part of the adherend is assumed to be a circular arc of radius of ρ(cf. Fig.12), (ii) the plastic strain energy absorbed during bending is assumed to be equal to that absorbed during rebending, (iii) aluminum foil is taken to be elastic-perfect plastic body.

Under these assumptions, the process of peeling from (i) to (ii) of Fig.12 is described now. The part AB, which has been originally straight, is bent to AB'. The part BC suffers only rotation and becomes B'C'. The part of CD is straighten to C'D'. Let the plastic strain energy absorbed during the bending process from AB to AB' be denoted as dS_1. According to the theory of plasticity(9), the bending moment of the elasto-perfect plastic bar is:

$$M/M_E = (3/2)[1-(1/3)(\rho/\rho_E)^2] \tag{14}$$

where, the values with suffix E indicate the values at the elastic limit. They are(10):

$$M_E = bt^2Y/6, \qquad \rho_E = Et/2Y \tag{15}$$

where, Y is the yield stress. The radius of curvature at the peeling is very small comparing to ρ_E, that is, $\rho/\rho_E \ll 1$. Therefore,

$$M \simeq bt^2Y/4 \tag{16}$$

holds from eqs.(14) and (15). The plastic strain energy is the product of the bending moment and the bending angle, α:

$$dS_1 = M\alpha = (M/\rho)dl \tag{17}$$

According to assumption (ii),

$$dS = 2dS_1 = (2M/\rho)dl = (bt^2Y/2\rho)dl \tag{18}$$

Introducing eqs.(9), (10), (11) and (18) into eq.(13), we can get the following equation:

$$(K_\theta/2b)P^2 + (1-\cos\theta)P - (2\gamma b + bt^2Y/2\rho) = 0 \tag{19}$$

Solving eq.(19) with respect to P,

$$P = -\frac{b(1-\cos\theta)}{K_\theta} + \sqrt{\left[\frac{b(1-\cos\theta)}{K_\theta}\right]^2 + \frac{2b}{K_\theta}\left[2\gamma b + \frac{bt^2Y}{2\rho}\right]} \tag{20}$$

is obtained. At the limit of $\theta \to 0$, ρ becomes infinity and eq.(20) becomes to

$$P/b = \sqrt{4\gamma/K_\theta} \tag{21}$$

Equation (21) is identical to the result of the previous paper(8) for $\theta=0$. The above result is natural because no plastic bending occurs in the case of $\theta=0$. In the case of $\theta\neq 0$, the following expression is obtained by rearranging eq.(20):

$$P = \frac{b(1-\cos\theta)}{K_\theta} [-1 + \sqrt{1 + \frac{2K\theta}{(1-\cos\theta)^2}(2\gamma + \frac{t^2Y}{2\rho})}\,] \qquad (22)$$

In the case of $2K_\theta/(1-\cos\theta)(1+t^2Y/4\rho)<<1$, the following result is obtained by applying Taylor expansion to eq.(15) and by taking the first order term:

$$P/b = 2\gamma/(1-\cos\theta)(1+t^2Y/4\rho\gamma) \qquad (23)$$

The value of ρ is the function of the peel angle,θ. According to the author's experiments, ρ decreases with increasing θ, although the precise values are not measured. As a first approximation, we assumed ρ as:

$$\rho = s(\pi/\theta)^n t \qquad (24)$$

Thus eq.(23) becomes the function of γ, s n and Y. If we take the combination of γ=0.05Kg/mm, s=1, n=3 and Y=10Kg/mm^2, the theoretical value agrees fairly well with the experimental data, although the validity of these values are not confirmed. Comparison of this theory with experimental data is shown in Fig.13 together with the previous theory(8).

Conclusions

Using the peel tester developed on an experimental basis, some experiments were conducted to make clear the relationship (i) between the peel speed and the peel strength and (ii) between the peel angle and peel strength. It was made clear that the peel strength has the great dependency on the peel speed. There were two peeling patterns, that is, steady peeling and unsteady one. As for the steady peeling, the qualitative explanation is possible using the generalized Voigt model. As for the transition from steady peeling to unsteady one, however, we have not succeeded to explain yet. But it may be possible to explain the transition phenomenon by considering further the fundamental equations derived in the present paper. Concerning the second subject, it was made clear that the peel strength increases again at a peel angle near 180°. To explain the above phenomenon, the previous theory was modified by introducing the plastic strain energy of adherend. Although some unexplained points remain in the modified theory, it will be useful in predicting the effect of peel angle on the peel strength.

Acknowledgements. The authors want to express their sincere thanks to Dr. K.Nakao, Osaka Prefectural Industrial Research Institute and to Mr. M.Ono, Industrial Products Research Institute for their valuable discussions on adhesive strength testing.

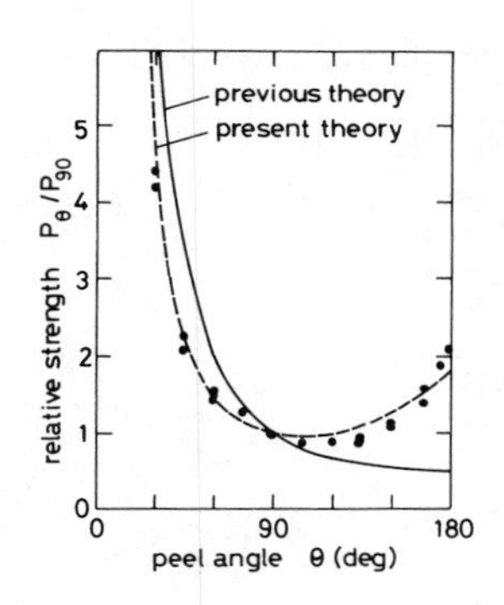

Figure 13. Relative strength of peeling

The authors also express their thanks to Mr. Y.Miyazaki, Mitsui Petrochemical Co., who offered us the test plates and to the members of the Work Shop of Institute of Space and Aeronautical Science, University of Tokyo, for their all-out cooperating in making the peel tester.

Literature Cited

1. Thrall,Jr.,E.W., J.Aircraft,1977,14,588.
2. ASTM Standard: D903(180° peel), D1781(Climbing drum peel test), D1876(T-peel test).
3. Hata,T; Gamo,M; Doi,Y., Kobunshi-Kagaku,1965,22,152(in Japanese)
4. Kaelble,D.H.; Lo,C.H. Trans. Soc. Rheol.,1974,18,219.
5. Kendall,K., J. Adhasion,1973,5,179.
6. Tobolsky,A.V.,"Properties and Structures of Polymers",John Wiley & Sons,1960.
7. Hata,T.,Kagaku,1947,17,16.(in Japanese)
8. Kawata,K.; Fukuda,H., Trans. Japan Soc. Compo. Mater.,1975, 1,36.
9. Kato,K.,"Plastic Processing of Metal",Maruzen,Tokyo,1971,292. (in Japanese)
10. Kudo,H.,"Plasticity",Morikita-Shuppan,Tokyo,1968,17(in Japanese)

RECEIVED January 8, 1980.

9

Waterborne Polymers for Aircraft Coatings

LOREN W. HILL

Department of Polymers and Coatings, North Dakota State University, Fargo, ND 58105

DANIEL E. PRINCE

Air Force Materials Labortory AFML/MBE, Wright–Patterson Air Force Base, OH 45433

Exterior aircraft coatings in current use contribute to air pollution by emission of volatile organic compounds that are present as solvents. The Air Force has begun a three phase program for reduction of emissions. The program consists of replacing air spray application by airless electrostatic spray, use of high solids coatings, and development of waterborne polymers for eventual use in aircraft coatings. The first phase is operational. Airless electrostatic spray is being used to apply solventborne topcoats and primers with substantial reductions in emissions resulting from elimination of most of the overspray. The high solids and waterborne coatings approaches are being pursued through contractural programs and an inhouse research program at the Air Force Materials Laboratory (AFML).

The purpose of the work reported here was to obtain a better understanding of the basic mechanical properties and curing behavior of current aircraft coatings so that evaluation of potential waterborne replacements could be performed on a more rigorous basis than using paint test methods alone. Methods chosen included characterization of unpigmented free-films by dynamic mechanical analysis (DMA) and use of Fourier transform infrared spectroscopy (FT-IR) for following the disappearance of isocyanate groups during polyurethane cure. Although characterization of the current topcoat is emphasized, preliminary results for several potential waterborne replacement systems are also presented.

Background

Current Solventborne Coatings. The topcoat is a two-package polyurethane consisting of an aliphatic isocyanate com-

0-8412-0567-1/80/47-132-083$05.50/0

ponent and a polyester polyol component. The impetus for finding a waterborne or high solids replacement is, of course, reduction of the solvent content. Unpigmented formulations contain approximately 65% solvent by weight. The gloss white pigmented formulation contains approximately 48% solvent by weight. The polyurethane system is normally applied in two coats, to a total dry film thickness of 51 $\pm$ 8 µm (2.0 $\pm$ 0.3 mils).

The current primer is a two-package epoxy/polyamide consisting of a bisphenol A type epoxy and an amine terminated dimer acid polyamide component. The solvent content of the mixed components is approximately 61% by weight in formulations containing corrosion inhibiting pigments. Normally one coat is used at a dry film thickness of 15 to 23 µm (0.6 to 0.9 mils).

Potential replacements will have to cure at ambient temperatures and be suitable for spray application, preferably airless electrostatic spray. The size and irregular shape of aircraft make other application methods impractical. The size also eliminates oven cure, and the irregular shape eliminates radiation cure, at least with currently available technology.

Aircraft Performance Requirements. The main function of aircraft coatings is to minimize corrosion. High level performance is required to insure that the coating system remains intact so that it can maintain its protective function in all aircraft environments. Flexibility at low temperatures and impact resistance are required. Resistance to softening in fluids used in aircraft, such as hydraulic fluids, lubricating oils, and hydrocarbons (fuel) is also required. The flexibility and fluid resistance requirements are often in conflict because the former is favored by a relatively low level of crosslinking and the latter by a high level of crosslinking. Thus, the combination of flexibility and fluid resistance, obtained with the current polyurethane topcoat, is a difficult challenge for potential replacements.

Paint test methods involving procedures well described in ASTM or federal test method publications are used to determine flexibility, fluid resistance and other coatings performance characteristics. These tests are used to establish aircraft coatings specifications. However, as is often true, the specifications tend to reflect the performance characteristics of the current system rather than actual in-use performance requirements. Therefore, evaluation of potential replacements using paint test methods and current specifications could be misleading. Use of DMA may permit more valid evaluation. DMA results may also suggest how modification of polymer structure might contribute to improved performance.

Polyurethane Cure. Paint test results carried out over many years for qualification purposes indicate that cure continues at least for several days after application of the current topcoat. Qualification tests are run after a minimum of seven days ambient cure. In this study, FT-IR has been used to follow the reaction of isocyanate groups in unpigmented films during this extended cure period. An attempt is made to relate isocyanate reaction with the development of film properties. The feasibility of using this FT-IR technique for studying cure of potential replacements is being investigated.

Experimental

Following Isocyanate Reaction by FT-IR. Clear coating films of the two-package polyurethane (Super Desothane, DeSoto, Inc.) were prepared for FT-IR analysis by air spray application onto either a polymeric substrate (clear polyethylene, i.e. commercial Glad Wrap, Union Carbide) or a tinfoil substrate. The polymeric substrate permitted recording of spectra immediately after spray application. Substrate absorption in the region of the isocyanate band at 4.4 μm was slight. The tinfoil substrate was used to obtain free films. After 17 hr. ambient cure, the films were sufficiently cured so that free films could be removed by mercury amalgamation.[1] The first spectrum for samples removed from tinfoil were recorded about 17 to 18 hr. after application. Both free films and films on polyethylene were cut (2.0 x 2.0 cm) and mounted on cardboard masks sized for insertion into the IR spectrophotometer sample holder. Mounted samples were stored under ambient conditions or in a dessicator, and IR spectra were recorded after various cure times. The extinction coefficient of the isocyanate band at 4.4 μm was 144 liter g^{-1} cm^{-1} as determined by n-butylacetate dilutions of the isocyanate component.

The FT-IR spectra were recorded using a Willey Model 318S Fourier Transform Spectrophotometer (Willey, Inc., see ref. 2). The main advantage of FT-IR over conventional IR to this study was greater accuracy in absorbance data associated with averaging multiple scans. The computer system also facilitated obtaining difference spectra.

Dynamic Mechanical Analysis (DMA). Samples for DMA were cut (3.0 x 0.4 cm) from tinfoil substrates approximately one hour before testing. After separation by amalgamation, the free films were mounted in the jaws (1.5 cm between jaws) of the DMA instrument. Film thickness ranged from 43 to 56 μm (1.7 to 2.2 mils). For cure studies the first property/temperature curves were determined after 17 hr. ambient cure, and subsequent determinations were made at approximately 24 hr. intervals with a different sample each time. The effect of ambient versus dessicator storage between determinations was investigated.

The effect of moisture on DMA plots was also studied using samples which had cured for 35 days or more. Moisture absorption was varied by storing samples at ambient relative humidity (RH), in a dessicator (RH = 0%), or in a dessicator containing liquid water and no dessicant (RH = 100%).

The tinfoil substrate sample preparation method was used for the current polyurethane topcoat, the current epoxy polyamide primer, and one potential waterborne topcoat replacement, which was a polyurethane aqueous dispersion (Polyvinyl Chemical's NeoRez R-960) cross-linked with 0.5 to 2.0 wt % of polyfunctional aziridine cross-linker (Cordova Chemical's XAMA 7). These samples were prepared at the AFML. Clear films of several potential waterborne replacement systems were supplied by DeSoto, Inc. Of the group of films supplied, DMA curves were obtained for the following: a second type of aqueous polyurethane dispersion (W. R. Grace's Hypol WB 4000), two coatings based on water soluble acrylic copolymers (B. F. Goodrich's Carboset 514 and Union Chemical's AMSCO Res 200) and another sample prepared with Polyvinyl Chemical's NeoRez R-960.

A Rheovibron Direct Reading Dynamic Viscoelastometer (Toyo-Baldwin, Ltd.) was used for DMA. This instrument and its operation have been described.[3] Briefly, the test consists of applying an oscillating strain to a rectangular sample under tension. The resulting stress and the lag of peak stress behind peak strain (phase difference) are obtained. In this study the data are presented in terms of storage modulus, E', and the tangent of the angle corresponding to the phase difference, tan δ. E' is a measure of stored energy or elastic response while tan δ is a measure of energy losses through viscous response of a viscoelastic sample that is subjected to an oscillating strain. The oscillating frequency was 11 Hz for all data reported here.

Results and Discussion

Polyurethane Cure, FT-IR. Comparison of pencil hardness results before and after exposure of the polyurethane topcoat to fluids indicates that cure continues for several days after application. The possibility that this change is related to continued curing involving isocyanate groups was investigated by FT-IR. As indicated in Figure 1A, the isocyanate band at 4.4 μm is still prominent after 18 hrs. ambient cure. Using the extinction coefficient and film thickness, the concentration of isocyanate groups remaining after 18 hrs. is calculated to be 1.56 equivalents/liter. After 41 hrs. and 65 hrs. the concentration had decreased to 0.35 eq./l and 0.09 eq./l, respectively. The ambient relative humidity (RH) during this period was high, ranging from 65% to 75%. Figure 1B shows that when the free film is stored in a dessicator following ambient cure for 18 hrs., the decrease in isocyanate absorbance is greatly reduced. After 185 hrs. in the dessicator, the 18 hr. concentration of 1.56 eq./l has only decreased to 0.87 eq./l.

The ambient exposure experiment was repeated under less humid conditions as shown in Figure 2. In this case absorbance is plotted on a linear scale so the relationship to concentration is more direct. The relative humidity was close to 50% throughout. After 17 hr. ambient cure the concentration of isocyanate remaining was 1.88 eq./l. After 45 hrs. and 65 hrs. the concentration had decreased to 1.38 and 1.03 eq./l, respectively. At all comparable cure times, the concentrations were considerably higher when the RH was lower (Figure 2).

The dominant curing reaction for the topcoat was previously assumed to be urethane formation as shown in simplified form in reaction (1). The dependence of rate of isocyanate

$$P\text{-}NCO + P'\text{-}OH \rightarrow P\text{-}\overset{H}{\underset{|}{N}}\text{-}\overset{O}{\overset{\|}{C}}\text{-}OP' \qquad \text{urethane} \qquad (1)$$

disappearance on RH and the slow loss of isocyanate in dessicated samples indicate that a moisture cure reaction is of much greater importance than previously realized. It is well known[4] that isocyanate containing systems can undergo moisture cure as shown in reaction (2). Berger[5] also noted a strong dependence of the rate of isocyanate loss on RH in polyester polyol/isocyanate systems. Reaction of isocyanate with water

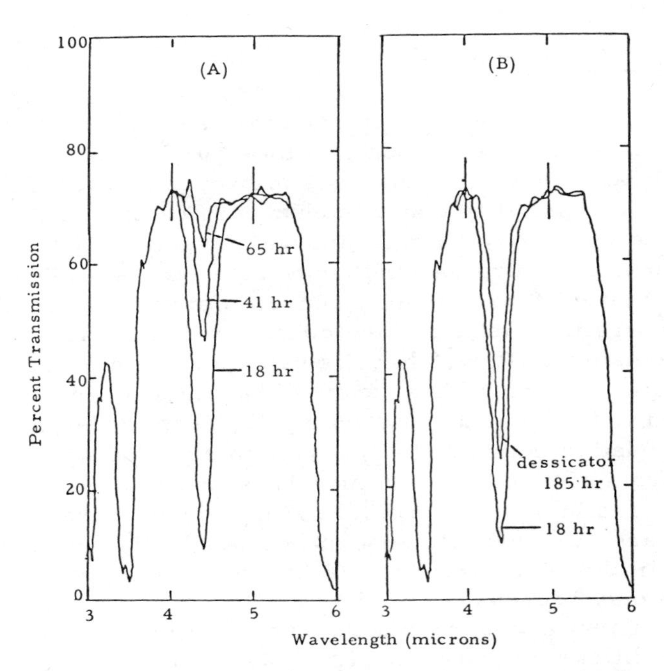

Figure 1. FTIR spectra of a two-package solventborne polyurethane unpigmented film: (A) ambient cure for the indicated times, (B) comparison of 18-hr ambient cure and a film cured for 18 hr under ambient conditions and then stored in a dessicator until 185 hr after application; film thickness = 38 μm (1.5 mils); RH = 65–75%

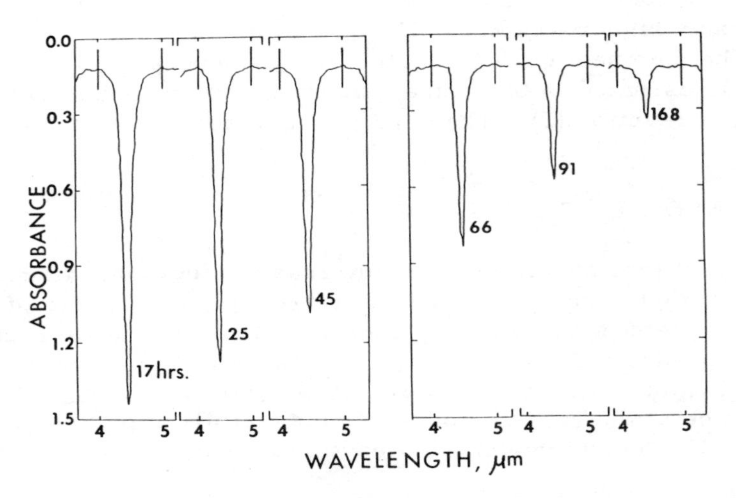

Figure 2. FTIR spectra of a two-package solventborne polyurethane unpigmented film after various cure times under ambient conditions at an RH of approximately 50%; film thickness = 48 μm (1.9 mils)

$$P\text{-}NCO + H_2O \rightarrow \left[P\text{-}\overset{H}{N}\text{-}\overset{O}{\overset{\|}{C}}\text{-}OH\right] \rightarrow P\text{-}NH_2 + CO_2 \quad (2)$$

$$P\text{-}NH_2 + P\text{-}NCO \rightarrow \underset{\text{urea}}{P\text{-}\overset{H}{N}\text{-}\overset{O}{\overset{\|}{C}}\text{-}\overset{H}{N}\text{-}P}$$

produces a carbamic acid intermediate which decomposes to form an amine and carbon dioxide. In a second step the amine group reacts with another isocyanate group to form a substituted urea crosslink. Of course, the actual cure is more complex than indicated here because both P and P' are multifunctional and because either the urea or the urethane can undergo secondary reactions at the remaining -NH- sites.[5] Recommended mixing volumes are reported[6] to result in an excess of -NCO groups which could contribute to moisture cure as shown in reaction (2).

An estimate of the fraction of isocyanate that reacts by moisture cure (reaction 2) can be obtained from absorbance determinations for dessicated samples. For example, the dessicated sample considered in Figure 1 had a concentration of 0.87 eq./l after 185 hrs. Since it is likely that a very high percentage of the hydroxyl groups originally present have reacted after 185 hrs. (7.7 days), the isocyanate remaining in the dessicated sample must represent that which normally is lost through reaction with ambient moisture. By using the resin characteristics provided by the supplier, one can calculate that the initial concentration of isocyanate in the mixed components, corrected for solvent loss, is approximately 2.7 eq./l. Therefore, the minimum percentage of isocyanate that reacts by moisture cure is approximately 32%. This is a lower limit for moisture cure because the sample was cured under ambient conditions for 18 hrs. before removal from the tinfoil and was only dessicated thereafter.

To follow isocyanate loss by FT-IR beginning immediately after spraying, a polymeric film was used as the substrate in place of tinfoil. As shown in Figure 3, the absorbance of the 4.4 μm band decreased linearly over the 6 hr. period. The initial rate of disappearance of isocyanate obtained from the slope was 0.12 eq. l^{-1} hr^{-1}. Linearity indicates that the rate is independent of isocyanate concentration. This result would be expected if the slow step is absorption and diffusion of water. At longer times the rate decreased possibly due to a combination of: a decrease in diffusion rate of water as crosslinking increases, a decrease in segmental mobility so that the second step of reaction (2) is retarded, and depletion of isocyanate groups. After 24 hrs. the isocyanate concentration remaining was 0.20 eq./l. The concentration of 1.8 eq./l obtained at zero time in Figure 3 is lower than 2.7 eq./l calculated from resin characteristics because some isocyanate reacts during a one hour "digestion period" between mixing and application.

Polyurethane Cure, DMA. Since we felt that the moisture related crosslinking would be indicated by changes in dynamic mechanical properties, DMA determinations were made at approximately the same cure times as the FT-IR determinations in Figure 1A. Curing conditions were not identical however, because FT-IR samples were exposed to the atmosphere on both sides while the DMA samples were not removed from the tinfoil substrates until about 1 hr. before each DMA run. Nevertheless, DMA results, shown in Figure 4, are consistent with the occurrence of the crosslinking reaction. The change in temperature of the onset of the main transition from about -5°C to about 10°C for samples cured for 17 hr. as compared to those cured for 41 hr., and the shift in position and height of the tan δ peak indicate that the additional cure which occurs after 17 hr. has a significant effect on properties. Samples cured for 41 and 65 hr. gave an E' plot with a rubbery plateau at about 1 x 10^8 dynes/cm^2 whereas the 17 hr. cure sample did not exhibit a plateau. Since the lower limit of measurement for samples of the dimensions used was about 5 x 10^7 dynes/cm^2, it is not known whether the 17 hr sample was sufficiently crosslinked to produce a rubbery plateau below this detection limit. Since it is well established that fluid resistance is related to crosslink density[7] and that the storage modulus in the rubbery plateau region increases with increasing crosslink density,[8] the improvement in fluid resistance of the topcoat over several days is very likely a result of the moisture related crosslinking indicated by the appearance and increase in modulus in the rubbery region.

Effect of Absorbed Water on DMA Curves. In addition to the effect of ambient moisture on cure, there is also a second moisture effect which was observed for fully cured polyurethane films (35 days or more ambient cure). In Figure 5A results are given for a sample exposed to 60% RH (ambient) while Figure 5B gives results for a sample obtained from the same substrate

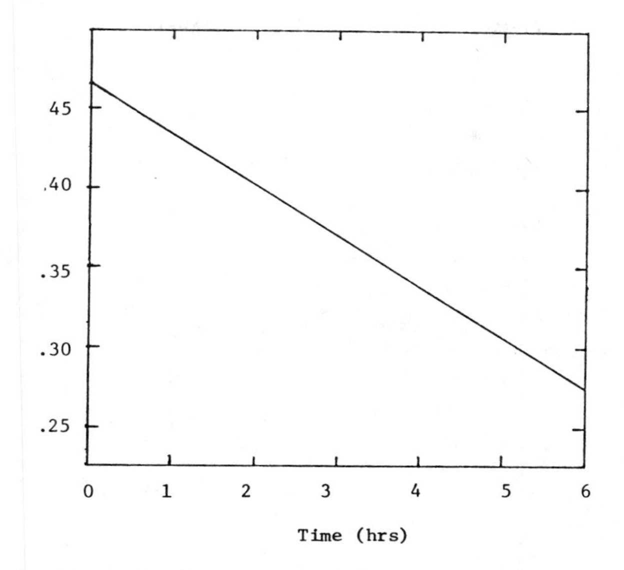

Figure 3. Isocyanate absorbance, corrected for baseline absorbance, as a function of time after spray application for a two-package solventborne polyurethane applied as a clear on a polyethylene substrate; film thickness = 18 μm (0.7 mil); RH = 75–80%

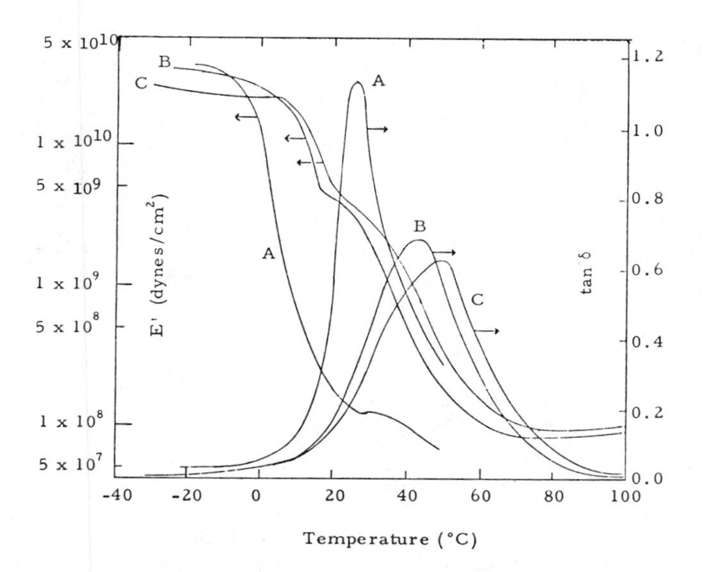

Figure 4. DMA properties of a two-package solventborne polyurethane unpigmented film cured under ambient conditions for: (A) 17 hr, (B) 41 hr, and (C) 65 hr; storage modulus (E′) on left ordinate, loss tangent (tan δ) on right ordinate; film thickness = 38 μm (1.5 mils); RH = 65–75%

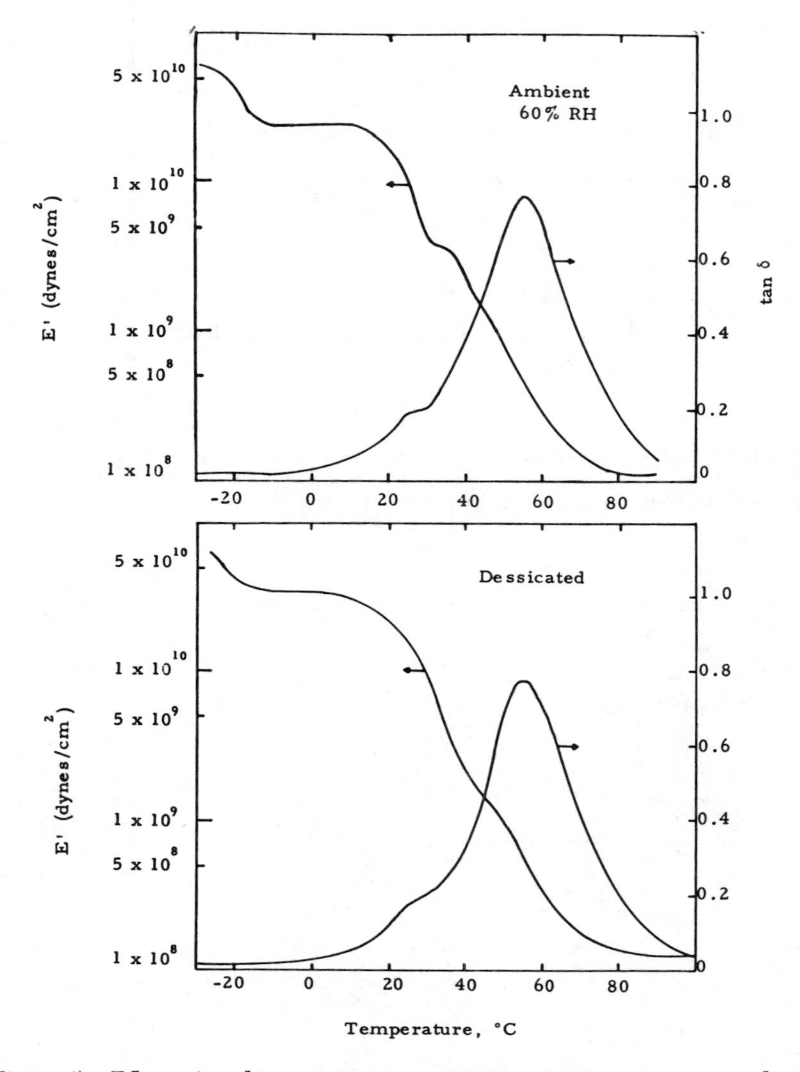

Figure 5. Effect of ambient moisture on DMA properties of a two-package solventborne polyurethane unpigmented film: (A) conditioned at 60% RH, (B) stored in a dessicator for 20 hr; film thickness = 58 μm (2.3 mil)

after storage for 20 hrs. in a dessicator. The "plateau" that appears in 5A at about 30°C is eliminated by dessication (5B). Dessication also shifted the onset of the glass transition to a slightly higher temperature.

Figure 6 also shows the effect of absorbed moisture. The curve labeled "humidified" was obtained after exposure to 100% RH for 20 hrs. In this case two plateaus were observed in the first determination. The second determination was made with the same sample immediately after cooling from 100°C to -30°C. A stream of dry nitrogen gas was used to prevent water absorption during cooling. Differences between the first and second determinations are attributed to loss of water at elevated temperatures encountered near the end of the first run. Further temperature cycling after the second determination caused no further change. Rehumidification of the same sample caused the plateau regions to reappear (curve not shown).

The occurrence of plateaus in DMA curves for samples containing absorbed water is thought to result from plasticization. Deanin and Nalepa[9] observed somewhat similar plateau regions in storage modulus versus temperature plots for plasticized polyvinylchloride. Bolon,[10] et al, reported T_g differences of 20°C or more for UV-cured polyurethane films under "wet" versus "dry" conditions. Although one would expect absorbed water to have some effect on properties, the magnitude of the effect reported by Bolon and the differences in storage modulus for the first and second runs shown in Figure 6A are remarkable. For example, at 30°C the humidified sample has a storage modulus of about 3×10^9 dynes/cm^2 while the "dry" second run gives a modulus of about 1.5×10^{10} dynes/cm^2 at 30°C. Thus, removing moisture causes the modulus to increase by a factor of five at 30°C. The strong dependence of polyurethane properties on absorbed water may be one of the causes of poor reproducibility in certain paint tests such as mandrel bend tests.

For polyurethanes the combination of low temperature flexibility and fluid resistance has been attributed to interchain hydrogen bonding as in structure (3).

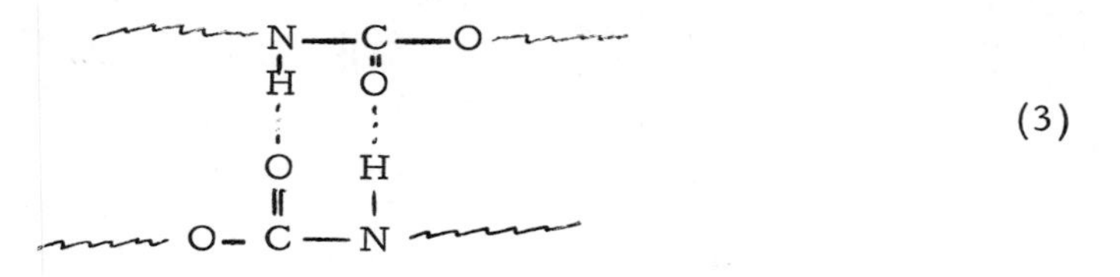

(3)

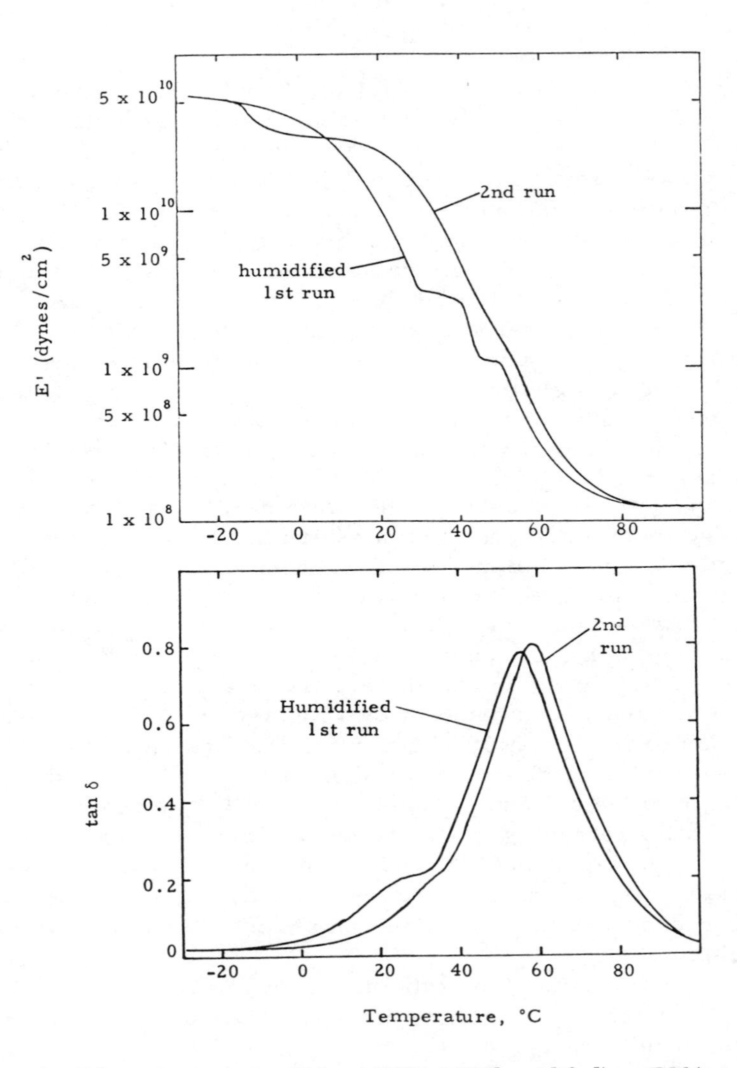

Figure 6. Effect of storage for 20 hr at 100% RH (humidified) on DMA results (first run). The second run was made on the same sample after cooling. The sample was the same as described in the caption of Figure 5.

The occurrence of two hydrogen bonds in proximity provides an effective interchain link, sometimes referred to as a "physical crosslink". Physical crosslinks may contribute to fluid resistance without decreasing flexibility to the same extent as covalent crosslinks. Water absorption could result in replacement of crosslinks like structure (3) by hydrogen bonds involving water, which would explain the strong dependence of properties on absorbed water observed in this study and in previous reports.[10] The amide links in nylon can form hydrogen bonds similar to (3), and it is well known that nylon properties depend strongly on absorbed moisture.[11] This dependence has been attributed[12] to intercatenary hydrogen bonds (i.e. a flexible link between polymer chains) involving several water molecules in place of a structure similar to (3).

Low Temperature Transitions. Figures 5 and 6 both show transitions at about -20°C, well below the glass transition which begins at about +20°C. These low temperature transitions may be related to impact resistance of polyurethane coatings.[13] Further study would be required to establish this relationship.

One cause of low temperature transitions is micro-phase separation.[14] Kim[15] et al report that crosslinked polyurethane is partially crystalline within hard segments, and Lagasse[16] proposes that crosslinked urethane elastomers exhibit a time-dependent room temperature modulus after thermal pretreatment that results from densification (but not crystallization) within hard segment domains. The occurrence of micro-phase separation in the polyurethanes used in coatings has not been established, however.

The storage modulus values (E') calculated at low temperatures were subject to variation depending on the "instrument constant correction" (called the "K value" in the Rheovibron manual). This correction was slight in the transition and rubbery plateau regions but large in the glassy region. Use of longer samples for the low temperature glassy region will eliminate this experimental difficulty, and work of this type is underway.

Potential Waterborne Replacements. Several clear films prepared using a polyurethane aqueous dispersion (Polyvinyl Chemicals NeoRez R-960) are compared with the current solventborne clear topcoat in Figure 7. The dispersion was crosslinked with 0.5% (curve B), 1.0% (curve C) or 2% (curve D) by

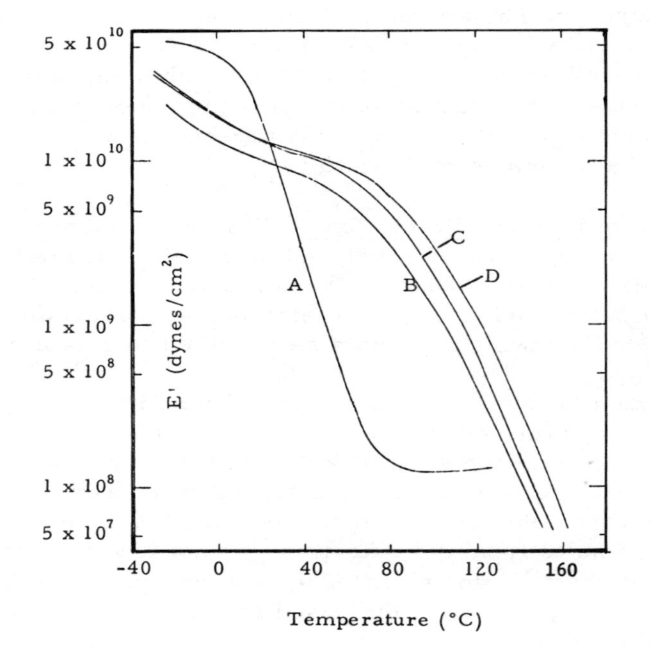

Figure 7. Comparison of DMA results for: (A) two-package solventborne polyurethane and an aqueous dispersion polyurethane (Polyvinyl Chemical's Neo Rez R-960) crosslinked with (B) 0.5%, (C) 1.0%, and (D) 2.0% of polyfunctional aziridine crosslinker (Cordova's XAMA 7); film thickness range = 40–66 μm (1.6–2.6 mil).

weight (based on dispersion solids) of polyfunctional aziridine crosslinkers. Cala and Lapkin[7] have reported improvement in fluid resistance when such crosslinkers are used even at these low levels. A shift in the transition to higher temperatures is observed as crosslinker content is increased. However, the dispersion curves do not exhibit a rubbery plateau. The absence of plateaus within the range of detection suggests that the dispersions have a lower crosslink density than the solventborne polyurethane (curve A).

It is interesting to note that at room temperature the E' values are not very different, whereas at most other temperatures the differences are large. Paint tests of flexibility or hardness carried out at room temperature and at a strain rate corresponding to the 11 Hz oscillating frequency used here would tend to give similar results for the two types of films. In contrast, the wide range of temperatures experienced by aircraft coatings is likely to produce conditions under which the response of the two systems to strain would be quite different.

It is evident in Figure 7 that the transition region is much broader for the aqueous dispersions. For acrylic latexes, Hoy[17] has shown that coatings having a broad transition give an advantageous combination of low film forming temperature and blocking resistance. Whether or not the broad transition of the aqueous dispersions provides any advantages for aircraft coatings has not been established, however.

Samples of several potential waterborne replacements were supplied by DeSoto, Inc. In order to check the validity of comparing samples prepared at different locations and on different substrates (tinfoil at the AFML and polyethylene at DeSoto), samples of NeoRez R-960 crosslinked with polyfunctional aziridines were prepared at both locations and analyzed by DMA with the results shown in Figure 8. Data points are shown at 10°C intervals for the AFML sample (-25°C, -15°C, etc.) and for the DeSoto sample (-30°C, -20°C, etc). A single curve could represent both sets of points between +10°C and 160°C. The scatter of data points from a smooth curve is minimal. These observations are offered as justification for comparing AFML and DeSoto film preparations and for showing other DMA curves without data points. The divergence below 10°C may reflect uncertainties in the "K value" correction noted previously. Figure 8 also shows results obtained with another polyurethane aqueous dispersion (W. R. Grace's Hypol WB 4000). In contrast to NeoRez R-960 which contains amine

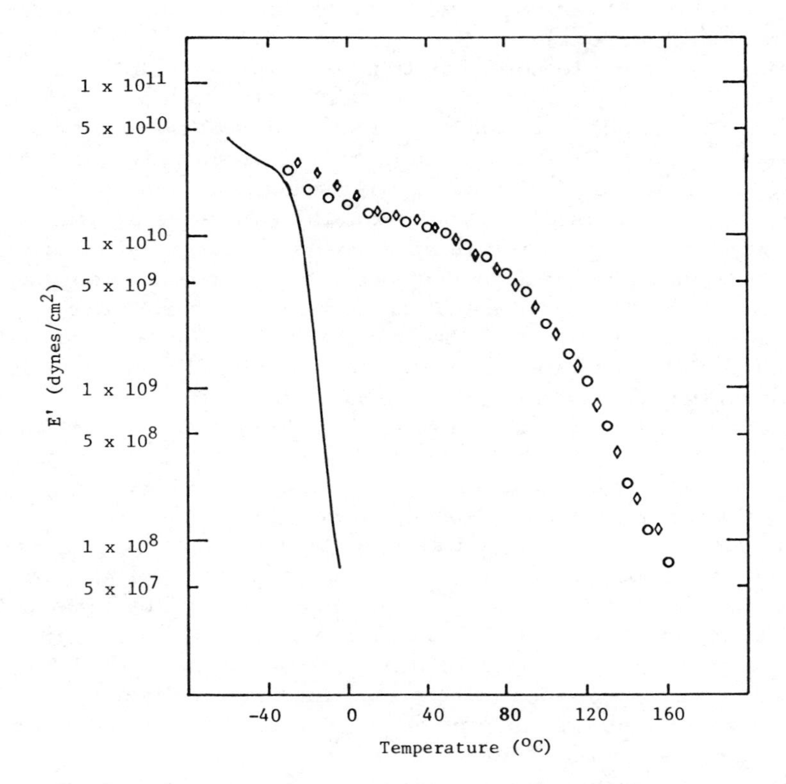

Figure 8. Comparison of DMA results for samples of Neo Rez R-960 prepared (◇) at the AFML and (○) by DeSoto, Inc. using polyfunctional aziridine crosslinkers, (——) another polyurethane aqueous dispersion (W. R. Grace's Hypol WB 4000) crosslinked with an epoxy resin (Dow Chemical's DER 732); film thickness range = 38–51 μm (1.5–2.0 mils)

neutralized carboxylic acid groups, Hypol WB 4000 is solubilized by acid neutralized pendant amine groups. Epoxy resins were used as crosslinkers for the latter. It is evident that the Hypol film has a much lower transition temperature. The rubbery nature of the film at room temperature was obvious during removal from the substrate and mounting in the Rheovibron jaws. This characteristic is not necessarily a major drawback for use as a primer. For certain aircraft applications where flexibility is particularly important, the epoxy polyamide primer has been replaced by a polysulfide primer which is very rubbery.

DMA results for two water soluble acrylic resins are given in Figure 9. These resins were crosslinked with low levels of polyfunctional aziridine crosslinkers (Cordova's XAMA 7). The absence of a rubbery plateau in the range of detection indicates that the crosslink density of both is low. These acrylics were not specifically designed for ambient crosslinking according to the suppliers literature. Therefore, modifications may be required to obtain optimum response to ambient cure with polyfunctional aziridines.

More generally it should be noted that all of the potential waterborne replacements described herein are still under study, and it should not be assumed that results presented represent the optimum achievable with the resins identified. Selection of these systems for study does not constitute endorsement by the AFML, nor does omission of other waterborne systems reflect an unfavorable preliminary evaluation. The high reactivity required for ambient cure raises the possibility of toxic effects. Since possible toxic effects are still being investigated, inclusion in this study should not be interpreted as having any implications regarding toxicity of crosslinkers or resins.

Paint Test Results and DMA. As noted previously, the storage modulus (E') in the rubbery plateau region is a measure of crosslink density. Unpigmented films of the current polyurethane have E' values in the plateau region of slightly over 1×10^8 dynes/cm^2 (Figures 5, 6 and 7). Unpigmented films of the current epoxy polyamide also have a well defined rubbery plateau at about 6.4×10^7 dynes/cm^2 (curve not included). In contrast, none of the potential waterborne replacements have a rubbery plateau within our limits of detection (Figures 7, 8 and 9). The low crosslink density of the waterborne systems, indicated by DMA, is also apparent in fluid resistance testing.

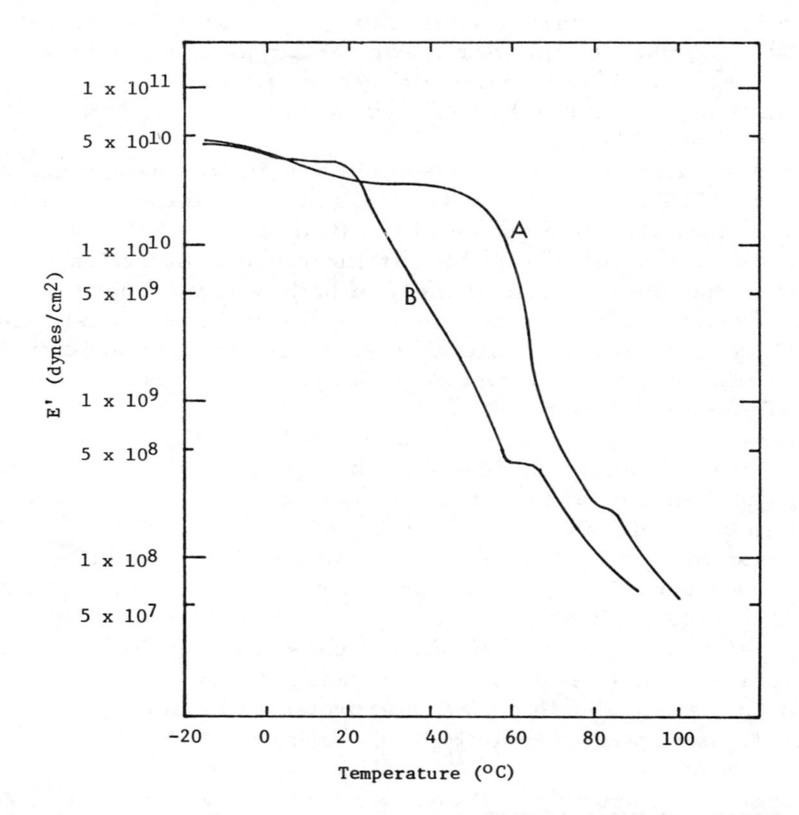

Figure 9. DMA results for water-soluble acrylic resins crosslinked with polyfunctional aziridine crosslinkers: (A) Union Chemical's AMSCO Res 200, film thickness = 36 μm (1.4 mils); (B) B. F. Goodrich's Carboset 514, film thickness = 38 μm (1.5 mils)

The test used for fluid resistance in Air Force specifications is reduction in pencil hardness following immersion. A comparison of the pencil hardness results of the current solventborne polyurethane with the two aqueous polyurethane dispersions indicates that the original hardness of the dispersions is slightly lower. The two water soluble acrylic systems have original hardnesses similar to that of the current solventborne polyurethane. Greater differences are noted after immersion. The waterborne systems all show more sensitivity to water immersion than the solventborne polyurethane. The waterborne systems all have extreme sensitivity to Skydrol 500B fluids with changes of greater than four pencil hardnesses following seven days immersion. It should be noted, however, that unpigmented formulations of the current solventborne polyurethane also undergo substantial softening in Skydrol 500B. No significant differences exist between any of the waterborne systems and the solvent borne polyurethane in sensitivity to other test fluids such as lubricating oil and hydraulic fluid.

If the fluid resistance tests are an accurate estimate of fluid resistance required in aircraft environments, it is evident that modifications directed at increasing crosslink density of waterborne systems are necessary. The contractural program has recently begun investgating such modifications.

Pigmented formulations of the current topcoat and primer produced coatings having substantially higher E' values in the rubbery plateau region than the corresponding clear films. The effect of pigment on the DMA curves of waterborne systems is currently being studied.

Although the aqueous polyurethane dispersions do not exhibit the same fluid resistance as the solventborne polyurethane, their impact resistance and low temperature flexibility are very good. In general, acrylics are not considered to have good flexibility or impact resistance; however, some of the formulations based on the two water soluble acrylic resins, mentioned previously, were better than expected in this regard.

Summary

Solventborne polyurethane topcoats and epoxy polyamide primers, currently used on Air Force aircraft, contribute to air pollution by emission of volatile organic compounds. The possibility of replacing these coatings with waterborne coatings is being investigated. A basic understanding of the curing reactions and properties of the current systems is required to provide a basis for evaluation of potential replacements.

The dependence of the rate of disappearance of the isocyanate IR band at 4.4 μm on relative humidity (RH) indicates that a substantial fraction (> 32%) of isocyanate loss results from moisture cure. The slow absorption and diffusion of water, especially at low RH, limits the cure rate and results in quite long term changes in properties. DMA results obtained over the same time interval (several days) as the FT-IR results indicate that isocyanate loss is accompanied by an increase in crosslink density.

A second moisture related effect was observed by DMA for fully cured (>35 days) films of the solventborne polyurethane. Films conditioned under ambient or high RH exhibited plateau regions in the middle of the glass transition. These plateaus were not observed for dry samples. The storage modulus of "wet" versus "dry" samples in some cases differed by a factor of five near room temperature. This effect of absorbed moisture on properties could be a cause of poor reproducibility of paint test results.

Several potential waterborne replacements have also been studied by DMA. None of the clear waterborne systems gave rubbery plateaus within the range of detection whereas clear films of both the current topcoat and current primer do exhibit such plateaus at 1×10^8 and 6.4×10^7 dynes/cm^2, respectively. This comparison indicates that the waterborne systems have a relatively low crosslink density. One paint test which is sensitive to crosslink density is fluid resistance. The waterborne systems are all more sensitive to softening during fluid immersion in Skydrol 500B and water than the solventborne polyurethane. Work directed at eliminating this deficiency is underway.

The relationship of DMA results to impact resistance and flexibility is still being investigated. Use of longer samples will permit obtaining more accurate data at low temperatures. DMA curves for several of the films gave indications of low temperature secondary transitions, which are often observed for materials having high impact resistance. Preliminary results with pigmented films indicate that pigmentation of the current topcoat and primer causes substantial increases in E' in the rubbery plateau region.

Acknowledgements

The authors wish to recognize and express appreciation to C. J. Hurley, R. L. Vissoc, and D. W. Hamilton of the University of Dayton Research Institute for their assistance in operation of the FT-IR and DMA instruments. Recognition and appreciation is also extended to DeSoto, Inc. for supplying some of the samples and paint test data.

Literature Cited

1. Yaseen, M. and Ashton, H.E., J. Coatings Technol., 1977, 49, (629), 50.
2. Willey, R.R., Applied Spectroscopy, 1973, 30, (6), 593.
3. Akay, M.; Bryan, S.J.; and White, E.F.T., J. Oil Col. Chem. Assoc., 1973, 56, 86.
4. Saunders, K.J., "Organic Polymer Chemistry," Chapman and Hall: London, 1973, p.342.
5. Berger, W., Federation d'Association de Techniciens des Industries des Peintures, Vernis, Emaux et Encre d'Imprimerie de l'Europe Continentale (FATIPEC) Congress, 1962, p.300.
6. Vanderhoff, J.W., "Water-Base Coatings," Technical Report AFML-TR-74-208, Air Force Materials Laboratory, Wright-Patterson AFB, 1976, part II, p. 85, (Available through National Technical Information Service).
7. Cala, J.A. and Lapkin, M., Am. Chem. Soc. Div. Org. Coat. Plast. Chem. Pap., 1976, 36, (2), 431.
8. Tobolsky, A. and Mark, H., "Polymer Science and Materials," Wiley-Interscience: New York, 1971, p.216.
9. Deanin, R.D. and Nalepa, S.M., Am. Chem. Soc. Div. Org. Plast. Chem. Pap., 1976, 36, (2), 811.
10. Bolon, B.A.; Olson, D.R.; Lucas, G.M. and Webb, K.K., Am. Chem. Soc. Div. Org. Coat. Plast. Chem. Pap., 1978, 39, 512 and 518.
11. Billmeyer, F.W., "Textbook of Polymer Science," Second Edition, Wiley-Interscience: New York, 1971, p.435.
12. Reimschuessel, H.K., J. Poly. Sci., Chem. Ed., 1978, 16, 1229.
13. Hill, L.W., Prog. Org. Coatings, 1977, 5, 277.
14. Senich, G.A. and MacKnight, W.J. in "Multiphase Polymers," R.F. Gould, Ed., Advances in Chem. Series 176, American Chemical Society: Washington, D.C., 1979, p.97.
15. Kim, S.C.; Klempner, D.; Frisch, K.C.; Radigan, W.; and Frisch, H.L., Macromolecules, 9, 1976, 258.

16. Lagasse, R. R., J. Applied Poly. Sci., 21, 1977, 2489.
17. Hoy, K. L., J. Coatings Technol., 1979, 51, (651), 27.

RECEIVED January 8, 1980.

10

The Chemistry of Low-Energy Curable Coatings for Aerospace Applications

A. FRANK LEO[1]

DeSoto, Incorporated, 1700 South Mt. Prospect, Des Plaines, IL 60018

The chemistry of Low Energy Curable Coatings for the Aircraft Industry is basically the chemistry of ambient cured epoxy-amine primers and isocyanate-polyol topcoats. Coatings based on these classes of raw materials are used by aircraft design engineers to retard corrosion of the high strength, light weight alloys used in aircraft construction. Unfortunately, these high strength alloys lose strength at a faster rate than the lower strength alloys because they are more susceptible to corrosion, particularly stress corrosion. (1)

The fact that aircraft travel further and faster between periods of maintenance than ground transportation equipment, through environments which are quite extreme puts these protective coatings to severe tests. Not only do these coatings have to endure cyclic and rapid temperature extremes, intense U.V. light, wind and rain abrasion, but they must do this in the presence of "stripper-like" fluids such as the phosphate ester hydraulic fluids, defrosting fluids and jet fuels. (2)

Before the aliphatic polyurethane topcoats became available in the mid 1960's, epoxies were popular both for topcoats and primers. (3) Until then no coatings equaled the epoxies in abrasion resistance, hardness, chemical resistance and adhesion. The introduction of aliphatic isocyanates, however, allowed for the development of even more abrasion resistant coatings, with better flexibility better resistance to chalking, discoloration on exposure and better acid resistance than amine cured epoxies. (4) Although the use of epoxies in topcoats is quite limited, amine cured epoxy vehicles make up the bulk of the aircraft primers in use today. These primers develop excellent adhesion to the aircraft substrates and allow the polyisocyanate cured polyol topcoats to adhere to the metal. Generally, aliphatic polyurethanes have poor adhesion to unprimed metal.

[1] Current address: Midland Division of Dexter, East Water Street, Waukegan, IL 60085

0-8412-0567-1/80/47-132-105$05.00/0

Epoxy Primers

Most epoxy primers are prepared from the diglycidyl ether of bisphenol A or one of its derivatives. Occasionally, other epoxies such as the diglycidyl ether of resorcinol and the Novalac epoxies are used as modifying resins.

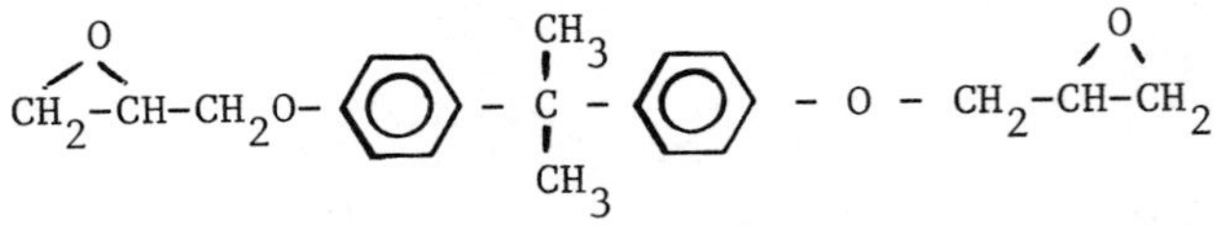

$$CH_2(O)CH-CH_2O-C_6H_4-C(CH_3)_2-C_6H_4-O-CH_2-CH(O)CH_2$$

Diglycidyl Ether of Bisphenol A

All of these epoxies are derived from aromatic monomers since these are very reactive with aliphatic amines at ambient temperatures. Normally, aromatic amines and carboxylic acids do not react with these epoxies at room temperature and are not used as curing agents. Some of the most alkaline amines such as diethylene triamine and triethylene tetraamine must be used in modified form, such as an amine functional amine epoxy adduct to prevent blushing (carbonate formation on the surface)[5] and improve compatibility of the admixed reactants. Other amines such as N-aminoethyl piperazine, dimethylaminopropylamine and polyamides are often used as partial or sole curing agents to reduce crosslink density and improve flexibility and impact resistance.[6,7]

$(CH_3)2\ N-CH_2-CH_2-CH_2-NH_2$

Dimethylamino Propylamine

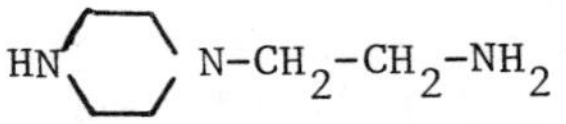

$HN(C_4H_8)N-CH_2-CH_2-NH_2$

n-Aminoethyl Piperazine

Military Specification Primers[8] require the use of amino hydrogen functional polyamides (derived from dimer fatty acids and multifunctional amines) and epoxy resins derived from bisphenol A with a molecular weight of about 1000.

The stoichometry of the epoxy-amine reactions is based on the theoretical number of equivalents of epoxy available for reacting with each amino hydrogen. So an amine such as ethylene diamine $H_2N-CH_2-CH_2-NH_2$ with a molecular weight of 60 would have a theoretical amino hydrogen equivalent weight of 15. A primary amine will react with an epoxy to form a secondary alcohol and secondary amine. This newly formed secondary amine, at least theoretically, is then capable of reacting with a second epoxy to form a tertiary amine and another secondary alcohol.[9] In actual practice this second reaction is quite difficult at room temperature; steric hinderance and poor mobility of the partially cured epoxy-amine systems brought about by increasing viscosity works against this secondary amine reaction from actually taking place.[10]

At room temperature alcohols and phenols are often used to promote the epoxy-amine reactions. It is thought that the hydrogen bonding effects of these hydroxyl groups assist in opening the epoxy ring. Phenols, because of the greater activity of the hydrogen speed the reaction rate of the amine-epoxy cure.[11]

Tertiary aliphatic amines are used both as catalysts for the epoxy amine reaction and as sole curing agents; these amines catalyze the epoxy-epoxy reaction which results in the generation of ethers and secondary alcohols.[11]

Some of the common tertiary amines are triethylamine, benzyldimethylamine, and tris 2,4,6 - (dimethylaminomethyl)phenol (DMP-30 from Rohm & Haas). DMP-30 is particularly reactive because of the presence of the three basic tertiary amine groups plus the phenolic hydrogen. It is often used in combination with primary and secondary aliphatic amines to assist the conversion of the unreacted epoxy groups remaining after the active hydrogen amine-epoxy reaction has progressed to the point where further reaction is prevented by the viscosity buildup of the partially cured reactants of the primer. In the industry, combinations of amines and amine adducts are commonly used to provide a convenient mix ratio and to obtain acceptable levels of water and hydraulic fluid resistance.

These epoxy primers pigmented with calcium, strontium, zinc and lead chromates are the staples of the industry. They yield films with low shrinkage, excellent adhesion, protect the substrate from corrosion and provide a tenacious bond for the intercoat adhesion of polyurethane topcoats.

Polyurethane Topcoats

The isocyanates used in coatings consist basically of two classes: the aromatic and the aliphatic isocyanates. The aromatic isocyanates tend to yield hard films with excellent acid and chemical resistance but poor resistance to yellowing and embrittlement on exterior exposure, consequently, they are rarely used in topcoats but sometimes find their way into speciality primers.

The aliphatic isocyanates used in exterior topcoat applications are non-yellowing, softer, more flexible, slower reacting and significantly more expensive than the aromatic isocyanates. Some of the aliphatic isocyanates that have been available and were used in the study described later are: Isophorone diisocyanate and its isocyanurate trimer, T1890, Veba-Chemie AG, Biuret of hexamethylene diisocyanate (Desmodure N Mobay) and a Linear difunctional aliphatic isocyanate, K5-2333, from Mobay.

Isocyanates react with active hydrogen atoms at ambient temperatures. Generally, the aliphatic isocyanates are somewhat sluggish at room temperature and require catalysis in the form of tertiary amines and/or tin compounds such as dibutyl tin dilaurate to effect an acceptable cure rate in paint films.

In paint films the most common reactions are these:(12) An isocyanate reacts with an active hydrogen to form an amide. When the active hydrogen is an alcohol, the product is a urethane. If the active hydrogen is an amine a urea is formed. If the active hydrogen is supplied by atmospheric moisture the isocyanate decomposes to an unstable carbamic acid which decomposes to form an amine (with the release of carbon dioxide) which ultimately reacts quickly to form a urea. The carboxylic acids will also react with isocyanates slowly, ultimately forming an amide with the loss of carbon dioxide. These "side reactions" with CO_2 release contribute to the haze sometimes observed in polyurethanes applied under humid conditions. This is somewhat different than the blush observed at times in the application of amine-epoxy films where CO_2 absorption by the alkaline amines cause carbonate formation.

The types of active hydrogen containing compounds differ considerably in their reactivity. Amines react many times faster with isocyanates than do alcohols and different amines will react at rates depending on their basicity and lack of steric hinderance.

Generally, the reaction rates of active hydrogen will follow this order:

$$\underset{1^{o}}{R\text{-}NH_2} > \underset{1^{o}}{AR\text{-}NH_2} > \underset{1^{o}}{R\text{-}OH} > \underset{2^{o}}{R\text{-}OH} > H_2O > \underset{3^{o}}{R\text{-}OH} > AR\text{-}OH > R\text{-}SH > R\text{-}COOH$$

Relative Reactivities of Amine with Phenyl Isocyanate in Diethyl Ether at 0°C (13)

Amine	Relative Reactivity
Ethyl amine	9.72
n-Butyl amine	9.17
Aniline	0.53

Approximate Relative Reactivity of Phenyl Isocyanate with Active Hydrogen Compounds in Toluene at 25°C (13)

Compound	Relative Reactivity
Aniline	10-20
1 -Butanol	2-4
2 -Butanol	1.0
Water	0.4
Tert -Butanol	0.01
Phenol	0.01
1-Butanethiol	0.005

Many of these functional groups react either too slowly (carboxylic acid) or too quickly (amines) to be used in practical topcoats. Consequently, the most widely used coreactants are the hydroxyl functional polyesters, polyethers and acrylics. These polyols offer wide latitude in the physical and chemical properties attainable in polyurethane topcoats, as well as offering a wide range in formulation for potlife and dry time.

Since polyethers do not resist U. V. exposure well, they are rarely used in aircraft topcoats. Hydroxyl functional acrylics and polyesters are the most common exterior topcoats. Acrylics are often favored by paint applicators because they are usually high molecular weight materials which yield fast apparent dry times. This minimizes dust collection on the wet film and allows for more rapid masking and color striping of the aircraft. However, aliphatic acrylics generally do not yield the high gloss that many customers demand. If a monomer of high refractive index [14] such as styrene is used in the preparation of the acrylic, the high gloss is attainable but flexibility and impact resistance suffer.

Polyesters of various types are probably the most commonly used polyols for aircraft exteriors. In our laboratory on the order of 100 different polyesters for use in polyurethane aircraft topcoats have been examined. Studies have been performed on the effect of acid monomer on the polyester urethane performance properties, as well as the effects of diol monomer, acid number, hydroxyl number, and crosslink density. These polyesterurethanes were examined for gloss, gloss retention, hydraulic fluid resistance, jet fuel resistance, water resistance, flexibility and impact resistance before and after exposure to U. V. and low temperature flexibility. [15, 16, 17, 18, 19]

The general trends observed in these coatings was an increase in gloss, gloss retention, chemical and water resistance with increasing aromatic acid content. If the aromatic acids were replaced with aliphatic acids, the coatings obtained had excellent impact flexibility both before and after exposure to ultraviolet light. Of course, the length of the diol monomer was very important also, but it followed the predictable course of increasing flexibility and decreasing chemical resistance with increasing monomer chain length. The aromatic acids were somewhat of an anomaly since coatings with excellent initial impact flexibility were obtainable with some combinations of aromatic acids and flexible diols but these same paint films, after exposure to artificial and natural weathering failed impact flexibility catastrophically.

The choice of isocyanate for the work described above was the trifunctional, biuret of hexathylene diisocyanate. The other commercially available aliphatic isocyanates examined, were the isocyanurate of isophorone diisocyanate, 4,4' dicyclohexylmethane diisocyanate, and a linear aliphatic difunctional isocyanate with an equivalent weight approximately equal to Desmodur N. Various aromatic isocyanates, were also examined but were found to be unsatisfactory because of poor flexibility and poor color retention on exterior exposure.

In the study to follow, three isocyanates were examined. The polyol was a proprietary polyester. The primer used was the Mil P-23377C type, epoxy-polyamide. The substrate was alclad 2024T3, aluminum blanks, 20 mils thick pretreated with a chromate conversion coating Alodine 1200S from Amchem Products, Incorporated. Table I attached summarizes our experimental findings in this study of isocyanates.

The isophorone isocyanurate triisocyanate cured system was found to have good gloss and gloss retention when compared to the biuret of hexamethylene diisocyanate, but reverse dart impact flexibility was less, particularly after exposure to U. V.; the paint films fractured at 5 in.-lb compared to 80 in.-lb with the biuret both before and after exposure to U. V.. Resistance to aircraft fluids was better with the isophorone isocyanurate triisocyanate, but the reaction rate was substantially less than that of the buiret triisocyanate. Because of the very slow reaction of the isophorone isocyanate, it is very difficult to obtain rapid dry, so dry to tape time in reasonably short intervals is difficult.

Replacing the biuret with a difunctional isocyanate of similar equivalent weight produced a coating with similar initial gloss but gloss retention was better. Reverse impact flexibility was as least as good, but the decrease in crosslink density resulted in inferior resistance to aircraft hydrolic fluid, Monsanto's Skydrol 500B.

It is apparent that obtaining all the film properties required by the industry is difficult. There are a number of areas where trade offs are necessary. Probably, the most notable are in these areas: gloss and gloss retention after weathering, flexibility and flexibility retention after weathering and resistance to aircraft fluids.

Weatherability, gloss, color and flexibility require the use of aliphatic isocyanates. However, aliphatic isocyanates are not as fluid resistant as aromatic isocyanates, although some of this loss can be compensated for by increasing the isocyanate content and/or the crosslink density of the final cured film. Of course, increasing the crosslink density may cause adverse effects in flexibility.

The polyester polyol component, which generally constitutes sixty to eighty percent of the resin content of the cured polyurethane film, also plays an important role in determining film properties. Generally, it is difficult to balance the monomer content of the polyester to obtain all the desirable film properties. With our particular proprietary polyester and the biuret of hexamethylene diisocyanate we have been able to meet most of the requirements.

TABLE I

	Biuret of Hexamethylene Diisocyanate	Isocyanurate of Isophorone Diisocyanate	Linear Aliphatic Diisocyanate
NCO Equi. Weight	190	240	183
NCO Functionality*	3	3	2
Gloss: Initial at 60°	93	94	91
Gloss: After 500 hrs.WOM	78	86	88
Gardner Reverse Impact:			
Initial Impact	80 in.-lb	5 in.-lb	80 in.-lb
After 500 hrs.WOM Twin Arc	80 in.-lb	5 in.-lb	80 in.-lb
Pencil Hardness:			
Initial	HB-F	HB-F	HB-F
7 Days in D.I.H_2O,25^oC	HB	HB	B
7 Days in Skydrol 500B, 25^oC	4B	HB	4B Blistered
7 Days in Type III, Reference Fluid, 25^oC	2B	2B	2B

*Theoretical number of functional groups per molecule.

However, it is not proper to report on the properties of the polyurethane topcoats without mentioning the contribution made to the total system by the primer.[20] Generally, the more tightly crosslinked epoxy primers will improve the fluid resistance of the topcoat. Unless this crosslink density is taken to the extreme or primer film thickness is excessive, there is no loss in reverse impact flexibility in the film thickness range normally tested.

In the isocyanate study described above, the Military Specification Primer was used. Although this primer is excellent for corrosion resistance, strippability and flexibility, it is not very resistant to the phosphate ester hydraulic fluids encountered on commercial aircraft. Consequently, that primer is not commonly used on commercial aircraft since any weakness in chemical resistance of the primer shows up in the total paint system. In order to obtain a total paint system (primer-topcoat) which will show no softening in hydraulic fluid or fuel, the crosslink density must be increased or the polyester must have increased aromatic character which would decrease flexibility - another trade off.

Summary and Conclusions

The chemistry of coatings for the aircraft industry is basically the chemistry of the epoxy-amine primer and polyisocyanate-polyol topcoats. These materials work synergistically, in that the primer has excellent adhesion to aircraft substrates and retards the corrosion of these metals. The polyurethane topcoat protects the epoxy primer from the affects of weathering and abrasion. Each component of the system has a vital function and would be almost useless without the other.

The polyurethane topcoats are generally prepared from polyester polyols and are cured with polyisocyanates; the biuret of hexamethylene diisocyanate having the best balance of properties obtainable with commercially available isocyanates.

Acknowledgement

The author expresses his appreciation to F. Hawker for helpful discussions in the preparation of this paper.

Literature Cited

1. Cole, H. G., "Corrosion and Protection of Aircraft", Proceedings of the First International Congress on Metallic Corrosion, 1962, London, (p. 642 - 5).

2. Evans, G. B., "Heavy Duty Coatings in Airframes", Industrial Finishing and Surface Coatings, September, 1972, (p. 24 - 30).

3. Catalyzed Coatings Protects Jet Fighter, Product Finishing, June, 1963.

4. "Urethane Coatings in the Transportation Industry", 12th Annual Symposium of the Washington Paint Technical Group, 1972.

5. Lee and Neville, Handbook of Epoxy Resins, 1967, McGraw Hill.

6. Bell, J. P., Reffner, J. A., and Petrie, S., "Amine-Cured Epoxy Resins: Adhesion Loss Due to Reaction with Air", J. App. Polymer Sci., 1977, 21, (p. 1095 - 1102).

7. Breslan, A. J. Epoxy Resins: Chemistry and Technology, edited by May, C. A., and Tanalsa, Y., Chapter 8, Mercel Dekker, N. Y., 1973 (p. 507 - 508).

8. MIL P-23377C, Military Specification, Primer Coating, Epoxy Polyamide, Chemical and Solvent Resistant.

9. Dusek, K. and Bleha, M., "Curing of Epoxide Resins: Model Reactions of Curing with Amines", J. Polymer Sci., 1977, 15, (p. 2393 - 2400).

10. Luvak, S., Vladyka, J., and Desek, K., "Effect of Diffusion Control in the Gloss Transition Region on Critical Conversion at the Gel Point During Curing of Epoxy Resins", Polymer, 1978, 19, Aug.

11. Lee and Neville, Handbook of Epoxy Resins, Chapter 9.

12. Saunders, J. H., and Frisch, K. C., Polyurethanes: Chemistry and Technology, Vol. XVI, Part I, Interscience, N. Y., Oct. 1965.

13. Saunders, J. H., and Frisch, K. C., Polyurethanes: Chemistry and Technology, Vol. I, Chemistry, Interscience, 1962, (p. 173, 208).

14. Simpson, L. A., Factors Controlling Gloss of Paint Films, Progress in Organic Coatings, 6, 1978.

15., 16., 17., 18., 19., Leo, A. F., Internal Reports, Unpublished Data: March 1973, Feb. 1974, Dec. 1975, March 1976 and August 1976.

20. Featherston, A. B., "The Significance of Processing Variables on the Adhesion of Sealants and Organic Coatings to Metallic Aircraft Surfaces, AD778019, 1974 April.

RECEIVED January 8, 1980.

11

High-Solids Coatings for Exterior Aircraft

R. E. WOLF, C. J. RAY, G. McKAY, and J. M. BUTLER

DeSoto Incorporated, 1700 South Mt. Prospect, Des Plaines, IL 60018

Recently, the effects of numerous industrial effluents and waste materials which ultimately enter the environment have become a major concern for everyone. Solvent emissions from the application of large volumes of coatings and paints contribute significantly to this problem. Additionally, economic parameters and the threat of energy shortages have put enormous political pressure on the coatings industry to reduce the total amount of solvent emissions.

The major generic system types available to the industry are water-borne, ultraviolet curable, and high solids coatings. High solids coatings systems can completely eliminate the use of photochemically reactive solvents, while concomitantly reducing the total solvent emissions required to deliver an equivalent amount of paint.

This paper reports partial results obtained with contract F-33615-77-C-5101 from the Air Force Materials Laboratory, Wright-Patterson Air Force Base.[1] The program is a continuing effort toward the development of a sprayable, high performance, 65% volume solids, two-package polyurethane exterior coating for military aircraft. Typically, two package, conventional urethane systems, sprayed at 35% volume solids are used for this application.

Theoretical Approaches:

A number of theoretical approaches are available to obtain the program objective. These include "simple" solvent reduction, solubility parameter studies, application equipment modification, reactive diluents, and polymer modification.

0-8412-0567-1/80/47-132-115$05.00/0

TABLE I

Four methods of measuring viscosity were employed.

1. #2 Zahn Cup	This useful, but unsophisticated instrument provides a constant low shear measurement.
2. Brookfield Viscometer	This instrument provides a low, non-uniform shear over a surface. Useful shear rate range is 0.52-49.2 sec^{-1}.
3. Wells-Brookfield Micro	This instrument provides a plate and cone, medium range shear rate, useful at 1.15 to 230 sec^{-1}.
4. Haake Rotovisco	A plate and cone attachment is used, which provides a shear rate range of 74-12,000 sec^{-1}. A Bob and cup attachment is also available and was utilized for pigmented formulations.

In order to achieve the 65% volume solids requirement, solvent reduction studies were conducted employing a number of individual polyol and isocyanate components of the polyurethane coating. In this study, various solvents were examined to test their ability to provide a low viscosity, high solids vehicle system for use in a high performance exterior aircraft enamel.

Viscosity-Concentration Profiles [2,3,4]

The ultimate application technique of the coating will be either conventional air spray or airless electrostatic spray. The effect of polymer-solvent interactions on solution viscosity at high polymer concentration is important for both application techniques. Accordingly, the viscosity-concentration profiles of several polymer-solvent combinations were made. This paper will report data obtained on one polyol component of the two-package urethane system.

Viscosity-concentration profiles were obtained

for Acryloid AU-568, an oxazolidine containing material from Rohm and Haas, supplied at 85% volume solids in ethoxy ethyl acetate. A brief discussion of this component is required in order to better understand the total system. The oxazolidine was originally supplied under the developmental code QR-568. The basic chemistry, while simple in nature, is actually quite complex when consideration is given to the extensive equilibra which accompany the entire reaction sequence producing the final cured film. The reaction sequence is initiated by atmospheric water. Without this water, the two-package system is stable almost indefinitely.

STEP	REAGENT	PRODUCT
1.	a. Oxazolidine b. Water	a. B-alkanolamine b. Carbonyl
2.	a. B-alkanolamine b. Isocyanate	"cured" polymer

The B-alkanolamine reacts with two equivalents of isocyanate to produce the cured film. Note should be made that an equivalent of carbonyl material is lost to the atmosphere during the cure cycle. The fate of these cure volatiles may be a major concern where large quantities of coating are applied in a closed structure. Consequently, proper ventilation must be provided during the early period of cure.

Over the range of interest of volume solids, the viscosity of the oligomer solutions ranges from 20 to over 100 sec. on a #2 Zahn Cup. This measurement yields a kinematic viscosity which is the ratio of the actual viscosity to the density of the fluid.

The data obtained is the efflux times, in seconds, and monographs are available for converting these efflux times of the viscosity cup to centipoise units. Typical spray viscosities reported for the #2 Zahn Cup range between 18 and 25 seconds.

Using the Zahn Cup, various solvents were tested to determine the effectiveness in reducing the viscosity of the oligomeric solution (Figure 1). Of the solvents tested, methyl ethyl ketone (MEK) appears to be the most effective solvent for reducing the viscosity of these solutions. Also, the individual

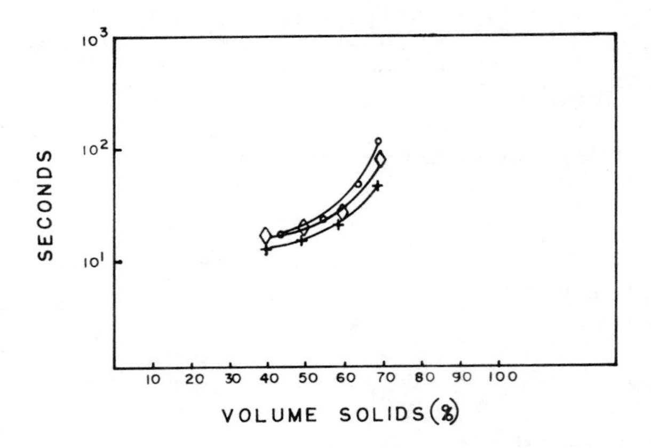

Figure 1. Viscosity–concentration profile: Rohm & Haas QR 568; 32 Zahn Cup; 77°F; (○) cellusolve acetate, (◇) MIBK, (+) MEK

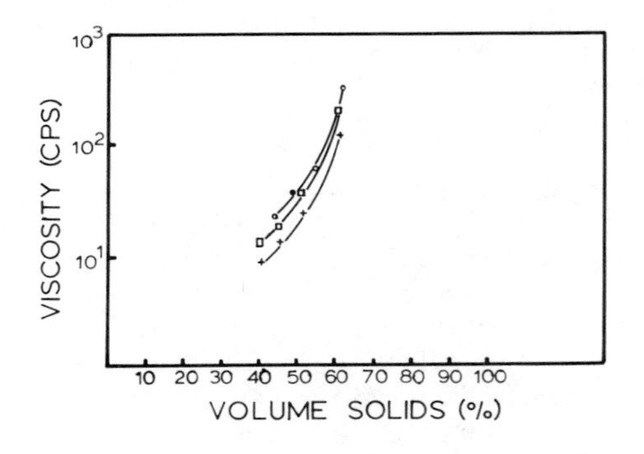

Figure 2. Viscosity–concentration profile: Rohm & Haas QR 568; #2 Brookfield Spindle; 20 rpm; 77°F; (○) cellusolve acetate, (□) MIBK, (+) MEK

solutions appear to increase in viscosity exponentially as the percent volume solids increases.

Similarly, a plot of the viscosity-concentration profile for the same solution measured on a Brookfield Viscometer is shown in Figure 2.

Again, the data indicates that MEK is the solvent of choice for this solution for the purpose of obtaining the highest solids at the lowest viscosity. These solutions also appear to exhibit an exponential increase in viscosity with increasing volume solids. These plots are displayed on a semi-log graph so the range in viscosity over the region of interest lies between 10 and 500 cps. Typically, the spray viscosity of spray applied materials ranges between 40 and 150 cps.

Viscosity is a function of the shear force exerted on the material in question divided by the shear rate. Additionally, the viscosity varies as a function of the shear rate. Newtonian fluids display a linear relationship on a shear rate versus shear stress plot, i.e., the viscosity is independent of the shear rate. Dilatant fluids exhibit an increase in viscosity with an increasing shear rate, while pseudoplastic and thixotropic fluids display a decrease in viscosity upon increasing shear rate. Therefore, emphasis should be placed on materials which exhibit shear thinning at high shear rates because the spray process imposes a shear rate ranging from 1000 to 40,000 sec^{-1} on these coatings. The coating should exhibit pseudoplasticity during the spraying stage, and dilatancy immediately upon contact with the surface of the object being coated. Obviously, a rapid increase in viscosity at rest is required because reduced sagging is desirable.

Viscosity-Concentration Profiles At Varying Shear Rates

The viscosity-concentration profiles of the AU-568 oligomer solutions in three different solvents at various shear rates are displayed in Figure 3. From this data, obtained with a Brookfield viscometer, the viscosity of the oligomer solutions at lower concentrations appear to increase as a function of an increase in shear rate. This apparent dilatant behavior is unusual since shear thinning would normally be expected to occur for these solutions. Additionally, the viscosity behavior at higher volume solids approaches Newtonian behavior at increasing shear rates. Methyl ethyl ketone is again the solvent

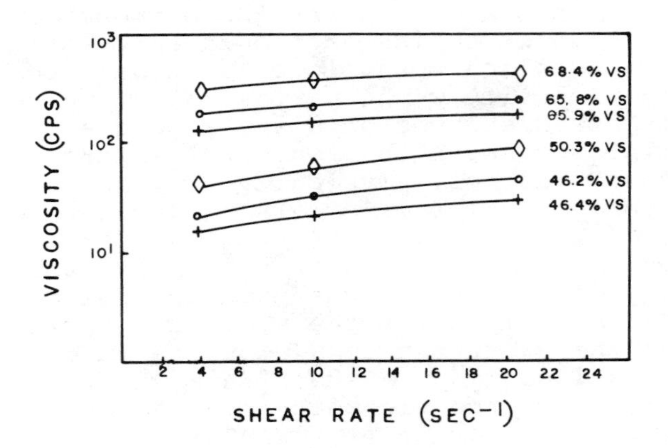

Figure 3. Viscosity shear rate: Rohm & Haas QR 568; #2 Brookfield Spindle; 77°F; (◇) cellusolve acetate, (○) MIBK, (+) MEK

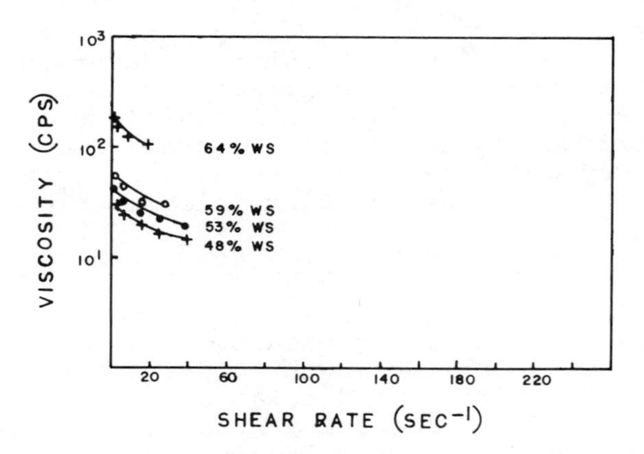

Figure 4. Viscosity shear rate: Rohm & Haas QR 568; cellusolve acetate; Wells Brookfield; 77°F

of choice.

Typically, Newtonian behavior is observed in the low volume solids concentration systems. But, as solvent is removed, these materials generally revert toward a pseudoplastic behavior. From these initial data, the lack of sufficient shear rate range in this study becomes apparent. Data from a Brookfield viscometer alone was not enough to be able to predict the Newtonian or pseudoplasticity characteristics of these solutions.

Examining the viscosity-concentration profiles of the AU-568 solutions at higher shear rates yields interesting results. Data obtained from the Wells-Brookfield viscometer show the expected behavior. In both cellosolve acetate and methyl ethyl ketone, shear thinning (pseudoplasticity) is observed over the extended shear range of the instrument. At the high solids concentration, AU-568 oligomer solutions exhibit a pseudoplastic behavior which at low volume solids the diluted solutions exhibit Newtonian behavior. See Figures 4 and 5.

To obtain data at even more extended shear rates, a Haake Rotovisco Viscometer was employed. See Figure 6. Because of the extended shear rate range of this instrument, and the particular plotting characteristics, the resulting pseudoplastic behavior does not appear as dramatic as did parallel behavior with the Wells-Brookfield instrument. However, even at the extended shear rate, considerable pseudoplasticity is actually observed. Again, the high volume solids appear more pseudoplastic than the lower volume solids solutions.

This type of information may be used advantageously in the formulation of spray application systems. Formulations that would normally be rejected on the basis of Zahn Cup or Brookfield viscosity data alone could possibly be found useful when examined with a larger shear rate instrument. In fact, partially as a result of this data, many systems involving the AU-568 oligomeric material have been fully formulated into a polyurethane enamel, and successfully spray applied at 65% volume solids.

Casson Plots

In 1969, Casson developed an equation to relate the viscosity of a polymer solution to a limiting high shear viscosity, e.g., the viscosity of the solution at infinite shear. See Figure No. 7.

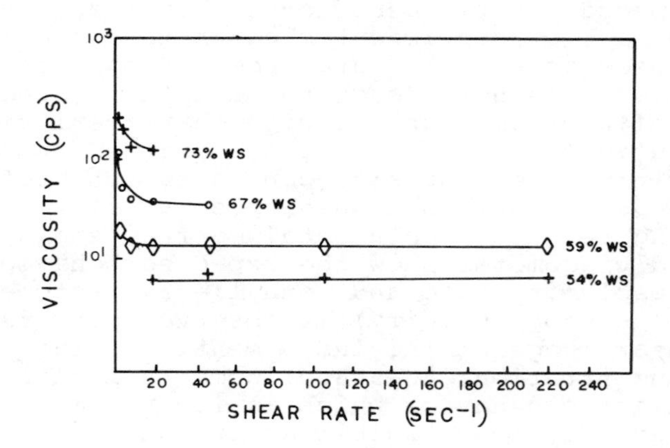

Figure 5. Viscosity shear rate: Rohm & Haas QR 568; MEK; Wells Brookfield; 77°F

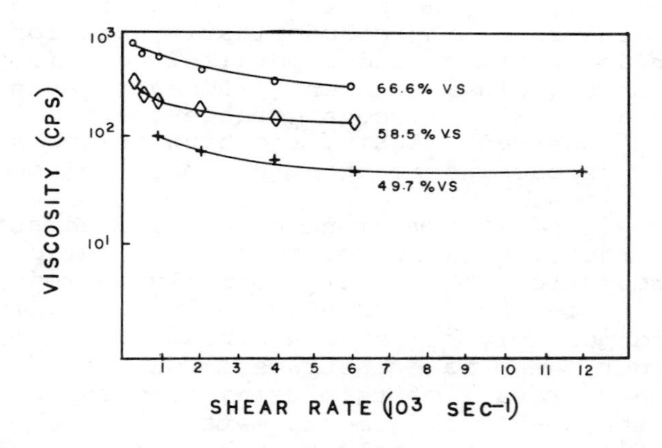

Figure 6. Viscosity shear rate: Rohm & Haas QR 568; cellusolve acetate; Haake Rotovisco; 67°F

$$\eta^{1/2} = \eta_{\infty}^{1/2} + \gamma_{o}^{1/2} \dot{\gamma}^{-1/2}$$

This equation states that a plot of the square root of the viscosity versus the reciprocal of the square root of the shear rate produces a straight line with an intercept equal to η_{∞}, the limiting high shear viscosity, and with the slope equal to $\gamma^{-1/2}$, the square root of the yield point. The limiting high shear parameter, η_{∞}, has been used as a measure of brushability or ease of coating application at high shear rates. In addition, the yield point parameter, γ_{o}, can be used as a measure of sag resistance. For example, this parameter may be employed to calculate the film thickness applicable to a vertical surface without the coating sagging.

The reported limitation of the Casson equation is that some coatings deviate from linearity at very high and low shear rates. However, despite this limitation, the Casson equation does represent a simple two parameter viscosity-shear rate equation with parameters which can be related to directly measurable properties of the coatings composition.

A family of Casson plots of AU-568 oligomer solutions in cellosolve acetate at various weight solids concentrations using data from the Wells-Brookfield viscometer was made (Figure No. 8)

As one would predict, the limiting high shear viscosity increases with increasing solids concentration. Of additional interest is the slight increase in T_{o} with increasing volume solids, indicating an improved probability for higher film build without sagging (Figure No. 9).

Casson Plots with Pigmented Systems

A number of Casson plots were generated using TiO_2 dispersions in various polyol components. A somewhat anomalous behavior was indicated with the viscosity data at 55% volume solids of an AU-568 dispersion. The mill charge yield in this dispersion at 60% volume solids was 96%, while, at 65% volume solids, the yield after grinding was 92%. This data is consistent with the obtained slope, T_{o}, indicating improved sag resistance. The high shear rate limiting viscosity, η_{∞}, is similar for both Cargill 5760 and AU-568, with the C-5760 showing a much greater shear thinning behavior. The yield point parameter is

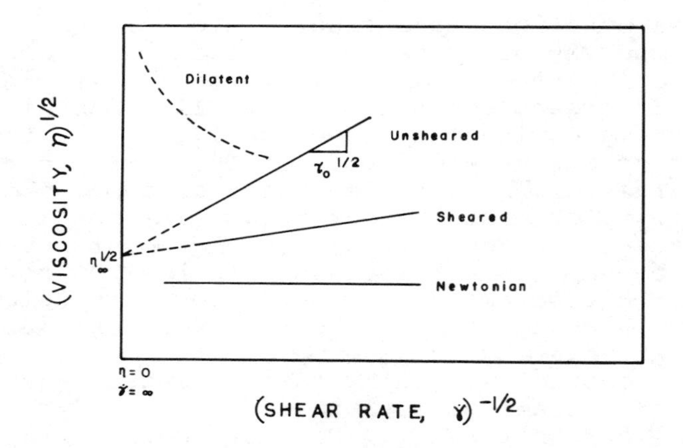

Figure 7. Viscosity shear rate: Casson plot

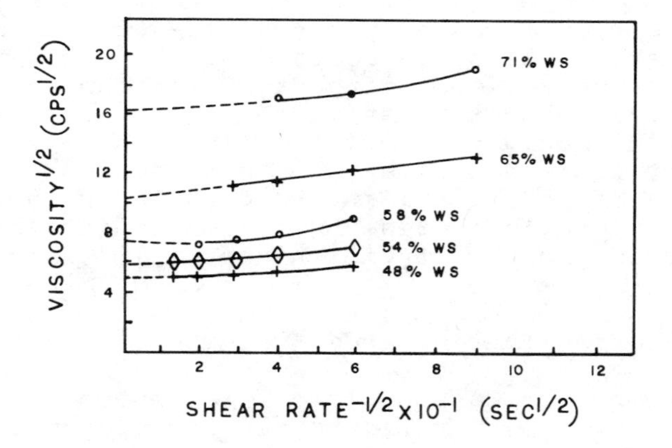

Figure 8. Viscosity shear rate: Rohm & Haas QR 568; cellusolve acetate; Wells Brookfield; 77°F; Casson plot

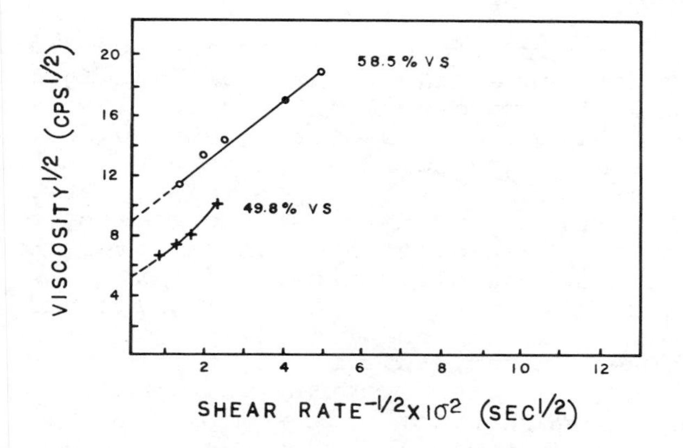

Figure 9. Viscosity shear rate: Rohm & Haas QR 568; cellusolve acetate; Haake Rotovisco; 67°F

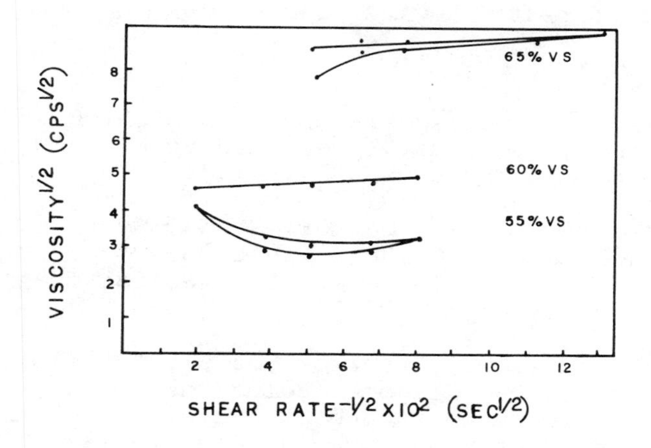

Figure 10. Viscosity shear rate: Rohm & Haas QR 568; TiO_2 dispersion; Casson plot

consistent with the fact that the mill yield was 65% at 60% volume solids. (See Figure 10).

Conclusions

As measured with high shear instrumentation, the viscosity of these high volume solids oligomer solutions exhibit a reversal in their rheological properties relative to the viscosities obtained at low shear rate and low volume solids. Although the possibility exists that a certain degree of this reversal may be due to an artifact of the instrumentation, a trend toward shear thinning at high shear rates for these high volume solids oligomers is clear.

The data from the Casson plots, specifically the yield point and high shear limiting viscosity, may be used to predict certain final film properties, i.e., sagging and indicate important information about materials processing of high solids systems. If the present data is truly indicative of the time behavior of these oligomer solutions, the rheological characteristics of high solids systems may be on a serpentine viscosity-concentration profile display, and the specific behavior for a desired application must be considered before an oligomer-solvent system is made.

REFERENCES

1. Wolf, R.E. "Viscosity Studies of High Solids Coatings Systems," presented at North Dakota State University, June 1, 1979.

2. Erickson, J. R., Garner, A. W., "Viscosity of Concentration Oligomer Solutions," presented at the 4th Water-Borne and High Solids Coatings Symposia, February, 1977.

3. Garner, A. W., Erickson, J. R., "Viscosity of Concentrated Oligomer Solutions For High Solids - II," presented at the 5th Water-Borne and High Solids Coatings Symposia, January, 1978, New Orleans, LA.

4. Erickson, J. R., "Viscosity of Oligomer Solutions For High Solids and UV Curable Coatings," Journal of Coatings Technology, 18 (620), September, 1976.

RECEIVED February 14, 1980.

12

Solvent-Removable Coatings for Electronic Applications

J. J. LICARI and B. L. WEIGAND

Rockwell International Corporation, Electronic Devices Division, P.O. Box 4192, Anaheim, CA 92803

A large variety of materials and processes are currently employed in the manufacture of hybrid microcircuits. Because of this, it is often difficult to completely remove the last traces of processing materials through normal cleaning prior to hermetic sealing. A single electrically-conductive particle, even in the micron size range, can cause an electrical short and result in a missile or spacecraft failure. In spite of the precautions taken by the electronics industry, it is extremely difficult to completely remove all particles, some being electrostatically attached to the surface of the circuit. One approach to resolving this problem — coating the entire circuit with a thin layer of organic polymeric coating in order to immobilize the particles — has been evaluated and has been successful. On a limited basis, coatings of high purity such as some semiconductor grade silicones and parapolyxylylenes have been used and are generally compatible with the active and passive devices and wire bonds of a high density circuit. However, these high molecular weight polymer coatings once applied and cured are difficult to remove for circuit rework. Because of the high cost of military and space grade microcircuits (ranging from several hundred to several thousand dollars), rework is important. Any particle immobilizing coating should therefore be easily removed to permit the repairs. The high molecular weight and highly cross-linked polymers are inherently difficult to remove. They are insoluble in the normal organic solvents. Removal by abrading, scraping, or cutting is operator dependent and often results in further damage to adjacent devices and interconnections. The objective of this work was therefore to evaluate and select a class of organic coatings that could easily be removed by a "hands-off" process by dissolving them in organic solvents that are commonly used in a manufacturing environment. This work was supported by the Air Force Materials Laboratory, Wright-Patterson Air Force Base, Ohio, Contract No. F33615-77-C-5141.

0-8412-0567-1/80/47-132-127$05.00/0

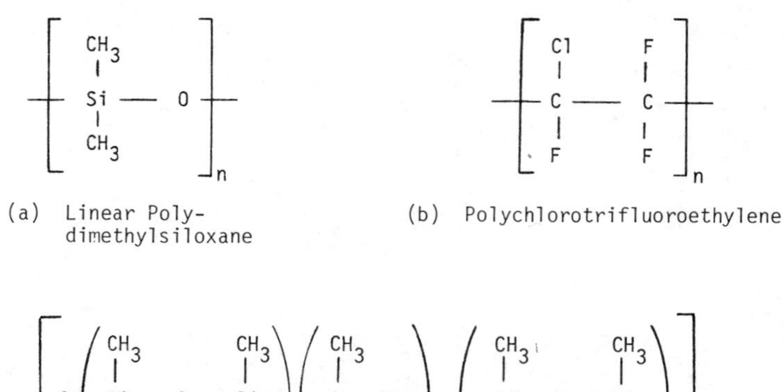

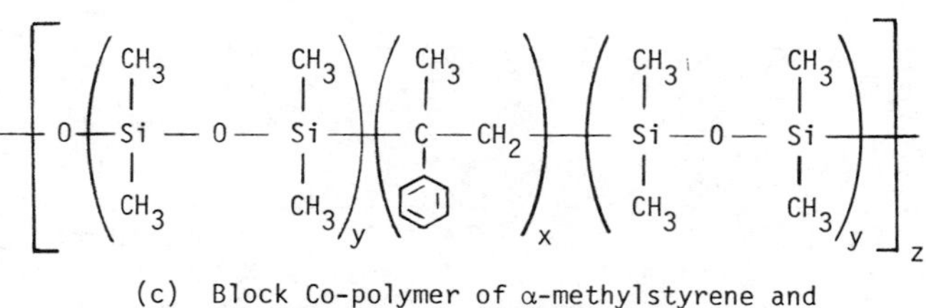

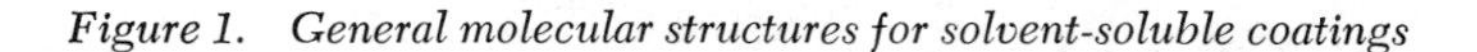
Figure 1. General molecular structures for solvent-soluble coatings

Table I. Description of Silicone Coatings Selected for Evaluation

Coating Type	Percent Solids (as rec'd)	Viscosity (Centistokes)	Film Description
Polydimethyl-siloxane I	27.1	49	Clear, waxy, does not melt or flow below 150°C, flows under pressure at 85°C.
Polydimethyl-siloxane II	16.0	570	Clear, tacky, flows under pressure at room temperature.
Block Co-polymer of α-methylstyrene & dimethylsiloxane	25.5	506	Clear, hard, tough coating, melting point 285-289°C.

Selection of Candidate Coatings

The five key requirements that the coating had to meet were:

(1) Easy application by spray, dip or flow coating in thin layers, 0.5 to 2 mils.

(2) High degree of purity containing low levels of sodium and chloride ions and no other corrosive constituents such as amines.

(3) Easy removal by solvent dissolution.

(4) Thermally stable to 150°C to withstand burn-in temperatures used for hybrid circuits.

(5) No stresses on fine-wire bonds and no deleterious effects on active devices.

Coatings were selected from a class of low-to-intermediate molecular weight uncrosslinked silicones and halocarbons which were soluble in Freon TF, xylene, or toluene — cleaning solvents widely used in the electronics industry. Specifically, three silicones formulated by Dow Corning, four halocarbon waxes from Halocarbon Corporation and mixtures of the silicones and halocarbons were evaluated. The general molecular structures and characteristics of these coatings are given in Figure 1 and Tables I and II. Early in the program, three of the four halocarbon waxes were eliminated from further consideration. Halocarbons 6-00 and 12-00 were too low in melting point to withstand the circuit burn-in temperatures of 125 to 150°C. Halocarbon 19-00 was insoluble in toluene and difficultly soluble in other solvents so that a coating solution could not be formulated. Two of the silicones, the linear polydimethylsiloxane types (I and II) were also eliminated after preliminary studies showed that residues remained on the circuit after Freon TF extraction, that one of the silicones remained tacky throughout processing, and that the other silicone had a high rate of outgassing.

Table II. Description of Halocarbon Waxes Selected for Evaluation

Halocarbon Corp. Serial No.	Melting Point or Pour Point	Solubility
6-00	60°C	Freon TF, Toluene
12-00	120°C	Freon TF, Toluene
15-00	144°C	Toluene
19-00	210-220°C	Insoluble

Table III. Chemical Analyses of Coating Materials

Method/Element	Silicone I	Silicone Co-Polymer	Silicone II	Halocarbon 12-00	Halocarbon 15-00
Emission Spectroscopy					
Silicon (wt %)	27	24	24	.046	.039
Magnesium	7.8	2.2	1.5	2.2	1.1
Iron	<9	23	20	20	11
Boron	<50	<50	<50	1.3	0.17
Aluminum	<3	8.3	<3	10	5.1
Copper } ppm/by wt	0.35	1.0	.32	0.70	0.33
Silver	<.5	<5	<.5	<0.10	<0.10
Titanium	17	<5	<5	<1.0	<1.0
Nickel	<3	<3	<3	<1.0	<1.0
Strontium	<10	<10	<10	<1.0	<1.0
Calcium	4.3	6.9	4.6	4.2	2.7
Loss on Ignition (wt %)	41.72	48.86	48.12	99.8896	99.9198
Atomic Absorption					
Sodium (ppm/by wt)	44.3	60.4	33.3	47.8	54.3
Potassium (ppm/by wt)	12.1	18.5	7.6	14.3	12.8
Titration					
Chloride (ppm/by wt)	60	209	110	153	62

Chemical and Thermal Characterization of Coatings

The silicone and halocarbon materials were characterized chemically and thermally by performing: infrared spectrographic analysis, atomic absorption spectroscopy for sodium and potassium ions, emission spectroscopy for other metal ions, mercuric nitrate titration for chloride ions, gel permeation chromatography for molecular weights, and thermal gravimetric analysis (TGA). Sodium ion concentration ranged from 33 to 60 ppm by weight while potassium ion concentrations were between 8 and 18 ppm. The measured chloride concentrations ranged from 60 to 209 ppm. Other metals were found in trace amounts (Table III) but are not considered significant. The molecular weight for the silicone co-polymer was found to be 100,000 while that for the soluble portion of Halocarbon 1500 was 1100. TGA curves for the three silicones are given in Figure 2. The silicone co-polymer and silicone II coatings were stable to 240°C. Silicone I showed loss of weight from 80°C. TGA curves for the halocarbons showed stability up to 220°C and essentially identical weight losses with increasing temperature (Figure 3). As a result of these characterizations, the silicone co-polymer, silicone II, and Halocarbon 15-00 were selected for further evaluation.

Electrical Characterization of Coatings

Electrical properties that are important to the performance of a coating used as an insulation or dielectric include insulation resistance (volume resistivity, surface resistance), dielectric strength, dielectric constant, and dissipation factor. Measurements of these electrical properties were made according to established ASTM and Military specifications at both 10^2 and 10^5 Hz, where possible. Electrical values for Dow Corning 6101 and DC-90-711 and G.E. EJC-261, representative of state-of-the-art semiconductor junction coatings, are given for comparison (Table IV). The results (Table IV) show that the halocarbon waxes had electrical properties of the same order of magnitude, except for dielectric strength, as the DC-6101. The dielectric strengths were similar to those for Dow Corning fluorosilicone coatings (340-380 volts/mil) which are recommended for electrical insulation for connectors. It was difficult to prepare dimensionally accurate test samples for silicone II because of its inherent tackiness. Samples had to be prepared and measured several times. The somewhat lower dielectric strength and volume resistivity may be attributed to these sample preparation problems. Generally, the electrical properties for the other coating materials were considered excellent and close to those of the best semiconductor junction and circuit coatings.

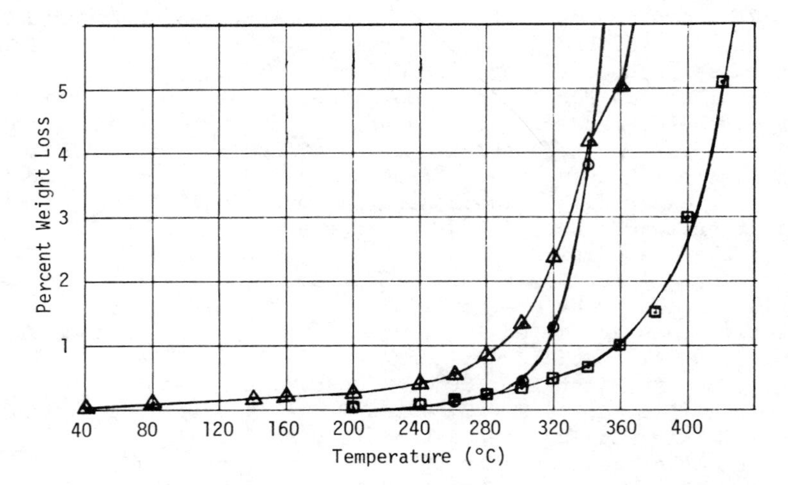

Figure 2. TGA curves for silicone coatings after vacuum bake-out for 4 hr at 135°C at < 1.0 torr pressure: (⊙) silicone co-polymer, (⊡) silicone II, (△) silicone I

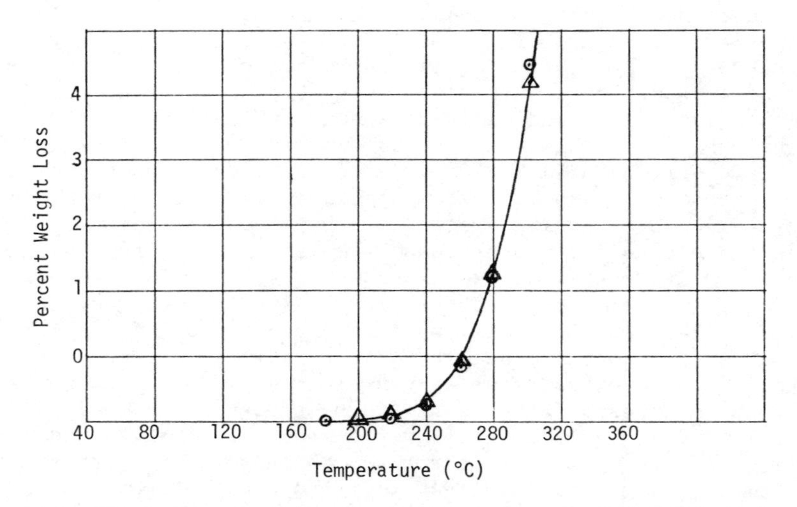

Figure 3. TGA curves for halocarbon waxes after vacuum bake-out for 4 hr at 135°C at < 1.0 torr pressure: (⊙) halocarbon 15-00, (△) halocarbon 12-00

Table IV. Electrical Properties of Coating Materials

MATERIAL	DIELECTRIC STRENGTH			VOL. RESISTIVITY (OHM-CM)		DIELECTRIC CONSTANT		DISSIPATION FACTOR	
	THICKNESS (MILS)	READING (VOLTS)	(VOLTS/MIL)	INITIAL	AFTER MOISTURE CONDITIONING	10^2Hz	10^5Hz	10^2Hz	10^5Hz
SILICONE I	105	40,725	388	1.2×10^{16}	4.4×10^{15}*	3.23	3.12	0.005	0.000023
SILICONE CO-POLYMER	105	46,170	469	1.4×10^{16}	2.0×10^{15}*	2.62	2.64	0.0016	0.000036
SILICONE II	115	26,180	228	5.1×10^{13}	1.9×10^{13}*	2.93	-	0.023	-
HALOCARBON 12-00	121	38,830	321	2.6×10^{15}	1.9×10^{15}**	3.76	3.11	0.016	0.0096
HALOCARBON 15-00	103	33,500	325	2.8×10^{15}	1.8×10^{15}**	3.39	3.05	0.0069	0.0064
DC6101 (DOW CORNING)	-	-	575	2.0×10^{15}	-	3.01	-	0.001	-
DC90-711 (DOW CORNING)	-	-	575	1.0×10^{14}	-	-	2.7	-	0.0004
G.E. EJC-261 (GENERAL ELECTRIC)	-	-	480	1.8×10^{15}	-	-	2.83	-	0.001

*10 Days at 70°C and 98% RH **10 Days at 23°C and 96% RH

Wire Bond Stress Evaluation

One of the main deterrents to the use of coatings on high density electronic circuits has been the stresses that coatings can impart to the fine interconnect wires and wire bonds. To measure these stress effects, a wire bond test matrix was designed which consisted of an aluminum metallized silicon die (0.2 x 0.2") to which one-mil diameter gold wires were ultrasonically bonded. Each die contained 42 bond pairs (84 bonds) and was mounted in a 42-lead ceramic package. This configuration provided for easy electrical measurement of each bond pair and diagnostic evaluation of the bonds. Eight packages were prepared for each coating. Bonds were nondestructively pull tested to two grams before coating, a test commonly used to assure their initial integrity. Electrical resistances were then measured for all bonds prior to coating, after coating, after 100 temperature cycles from -65 to 150°C, and after humidity exposure. The coatings were applied by both flow coating and spraying. The silicone co-polymer and Halocarbon 15-00 were selected as the best candidates to formulate mixtures and were formulated in 1:1, 2:1, and 3:1 ratios, respectively.

Results of Wire Bond Stress Tests

Considerable differences were found between the flow coated and spray coated packages. Flow coatings resulted in bridging between wires and between wires and the base of the package, whereas spray coating controlled the thickness better and bridging was avoided. After 100 temperature cycles, 24 wire bond failures out of 576 (~4%) were noted in the flow coated samples and all failures occurred in leads bridged by the coating. No failures were noted in the spray coated packages. Upon completion of the temperature cycling tests, the coated packages and uncoated controls were subjected to a 10-day cyclic humidity test and again bond resistances were measured. No changes in electrical resistances occurred, nor were there changes in the appearance of the coatings or evidence of corrosion.

Effects of Coatings on Devices

Because unpackaged active and passive devices in chip form are used in hybrid microcircuits, they are very sensitive to ionic contaminants, acidic or alkaline impurities, and moisture. Coatings applied to such surfaces must be of a higher purity than those for the more conventional electronic assemblies. It was therefore necessary to assess the effects of the coatings on the electrical parameters of a variety of electronic devices. For this purpose, a test circuit was designed which contained a variety of devices especially selected for their sensitivity to surface contamination (Figure 4). The circuit was designed to permit biasing and testing of the individual devices. In the

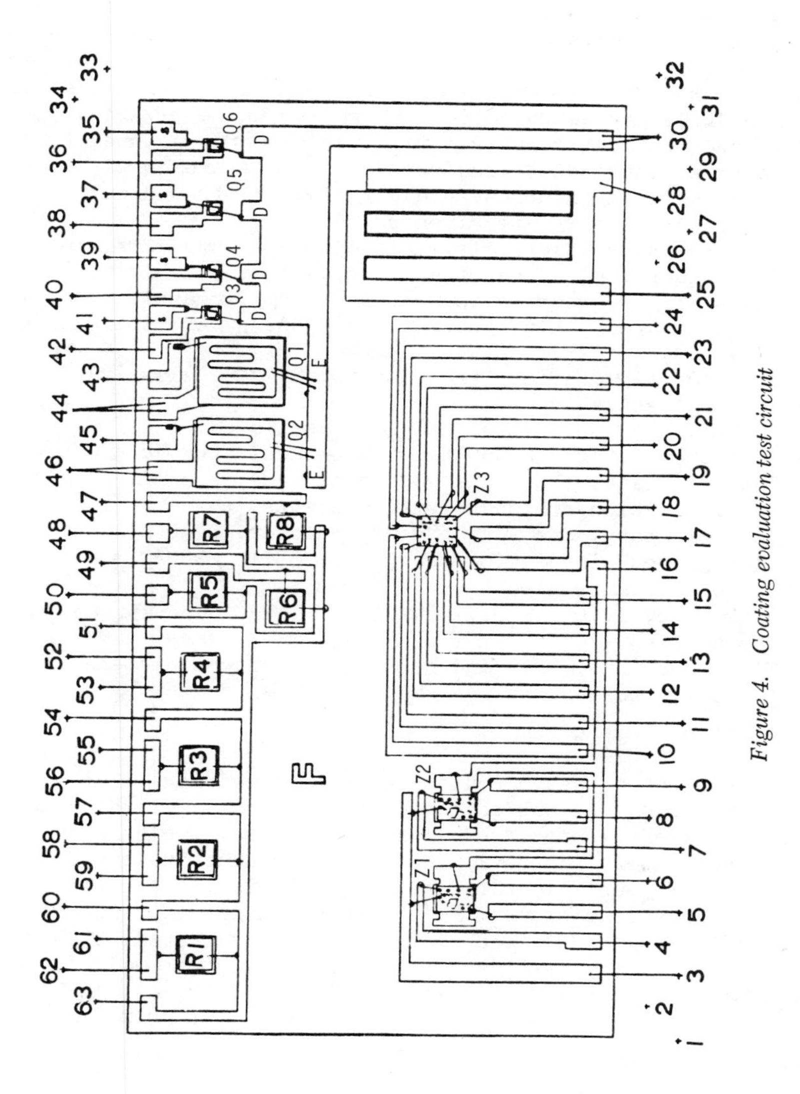

Figure 4. Coating evaluation test circuit

circuit, resistors R1 through R4 are PdAg-Pd oxide thick-film chip resistors. Two of the thick film resistors are glassivated and two are unglassivated. Resistors R5 through R8 are thin-film nichrome chip resistors. Two of these are also glassivated and two unglassivated. Transistors Q1 and Q2 are mesa transistors, 2NC5975, with unpassivated junctions. Transistors Q3 through Q6 are junction field effect transistors, 2N4391. Z1 and Z2 are operational amplifiers, LM741. The two op amps are glassivated. Two of the JFETs are glassivated and two are unglassivated. The interdigitated capacitor has 2.5 mil spacings and was used for insulation resistance and breakdown voltage measurements. Thin film metallization is nichrome/nickel/gold. Parts and devices used in the test circuit are given below:

- PdAg/Pd oxide thick film resistors (450 ohms ±20%, 500 ohms/square, glassivated and unglassivated, chip type).
- Thin-film nichrome resistors (500K ohms, glassivated and unglassivated, chip type, Hybrid Systems Part No. UHR-5E-500-3-N).
- Junction field effect transistors, 2N4391, glassivated and unglassivated.
- Operational amplifiers, LM741, glassivated.
- Mesa transistors, 2NC5975, with unpassivated junctions.
- Interdigitated thin film gold capacitor, 2.5 mil spacing—for insulation resistance measurements.
- Complementary metal oxide semiconductor, CD4001.
- Package, 63 pin, Tekform Part Nos. 35019, 35021 and 35029-2, seam sealed.
- Substrate, 1.600" x 0.775" x 0.027", MRC Superstrate, 99% alumina.
- Wire bonds, aluminum and gold wires, ultrasonic and/or thermally bonded.

High Stress Accelerated Tests

Sixteen coated test circuits (eight per coating) plus eight uncoated controls were subjected to high stress accelerated test conditions currently specified in MIL-STD-883B. One-half of the coated circuits were purposely seeded with conductive particles. The samples were serially subjected to mechanical shock (1500 g, 5 times), constant acceleration (10,000 g), high temperature storage (1000 hrs/150°C), and then electrically tested at 25°C and 125°C. Electrical measurements included: thick and thin film resistance; mesa transistors (beta, leakage); JFETs (zero gate drain current); op amps (V_{OFFSET}, I_B-); CMOS (quiescent current); insulation resistance; breakdown voltage.

Burn-In and Operating Life Tests

A second set of test specimens were subjected to the burn-in test per MIL-STD-883B, Method 1015.2, Condition A, for 240 hours at 125°C followed by the operating life test per MIL-STD-883B,

Method 1005.2, Condition A, 1000 hours at 125°C. All of the active devices were reverse biased during the tests and the thick and thin film resistors were subjected to a continuous voltage to evaluate electrolytic effects of the coatings. Electrical tests were performed after the burn-in test and at 168, 504, and 1000 hours for the operating life test. Leak tests and particle tests were also conducted after the burn-in and operating life tests.

Results of Electrical Tests

The electrical parameters for all devices measured after mechanical shock, constant acceleration, and high temperature storage were within specification requirements, except for the Pd/Ag/Pd oxide resistors. These resistors also changed in the uncoated control samples so changes could not be attributed to the coating. Fortunately, this type of resistor is no longer widely used in hybrid microcircuits. There was no evidence of particle induced shorts in any of the hybrids.

Summary of Results

1. Low-to-medium molecular weight, uncrosslinked silicones and halocarbons were evaluated as solvent soluble coatings to immobilize particles on hybrid microcircuits. Chemical, thermal, and electrical characterization of three silicones, four halocarbons, and three mixtures resulted in the selection of a block co-polymer of dimethylsiloxane with α-methylstyrene and a 3:1 mixture of this with a halocarbon wax as the best coatings for the application.
2. The coatings did not degrade fine wire bonds, provided that bridging between the wires or from the wires to the package or device did not occur. The coatings should be spray applied to a dried film thickness of 0.001 to 0.0038 cm (0.0004 to 0.0015 inch) to avoid bridging.
3. The coatings had no adverse effects when properly applied as coatings to the hybrid microcircuit devices tested in this study. Coated electronic devices after 1240 hours of burn-in at 125°C showed no electrical failures due to either the presence of mobile ions or other impurities in the coatings.
4. The coatings could be easily removed with trichlorotrifluoroethane (Freon TF) in a two-chamber, immersion degreaser or in a Soxhlet extractor. The use of these coatings for particle immobilization in microcircuits permits "hands-off" removal followed by standard rework and recoating procedures.
5. The coatings effectively immobilized particles introduced into packages either before or after coating application.

RECEIVED April 8, 1980.

13

Acetylene-Substituted Polyimides as Potential High-Temperature Coatings

N. BILOW

Hughes Aircraft Company, Culver City, CA 90230

Early in 1969, research was initiated, under U.S. Air Force Materials Laboratory sponsorship[1], aimed at the development of a new concept for chain extending and curing high temperature polymers, via an addition process. The need for such a process was apparent since until that time high temperature polymers such as polyimides, polyquinoxalines, polybenzimidazoles, etc., were all produced by condensation reactions which liberated large quantities of volatile byproducts during the polymerization, or cure. When prepolymers of these condensation polymers were used as molding compounds or laminating resins, they yielded porous structures with strengths and thermal oxidative stabilities well below that which would be expected from theoretical considerations and well below that which would have been observed with nonporous structures. Furthermore, high molecular weight polyheterocyclics invariably were too intractable to fabricate into thick structures.

After evaluating several potential cure concepts, it was found that uncatalyzed acetylene-terminated polyimide prepolymers could be chain extended and cured at temperatures of 200°C or above. It was furthermore discovered that high strength polyimides could be produced having thermal stabilities at least equivalent to those of conventional condensation type polyimides; thus, the polymerized acetylene groups had a high degree of thermal stability. This suggested that aromatic moieties were being produced. Support for this conclusion was provided at Hughes in a cursory study of one model compound which trimerized when heated in the absence of catalysts, producing a benzenoid structure. However, an in-depth study of this homopolymerization was first conducted by P. Hergenrother.[2]

0-8412-0567-1/80/47-132-139$05.00/0

Having made this discovery, various types of ethynyl-substituted polyimide prepolymers were synthesized, molded into void-free neat resin specimens, and evaluated as to their mechanical properties. Glass and graphite-fabric reinforced composites were also fabricated and tested and shown to retain high strengths at temperatures up to 370°C. In graphite fiber reinforced composites, up to 75% of the flexural strength was retained after aging in air for 1000 hours at 320°C.

Early descriptions of this work were first published in a series of limited distribution U.S. Air Force contract summary reports,[1] but the first public disclosure did not occur until 1974.[3] Subsequent papers were presented at meetings of the Society for the Advancement of Materials and Process Engineering (SAMPE).[4,5,6] Subsequently updated papers were presented at the American Chemical Society Meeting in Miami, Florida[7] and Honolulu, Hawaii. These papers provided extensive mechanical property data. Various U.S. and foreign patents have also been issued[8] and subsequently have been licensed to the Gulf Oil Chemical Company, Houston, Texas.

Discussion

Two series of polyimide prepolymers with terminal acetylene groups were synthesized from monomers such as those illustrated in Charts I and II. Both imidized prepolymers and their amic acid precursors are soluble in dimethylformamide or n-methylpyrrolidinone and the solutions can be used as varnishes for the preparation of coatings, adhesives, composite solid lubricants, and glass or graphite fabric reinforced laminated structures. Solvents such as tetrahydrofuran and acetone can often be used when the parent dianhydride contains central groups such as O, CH_2, S, $C(CF_3)_2$ and certain other moieties (see Charts I and II). Prepolymer of the general type, illustrated in Figure 1 (wherein n=1), melts at 195°C -198°C, and when cured at 250°C, and subsequently postcured for 10 hours at 370°C, has a glass transition temperature (Tg) as high as 370°C. Less severe postcures give lower Tg resins as shown in Table I. Mechanical properties of the cured polymer I (wherein n=1) are shown in Table II. [Compound I (wherein n=1) originally was designated HR600 by the author but the polymer is currently marketed under the Gulf Oil Chemical Co. trade name "Thermid 600". When n=2, the oligomer is designated HR602. Higher telomers have been designated in an analogous manner.]

Thermogravimetric analysis in air indicates that the cured polymer is stable to over 400°C although long term (1000 hour) air exposure of graphite fiber reinforced laminates at 316°C shows that strength retention is at best 75%, depending upon the specific type of graphite fiber reinforcement used. This is evidenced in Figure 3.

CHART I. Type I oligomers

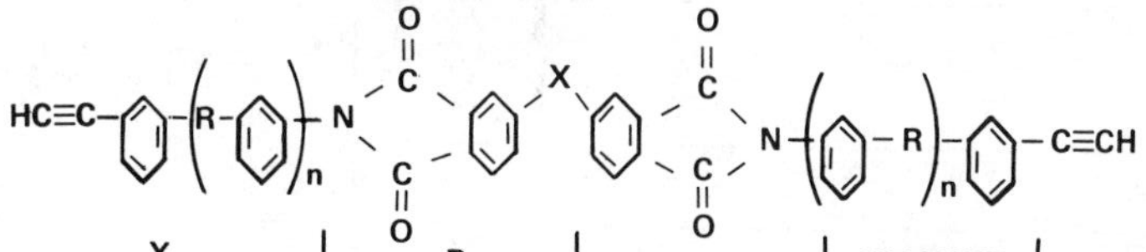

HC≡C–C₆H₄–(R–C₆H₄)ₙ–N(CO)₂C₆H₃–X–C₆H₃(CO)₂N–(C₆H₄–R)ₙ–C₆H₄–C≡CH

X	R	n	SOLVENT	M.P., °C	Tg, °C OF CURED POLYMER *
>C=O	–	0	NMP DMF	>250	>400
>C(CF_3)$_2$	–	0	ACETONE	184 - 5 190 - 4	410
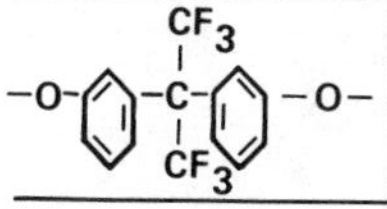 –O–C₆H₄–C(CF_3)$_2$–C₆H₄–O–	–	0	ACETONE	122 ± 2	312* 372**
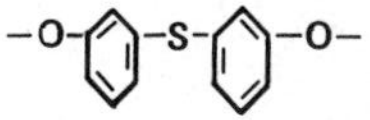 –O–C₆H₄–S–C₆H₄–O–	–	0	THF (ACETONE)	140	302
>C=O	–O–	1	ACETONE $CHCl_3$	175 - 185	>400

*AFTER 8 hr 370°C POSTCURE
**AFTER 24 hr 370°C POSTCURE

CHART II. Type II oligomers

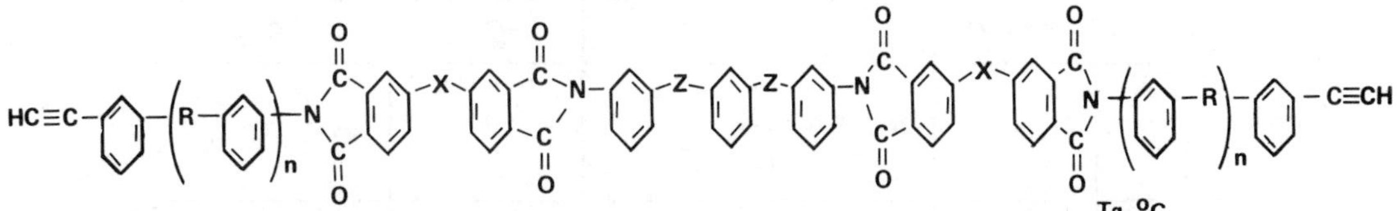

X	Z	R	n	SOLVENT	M.P., °C	Tg, °C OF CURED POLYMER	NOTES
$>C(CF_3)_2$	–O–	–	0	ACETONE	168 - 178	324 294	
$-O-C_6H_4-C(CF_3)_2-C_6H_4-O-$	–O–	–	0	ACETONE	182 - 185	296	
$>C=O$	–O–	–	0	NMP, DMF	195 - 200	370	HR600
$-CH_2-$	–O–	–	0	THF	160	263	
–O–	–O–	–	0	THF (ACETONE)	150	253	
$-O-C_6H_4-S-C_6H_4-O-$	–O–	–	0	THF (ACETONE)	134	212	
$>C=O$	–O–	–	0	NMP, DMF	200 - 210	CURE INCOMPLETE	PARA ETHYNYL
$>C=O$	–S–	–	0		175		
$>C=O$	–O–	O	1				HR650

TABLE I. GLASS TRANSITION TEMPERATURES OF POLYIMIDE (FIG 1, n =1) CURED UNDER VARIOUS CONDITIONS

Postcure Conditions		
Time, hr	Temp., °C	Tg, °C
40	290	285
40	320	295
40	340	310
40	370	320
40	400	340

These values, obtained by thermomechanical analysis, were somewhat lower (20-25°C) than those obtained by dynamic methods.

TABLE II. MECHANICAL PROPERTIES OF CURED POLYIMIDE I (n=1) (5)

Property	
Tensile Strength	14,000 psi
Tensile Modulus	550,000 psi
Elongation	2.6%
Flexural Strength	18,000-21,000 psi
Flexural Modulus	650,000 psi
Compressive Strength	Up to 66,000 psi
Hardness (Barcol)	85

SAMPE

Differential thermal analysis indicates a two step polymerization, since a small exotherm is observed at about 200°C and a major exotherm at 240-260°C. This conclusion is supported by polymerization mechanism studies of E.G. Jones, et.al.[9] Prepolymers have generally been molded and cured at the latter temperature, and subsequently postcured in air for 8-10 hours at 370°C. The latter treatment has been found desirable when optimum long term high temperature stability is to be realized.

Higher degree of polymerization prepolymers of the type illustrated in Figure 1 (to n=14) have Tg's which are progressively lower as the degree of polymerization increases (see Table III); thus, their maximum potential high temperature capability is compromised depending upon the application. Polymers with the higher n values (e.g., n=5) have provided films in studies conducted to date.

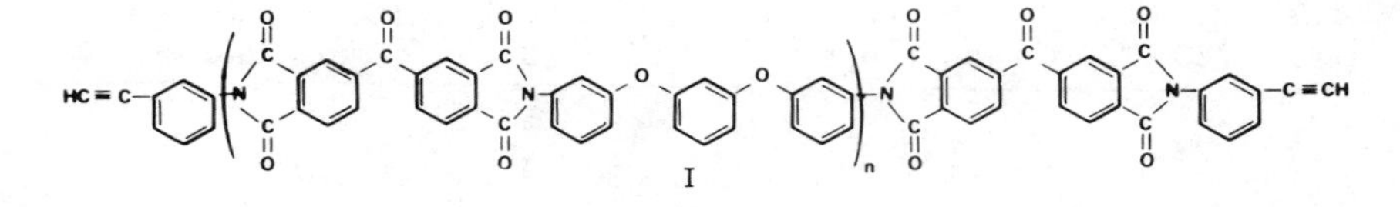

(Polymide I)

Figure 1. Prepolymer of the general type

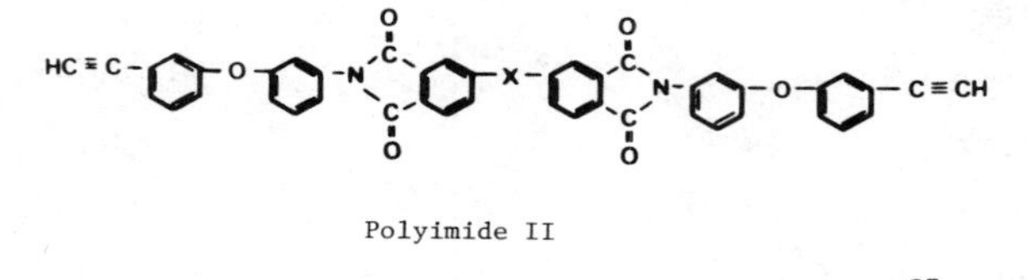

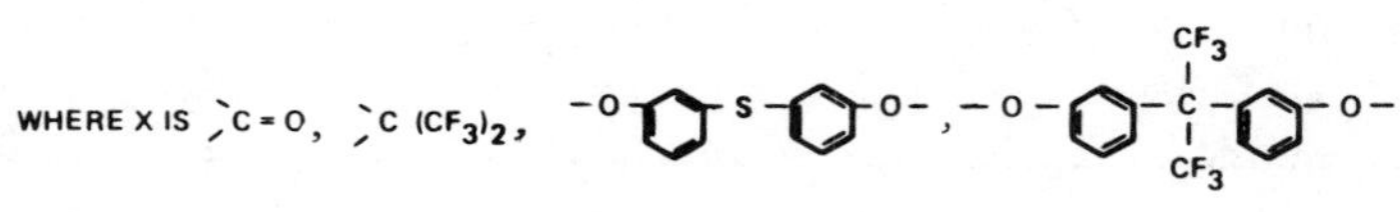

Figure 2. Four acetylene-substituted polyimide oligomers

TABLE III. Tg's AS A FUNCTION OF n
(HR 600 TYPE PREPOLYMERS)

n	Tg, °C
1	320
2	268
3	254
5	240
Postcures at 340°C in air	

The acetylene-substituted polyimides having structures analogous to that of Figure 1 (n=1) were made from various combinations of 4 diamines, 6 dianhydrides, and 2 aminoarylacetylenes. Structures of the monomers used are illustrated in Charts I and II. Several of the acetylene-substituted polyimide oligomers (Chart I) were made without a diamine. Four of these are illustrated in Figure 2. Prepolymers of this latter type are soluble in acetone and tetrahydrofuran, thus allowing them to be used in acetone-based varnishes or coatings.

Dielectric properties of the polyimide of Figure 1(n=1) were also measured and found to be virtually constant over the temperature range of 20-320°C and frequency range of 9.0-12 GHZ. The dielectric constant was 3.13 ±0.01 and the dissipation factor was 0.5 ±0.1.

Prepolymers of structure I were also found useful as adhesives for titanium. Typical Ti-Ti lap shear strengths depended upon the specific structure, but typical values were as follows:

	Temperature Range, °C	Lap Shear Strengths, psi
n=1	20-260°C	1200-2100
n=2	20-260°C	1600-3600
*	20-260°C	2700-4200

[*Figure 1, n=1, with the terminal ethynylpheny groups replaced by 3,3'-ethynylphenoxyphenyl.]

A comprehensive review of the adhesive research results has recently been submitted for publication.[10]

The strength range indicated includes initial ambient temperature values as well as values obtained after 1000 hrs. aging at 260°C. Long term use at the latter temperature thus is highly practical.

Flexural strengths of graphite fiber reinforced composites were measured before, during, and after prolonged thermal aging in air at 316°C. Results of these tests are shown in Figures 3 and 4. The study demonstrated that useful properties are maintained over the long term.

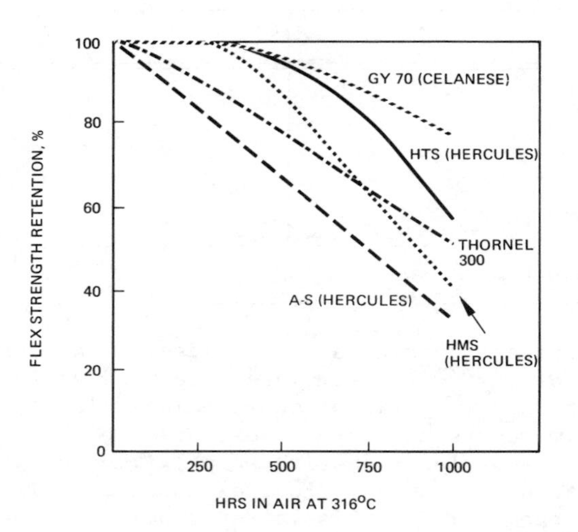

SAMPE

Figure 3. Flexural strength retention of graphite fiber-reinforced laminates made from prepolymer I (n = 1) (6). Refer to Figure 1.

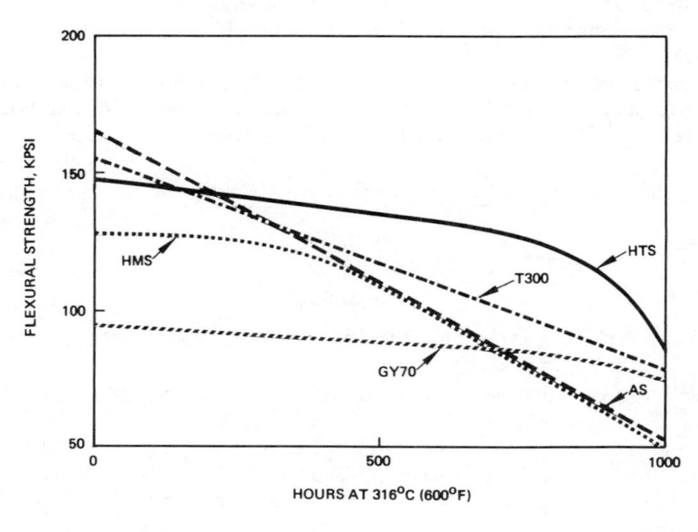

SAMPE

Figure 4. Flexural strengths of graphite fiber-reinforced laminates made from prepolymer I (n = 1) (6). Refer to Figure 1.

Polyimide I (n=1) not only has been used to prepare high strength glass and graphite fiber reinforced laminates, but also has provided very high compressive strength moldings, and high performance composite solid lubricants.

Interlaminar shear strengths of Celion 3000 graphite fiber reinforced laminates made from Polyimide I (n=1) were measured as a function of both thermal aging and humidity exposure at 70°C and 95% R.H. Results of these tests are shown in Figure 5. It is evident from this study that even after 1000 hours of continuous exposure, over 50% of the initial strength is retained.

Mechanical properties of other cured polyimides of the type illustrated in Figure 1 vary with the degree of oligomerization as shown in Tables IV, V, and VI. Table IV lists mechanical properties, Table V shows the various glass transition temperatures, and Table VI shows typical gel times. For applications such as in coatings, it is anticipated that polymers with the higher values of n will be most satisfactory, since such materials have a lower crosslink density and tend to be tougher.

TABLE IV. MECHANICAL PROPERTIES AS A FUNCTION OF n

HR600 TYPE PREPOLYMERS (STRUCTURE I)

	Tensile Strength, KPSI		Modulus, MPSI		Elongation, %		
n	20°C	200°C	20°C	200°C	20°C	200°C	250°C
1	14	14.0	0.55	0.5	2.6		
2	17	6.6	0.55	0.3	3.6	3.5	8.6
3	14	6.1	0.55	0.3	2.8	3.0	10.4
4	19	6.0	0.55	0.3	4.8	4.8	8.4

TABLE V. Tg's AS A FUNCTION OF n

n	Tg, °C
1	320
2	268
3	254
5	240
Postcures at 340°C, Structure I	

TABLE VI. GEL TIME AS A FUNCTION OF n

n	Time, Sec
1	180
3	275
5	399
14	288
Gel time at 250°C, Structure I	

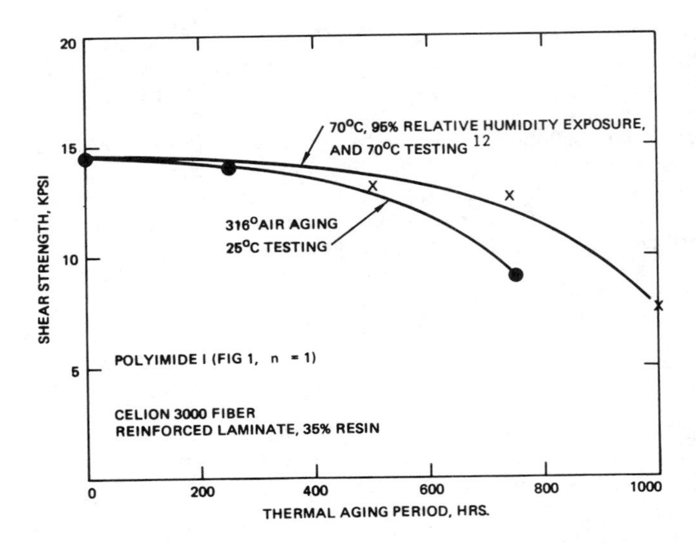

Figure 5. Interlaminar shear strength as a function of air and humidity aging

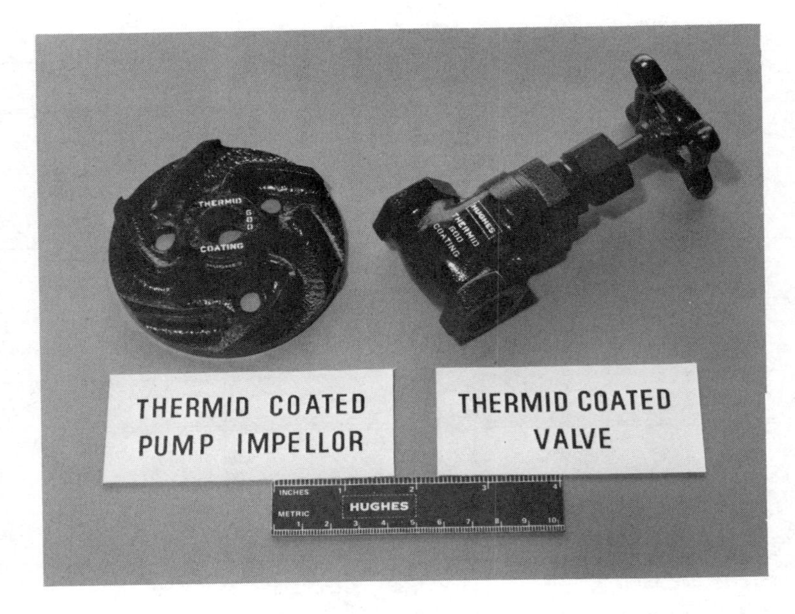

Figure 6. Structure I polymer adherent coatings on selected equipment

Research conducted to date also has included an investigation of fluidized bed coating of aluminum wire. This investigation was performed with Polymer I (n=1) by drawing heated wire through a Teflon TFE die while heating the prepolymer at 188°C. The coating produced was of uniform thickness and adhered well to the aluminum wire. Another study involved the coating of iron pipe with a 5% solution of polymer I (n=1) in dimethylformamide. Coatings were air dried, then cured at 250°C, and subsequently postcured up to 320°C. Adherent coatings produced from the polymer of structure I (n=1) have also been applied to both control valves and pump impellers as shown in Figure 6; however, the evaluation of these coatings has not yet been completed.

Synthesis Procedure

Imidized prepolymers were prepared by reacting a 10% by weight solution of dianhydride in N-methylpyrrolidinone (NMP) with a 10% solution of the diamine in the same solvent. After stirring for 1/2-1 hour at ambient temperature, the ethynylated amine in NMP was added, and the reaction was allowed to continue for 1/2 hour. The mixtures were heated and sufficient benzene was added to adjust the boiling point to 150°C. When water evolution ceased (4-6 hours), the mixtures were cooled, solvent was removed on a rotary evaporator, and the residual oligomer was subsequently triturated in ethanol, then dried.

Summary

Fourteen thermosetting acetylene-substituted polyimide prepolymers were synthesized and evaluated as to their thermal properties, solubility, and in some cases mechanical properties. Cured resin Tg's ranged from 212-410°C. Low boiling solvents such as acetone and tetrahydrofuran were useable in some cases, opening up the possibility of producing solvent based coatings for 300-370°C applications. Coatings prepared to date have shown considerable promise although the first comprehensive paper on their properties will not be published until mid 1980.

Aknowledgements

The author wishes to express his appreciation to Mr. A.A. Castillo, Mr. S. Goodman, Dr. A.L. Landis, Mr. W.H. Fossey and Mr. J. Tedesco for their valuable contributions to this research.

Literature Cited

1. Air Force Contracts F33615-69-C-1463; F33615-71-C-1458; F33615-71-C-1228.
2. P. Hergenrother, Polymer Preprints, Amer. Chem. Soc. Meeting, Honolulu, Hawaii, April 1979.

3. A.L. Landis, N. Bilow, et al, Polymer Preprints, Amer. Chem. Soc. Meeting, Atlantic City, N.J., 15, 533, 1974.
4. N. Bilow et al, SAMPE Symposium Preprints, San Diego, CA, Apr. 28-30, 1975.
5. N. Bilow & A.L. Landis, "Recent Advances in Acetylene-substituted Polyimides," Natl. SAMPE Technical Conference, Vol. 8, p. 94, Seattle, WA, Oct. 12-14, 1976.
6. N. Bilow, et.al., "New Developments in Acetylene-Terminated Polymides," 23rd Natl. SAMPE Technical Conference, Anaheim, CA, May 2-4, 1978.
7. N. Bilow & A.L. Landis, Polymer Preprints, Amer. Chem. Soc. Meeting, Miami, Fla., 1978, pps 23-28.
8. N. Bilow, A.L. Landis & L.J. Miller, U.S. patents 3,845,018; 3,879,349; 4,098,767; 3,928,450; 3,864,309; 4,075,111; 4,108,836.
9. E.G. Jones, J.M. Pickard, D.L. Pedrick, Polymer Curing and Degradation," Report No. TR-78-162, Air Force Materials Laboratory, Nov. 1978.
10. N. Bilow, et.al., submitted to Journal of Applied Polymer Science, June 1979.

RECEIVED January 8, 1980.

14

Development of Launch-Tube-Mounted Polyurethane Seals for Missile Launch Systems

JOSEPH F. MEIER, GEORGE E. RUDD, ALBERT J. MOLNAR, DONALD D. JERSON, and MORRIS A. MENDELSOHN

Westinghouse Research and Development Center, Pittsburgh, PA 15235

GIRARD B. ROSENBLATT

Marine Division, Westinghouse Electric Corporation, Sunnyvale, CA 94086

Over the last 15 years, scientists and engineers at Westinghouse Electric Corporation have been involved with polymer developments for missile launch systems. The subject areas investigated are flexible urethane foam (1,2), rigid-ductile polyurethane foam (3), rigid-brittle phenolic foam (4-8), neoprene launch tube liner pads (9,10,11,12), cast polyurethane launch tube liner pads (13,14,15,16), cast polyurethane launch seals (17,18,19), and creep resistant EPDM missile support pads (20). A schematic of a typical launch system giving the location of some of these materials is shown in Fig. 1.

A recent area of investigation deals with both launch tube-mounted and missile-mounted seals. These seals span the annular space between missile and launch tube and retain eject gases during launch. The eject system may be either a "cold launch" in which a gas generator or rocket motor accelerates the missile to launch velocity.

This paper concerns launch tube-mounted seals and defines seal performance in terms of the EI characteristics of candidate materials (E = modulus and I = moment of inertia). We also introduce the concept of push-through testing of nominal one-in. wide radial strips of seals, describe pressure testing of two-ft. diameter seals of various cross section, discuss the importance of seal cross sectional shape to include the humped-notched design, and show the interrelationships of theory and experimental results. The preparation of full scale 77-in. diameter seals for simulated launch tests is also described.

0-8412-0567-1/80/47-132-151$06.25/0

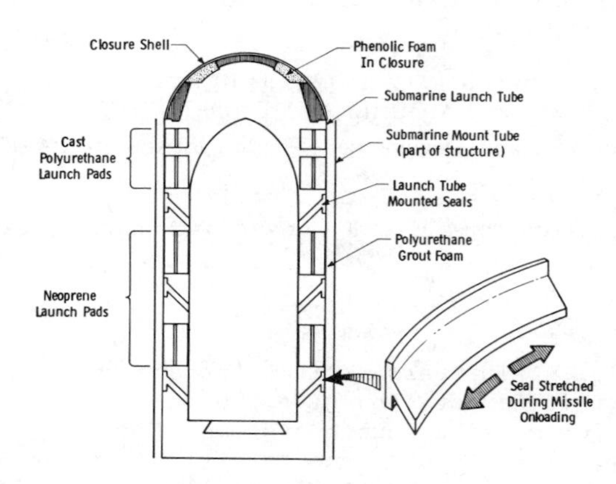

Figure 1. Location of polymeric components in the launch tube environment

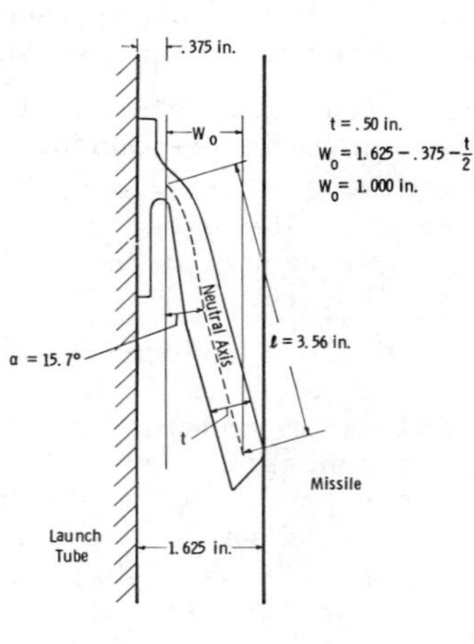

Figure 2. Nominal configuration of present lip-type launch seals without pressure (seal cross-section A)

Experimental Procedures

Functional Requirements of Seals

1. During launch the seal must not invert (i.e., flip) until the missile passes the seal.

2. The seal must not mechanically overload the missile skin.

3. The seal must not bottom before the shock pads (i.e., the bottomed thickness of the shock pads in the annular gap between missile and launch tube must be $\geq$ than the bottomed seal thickness).

4. Excessive friction must not be exerted on the missile by the seal.

5. Launch pressure of the existing system is ~100 psi, but a theory should be capable of predicting seal behavior at various launch pressures and gaps.

Theoretical Developments. In considering a seal of the cross section shown in Fig. 2, molded from the neoprene formulation given in Table I, two fundamental dimensions are defined, W_o and ℓ. W_o is the annular gap between missile and launch tube minus the distance from the launch tube bonding surface to the hinge point in the throat of the seal (i.e., 0.375-in.) minus the thickness of the beam (0.5-in.) divided by 2, or:

$$W_o = 1.625 - 0.375 - \frac{0.5}{2} = 1.0 \text{ in.}$$

For the system of interest, the nominal W_o is one-inch. ℓ is defined as the effective length of the lip seal neutral axis in terms of bending (OT in Fig. A of Appendix 1). Note that ℓ is considerably greater than the annular gap before pressurization. After pressurization ℓ becomes smaller, the seal deforms, takes on a humped n shape, and approaches W_o.

Uniform Beam Analysis. A unit radial strip of seal was analyzed as a uniform beam with necessary equilibrium forces at the point of tangency where the seal is assumed to be parallel to the missile. The derivation is given in Appendix 1. These conditions are met when a numerically or graphically solved transcendental function $g(\ell)$ equals zero, where ℓ is treated as the independent variable.

Solutions obtained by simultaneously satisfying the above conditions (i.e., when $g(\ell) = 0$) are shown in Fig. 3. It is interesting to note from Fig. 3 how theory predicts, for a typical set of data, and a stationary missile, that between 160 and 165 psi launch pressure, the no solution case (i.e., a flip condition)

TABLE I

NEOPRENE LIP SEAL FORMULATION

Ingredient	Parts by Weight	Weight %
Neoprene W	100	49.2
Stearic Acid	1.0	0.49
Akroflex CD	2.0	0.98
MgO	4.0	1.96
Andrez 8000AE	40.0	19.56
HiSil 233	40.0	19.56
ZnO	5.0	2.45
Carbon Black N-326	5.0	.245
Circo Light Process Oil	2.5	1.23
Cumar P-10	2.5	1.23
Benzothiazyl Disulfide	0.5	0.25
Tetramethylthiuram Monosulfide	0.5	0.25
Sulfur	1.0	0.49
	204.0	100.1%

Production Cure Conditions 90 mins @ 302°F (150°C)

is reached prior to the theoretical buckling load. (For a moving missile, the value would be 1/2 of these values.)

Taking solutions from Fig. 3, it is possible to construct a plot of launch pressure, p_f vs. W_o, as shown in Fig. 4. It should be noted that p_f is very dependent on the "effective EI characteristic". Since modulus varies with strain and strain rate, some uncertainty existed as to which modulus is appropriate. However, the curves of Fig. 4 were constructed using an effective E of 3,000 psi as a lower limit and E of 27,000 psi as an upper limit. Tests indicate the effective modulus is very near the lower value for this analysis to compare favorably with test results at slightly greater than nominal gaps ($W_o \approx 1.25$ in.). Furthermore, the analysis indicated a sharper rise in the invert pressure characteristics than test results at smaller missile-seal gaps ($W_o \approx 1.0$ in).

Non-uniform Beam Analysis. More recent work has been completed using a large bending analysis which includes equation 12 of Appendix 1 with a variable I (non-uniform beam) to reflect the actual seal cross-sectional geometry. This procedure involves a double numerical integration of equation 12 along with an iterative scheme to satisfy the external force equilibrium and slope constraint at the tangent point, T, on the missile. Results from this analysis are also compared to the stationary missile test results (on the two-ft. diameter pressurized seal) in Fig. 5. Also, the corresponding invert pressure for a launch condition where the friction force (μN in Fig. A of Appendix 1) between the missile and seal reverses direction and reduces the invert pressure is shown. This improved analysis predicts the seal invert pressure throughout the functional range of the missile-seal gap without any variation of the modulus, E (i.e., an effective or variation of modulus is not necessary as in the uniform beam method). The required material modulus used in this analysis is very near to that which would be obtained from standard procedures (i.e., from tensile or bending test specimen).

Two-Foot Diameter Seal Testing. To obtain pressure test data on two-ft. diameter seals, it was necessary to construct special seal test equipment. This apparatus is shown in Figs. 6 and 7.

The tester consists of two cylinders separated by an annular gap. The outside cylinder corresponds to the launch tube and the inner cylinder corresponds to the missile. The test seal is bonded to the launch tube using Chemlok 304 or 305 adhesive (Hughson Chemical Co.). High pressure N_2 gas from an accumulator is dumped by a quick release valve through a manifold to the annular gap between the simulated missile and launch tube to pressurize the seal. Unlike a true launch, the test missile remains stationary during the test. Different diameter "missiles" are available to vary the annular gap in a concentric manner, and the "missile" can be lo-

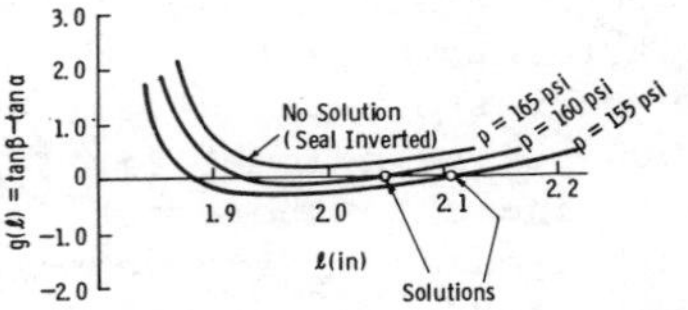

Figure 3. Graphical illustration as the launch pressure is increased and the no solution condition occurs

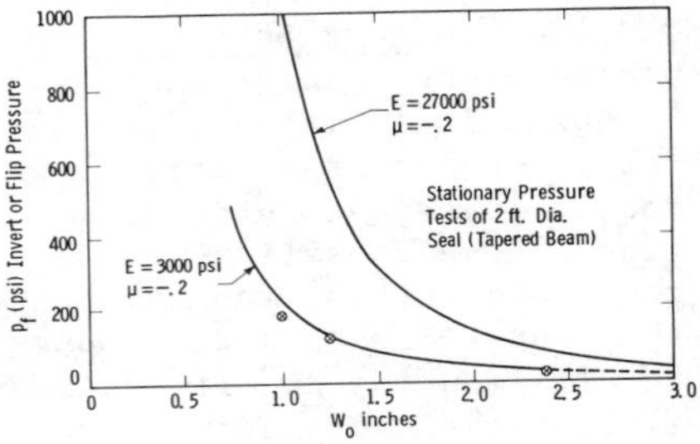

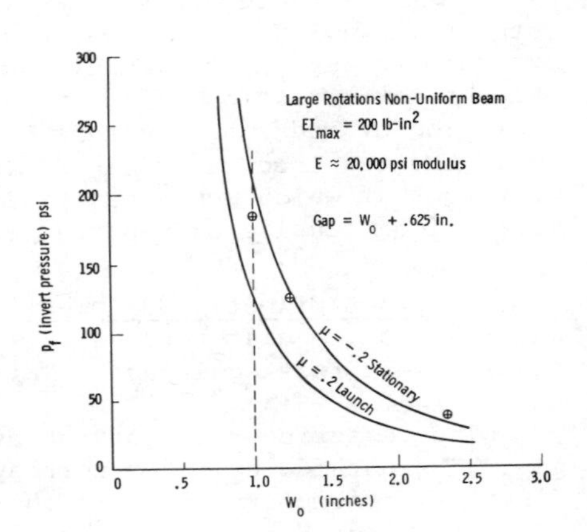

Figure 4. Summary of seal invert conditions for a stationary missile using force equilibrium and simple uniform beam theory dy/dx = *0*

Figure 5. Theoretical pressure invert characteristics using large rotations and nonuniform beam theory for tapered beam launch seal: (⊕) pressure tests of 2-ft-diameter urethane seal (tapered beam) with stationary missile

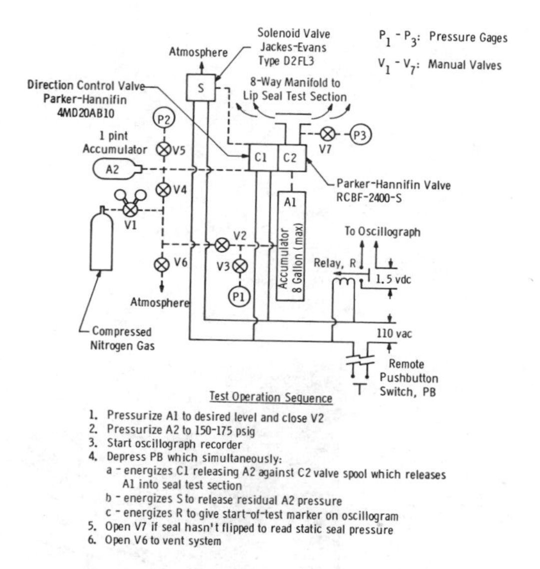

Figure 6. Schematic of the 2-ft lip seal pressure test facility

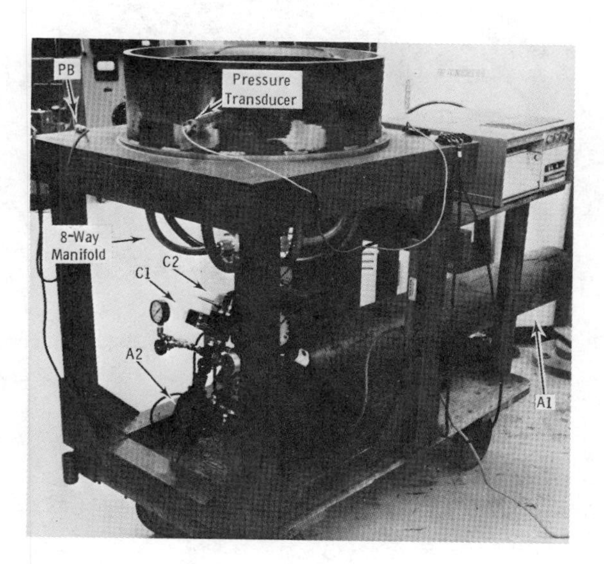

Figure 7. Two-foot-diameter seal test facility

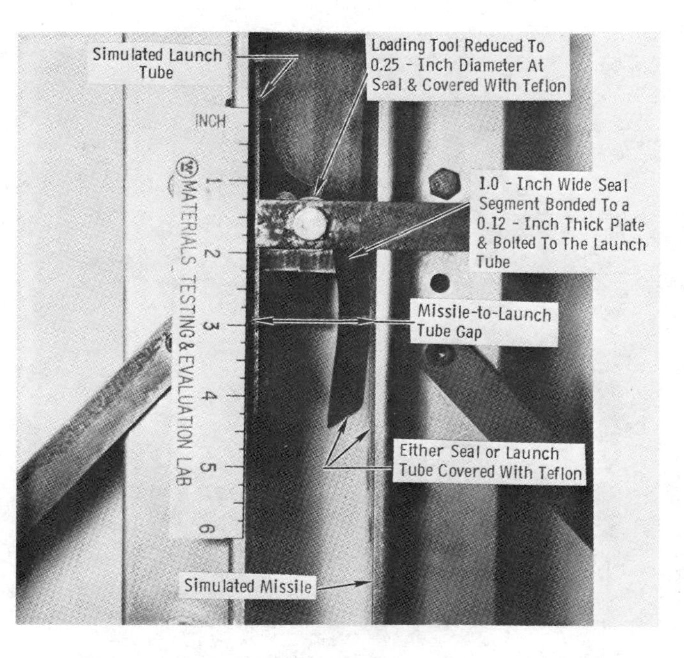

Figure 8. Push-through test fixture mounted in an Instron Universal testing machine

cated eccentrically to give a non-uniform gap. Seal vertical deflection was measured by an LVDT mounted in contact with the top of the seal.

Push-Through Testing. Although good agreement was obtained between theory and two-ft. diameter tests, the expense of the mold and time requirements necessitated the development of a simpler test method. To serve this need, the push-through apparatus shown in Fig. 8 was developed. Push-through tests were conducted on nominal one-in. wide radial strips of neoprene lip seals bonded to a steel mounting plate and deflected at 2-5 in./mil at any desired gap. From these experiments, load-deflection data were recorded at various gaps, the peak loads converted to pressure by the formula $p = \frac{2P}{W_0}$ (see Results and Discussion section) and compared to the theory which predicts seal invert pressure. Push-through tests were also conducted on one-in. wide strips of cast polyurethanes A and B, the formulations of which are given in Table II. This material was selected as the primary candidate material because of its outstanding performance characteristics and ease of procurement and processing.

Neoprene Seal Material. The neoprene seal material used in much of the early seal work was molded from the formulation given in Table I. Its properties are listed in Table III.

Cast Polyurethane Seal Materials. Due to the expense related to molding two-ft. diameter neoprene seals, we decided to prepare both seal segments and two-ft. diameter seals from cast polyurethane. Although many polyruethane formulations were evaluated, most work centered on the three polyurethane formulations given in Table II. These represent a very soft formulation, A, used primarily to check out the seal tester at low pressures; B, a material with a much higher modulus, close to the neoprene material and C, a commercially available polyether prepolymer and cure system from DuPont - Adiprene L-167/MOCA.

Properties of these polyurethane materials are contrasted with the neoprene properties in Table III.

Seal Bottoming Characteristics. An important consideration in seal design is the force that the seal exerts on the missile skin as the missile-launch tube gap changes during launch, high seas, docking, etc. To measure this force, a lateral excursion test was employed. In this test, a one-in. wide strip of seal material is bonded to a plate and deflected in an Instron as shown in Fig. 9. Although conducted at a low rate (2-5 in./min.), the load vs. displacement data collected are useful in comparing materials and geometries. More detailed experiments have been conducted where the distribution of the total lateral force is determined, but these measurements are not described in this paper.

TABLE II

CAST POLYURETHANE FORMULATIONS

Sample Code	Prepolymer Preparation				Amount of Prepol or Isocyanate Component, gms	Reactive Additive, Diol, Triol, etc.	Amount of Reactive Additive		Cure Conditions, Time, Hr/Temp., °C
	Polyol	Equivalents of Polyol	Isocyanate Component	Equivalents of Isocyanate			Eq.	gm	
A			Adiprene L-100	1.0	1023	HQDBHEE[a]	0.95	94	19/110
B	Polymeg 1000	1.0	Mondur HX	2.2	591	1,4 butanediol	0.95	42.75	24/85
C			Adiprene L-167			MOCA[b]			2/70

(a) hydroquinone di(β-hydroxyethyl ether)

(b) 4,4´-methylene bis(2-chloroaniline)

TABLE III

PHYSICAL PROPERTIES OF NEOPRENE AND POLYURETHANE ELASTOMERS

Sample Code	Stress at 100% Strain, psi	Stress at 200% Strain, psi	Tensile Strength,* psi	% Elongation	Shore A Hardness
Neoprene	770-870	960-1090	2810-3285	575-780	91-95
Polyurethane A	--	300	1715	>1000	62
Polyurethane B	--	1290	3150	>1000	95
Polyurethane C L-167/MOCA	1831-1921	2441-2595	4190-4883	340-370	98

*Tensile data measured at 20 in./min.

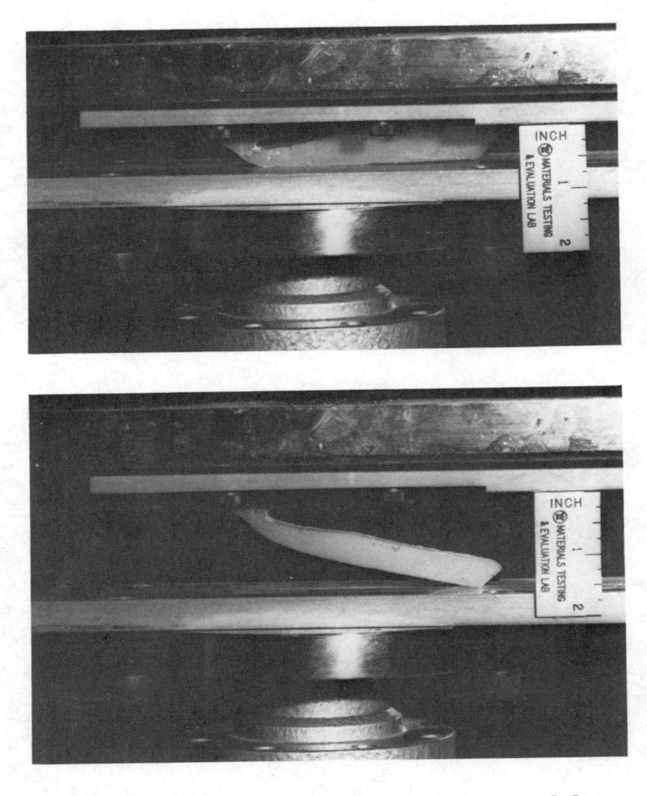

Figure 9. Seal bottoming test: top, seal completely bottomed; bottom, seal position at nominal gap of 1.62 in.

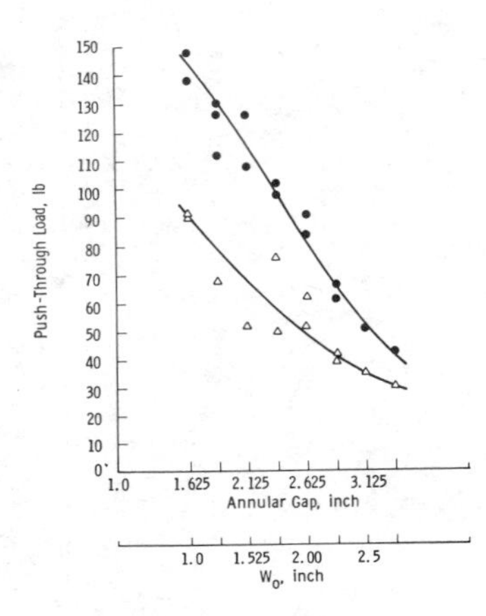

Figure 10. Peak push-through load as a function of annular gap for neoprene seal segments: (●) first cycle data; (△) second cycle data

Teflon Bonding Procedures. Ten mil thick Na etched, skived Teflon tape is bonded to the upper polyurethane seal surface using the proprietary adhesive DP8363 from Conap, Inc., Olean, NY. The purpose of the Teflon is to reduce the coefficient of friction between the seal and the missile skin during missile onloading and launch.

Results and Discussion

Push-Through Testing

Influence of Gap. Utilizing the push-through tester shown in Fig. 8 and radial strips of neoprene seals with the cross-section shown in Fig. 2, multiple push-through cycles were conducted as a function of gap. A virgin specimen was used at each gap. Selected data are shown in Fig. 10 for the neoprene seal material, and typical push-through curves showing actual seal deformation are presented in Fig. 11. From Fig. 10 several effects are obvious. Namely, the push-through load exhibits a large decrease from first to second cycle; in some cases there is a large variation in push-through load; at large gaps, the push-through values converge. It is evident that launch seal performance would be degraded by multiple pressure cycling and that the launch pressure retained by a seal would also be expected to decrease as the annular gap increases.

Influence of Test Temperature. The neoprene lip seal formulation given in Table I contains a thermoplastic material - Andrez 8000AE (Anderson Chemical Co.) - which is a high styrene content copolymer of styrene and butadiene. Because of this ingredient, it could be anticipated that seal performance would be degraded as a function of temperature. This is shown in Fig. 12. It should be noted that the data in Fig. 12 were obtained on seal segments that had a .010 in. Teflon film bonded to the upper surface. We have observed that the Teflon film increases the push-through load by about 10%.

In a similar manner, the push-through load of cast polyurethane seals, specifically formulation C, have been determined over the same temperature range and do not show nearly the temperature range and do not show nearly the temperature sensitivity that the thermoplastic containing neoprene formulation does.

Influence of Polyurethane Formulation. Similarly, two polyurethanes (A and B of Table II) were cast into one-in. wide seal segments of the cross-section shown in Fig. 2 and push-through tested. Polyurethane A was formulated to be a relatively flexible material that could be used in the two-ft. diameter seal tester to check out the equipment at low pressures. Polyurethane B was developed to be similar in strength to the neoprene material whose formulation is given in Table I. Push-through data for the poly-

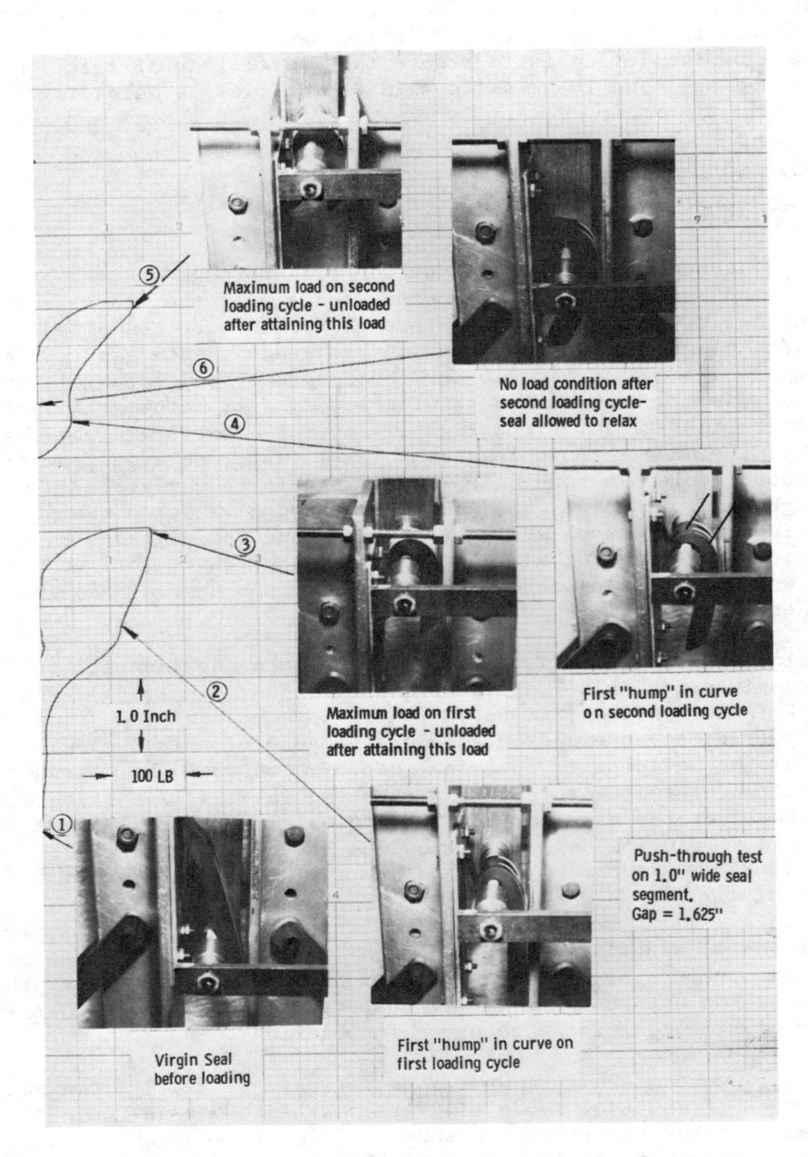

Figure 11. Correlation of push-through load-deflection curve characteristics with neoprene seal segment distortions

urethane materials are summarized in Fig. 13. Also shown in Fig. 13 are pressure invert data calculated by converting peak push-through loads to pressure loads using the equation $P_f = \frac{2P}{W_o}$ for (formula discussed in Section B).

Two-Foot Diameter Seal Tests. Subsequent to developing the push-through data in Fig. 13, two-ft. diameter seals of polyurethane A and B were fabricated and tested in the seal test facility shown in Fig. 7. The relationship between actual pressure tests and the push-through values for radial strips of polyurethane, also shown as the numbered points in Fig. 13 is excellent. After several of these tests were completed, additional pressure tests were conducted on two-ft. diameter seals of polyurethane B which is of comparable stiffness to the neoprene seal material. These results are also plotted in Fig. 13. Data from pressure tests with polyurethane B did not agree well with the predicted values computed from the push-through loads. It should be kept in mind, however, that in push-through testing, the rate is 2-5 in/min and in the pressure tests, the seal goes from a no pressure situation to a flipped position in 0.5 sec.

As a result of these tests, it was found that better agreement between push-through load and pressure invert tests could be obtained by using $P_f = \frac{1.75\ P}{W_o}$ and since we were interested in seals with high pressure capabilities, this relationship was adopted for further experimental work. If the integrated moments of a uniformly (pressure) loaded and point loaded beam are equated the result is $P_f = \frac{1.5\ P}{W_o}$. If the maximum moments are equated for the two conditions, the result is $P_f = \frac{2\ P}{W_o}$. Test results are between the two.

Scale-up to 77-In. Diameter Seals. One significant difference between the two-ft. diameter seal testing and actual missile launch tests is that the missile is moving in a launch. This could not be tested experimentally in our laboratory experiments with polyurethane seals. However, launch tests using the neoprene seal had been conducted previously and it was possible to correlate actual launch data and seal pressure performance with push-through values of segments cut from large seals. This was done for a number of experiments and it was found that the actual invert pressure of neoprene seals was about 90-100 psi at a nominal 1.6-in. gap which was equal to about half of the value predicted from push-through testing. This makes qualitative sense as the pressure required to invert a seal under launch conditions would be expected to be less than under stationary conditions due to the influence of friction between the seal and the missile. That is, as the missile begins to move, the seal tends to be dragged with the missile and the pressure required to invert the seal would be less.

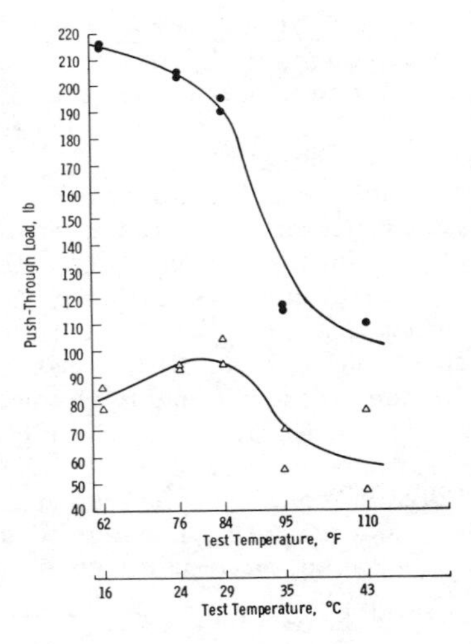

Figure 12. Peak push-through load of neoprene seal segments as a function of test temperature at a constant annular gap of 1.625 inch (●) first cycle data, (△) fifth cycle data

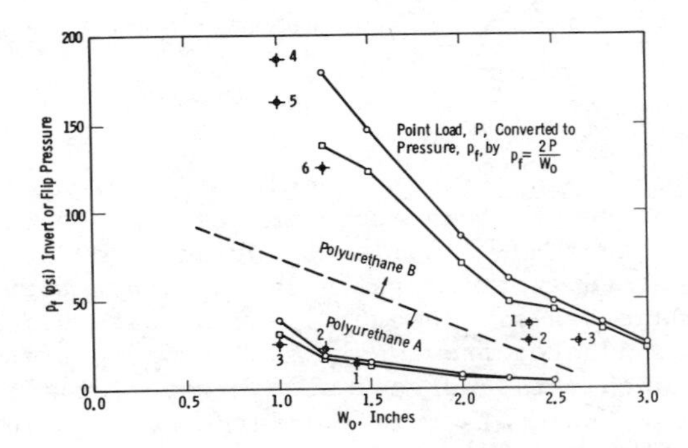

Figure 13. Summary of point load invert tests of 1-in. radial seal strips and actual pressure invert tests of 2-ft-diameter seals for both polyurethane A and B: (○) first cycle push-through; (□) second cycle push-through; (⊕) n—pressure invert tests of a complete 2-ft-diameter seal, n-cycle number

The theoretical influence of friction for both the stationary and the moving missile on sealing pressure is shown in Fig. 14. Superposed on the figure are experimental data for both two-ft. diameter urethane seals, urethane segments, and full scale neoprene seals.

Seal Geometry Considerations. Since it appears that the maximum launch pressure with tapered beam seals is limited both with the neoprene materials and polyurethane B, it was decided to evaluate seal geometry as a means of attaining higher sealing pressure.

To approach this goal, it is possible to develop materials with higher modulus (i.e., higher push-through load), but one is limited in this approach due to other considerations such as the load transmitted to the missile skin by the seal and the ability to stretch the seal during missile onloading.

As a result of numerous push-through tests on a variety of materials, we selected a commercially available material (Adiprene L-167/MOCA), designated polyurethane C, as our material of choice and varied seal geometry. A number of seal shapes and thickness were considered as shown in Fig. 15, but basically a seal cross-section shown as B or C in Fig. 15 was selected as being the most viable shape (19).

To demonstrate the outstanding performance of the humped-notched (H-N) seal, two molds were prepared and 10-in. long segments of polyurethane C were cast and cured to the shape shown in Fig. 2 and H-N configuration B of Fig. 15. In both cases, the seal beam was a nominal 0.5 in. thick. Both molds were heated to 70°C, and a single batch of polyurethane C was mixed, cast sequentially, and cured in the preheated molds for two hours at 70°C. After demolding, the samples were cut into one-in. wide segments and allowed to rest at room temperature for approximately one week. During this period the samples were bonded to steel plates and subsequently push-through tested with the results shown in Fig. 16. These data show that the first and fifth cycle H-N push-through loads are significantly higher than the tapered beam shape at all gaps tested. It follows that full scale seals with cross-section C fabricated from polyurethane C should give significantly higher launch pressure results than similar seals fabricated to the tapered beam shape shown in Fig. 2.

Full scale Adiprene L-167/MOCA seals of cross-section C have been fabricated and are scheduled for launch tests in 1979.

Seal Bottoming Characteristics. In addition to sealing characteristics, the seal must also meet a certain bottoming requirement - as the missile moves toward the launch tube, the seal must not bottom or compress fully before the shock pads bottom or it will place a high and concentrated load on the missile and damage the missile skin. The degree of deflection that a seal will undergo before it bottoms can be measured by the test apparatus

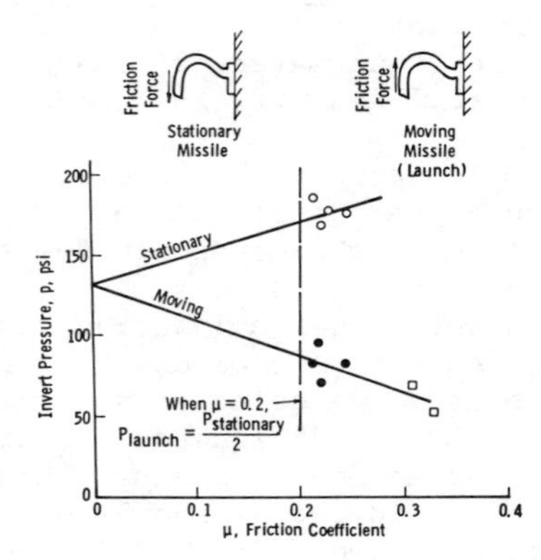

Figure 14. Seal invert pressure as a function of friction: (——) theory; test data: (□) full scale neoprene, (○) 2-ft-diameter urethane, (●) urethane segments

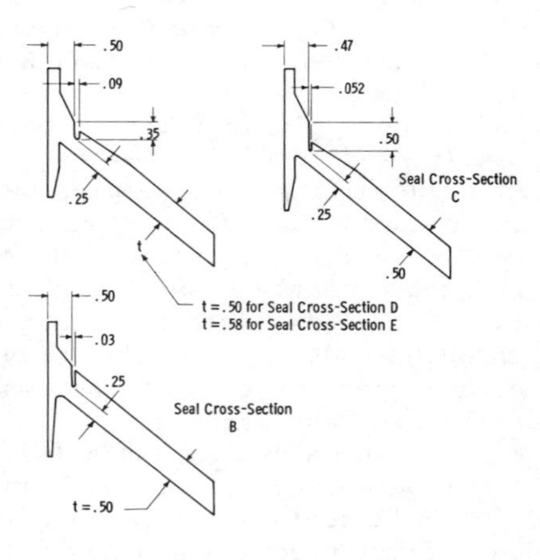

Figure 15. Various seal configurations

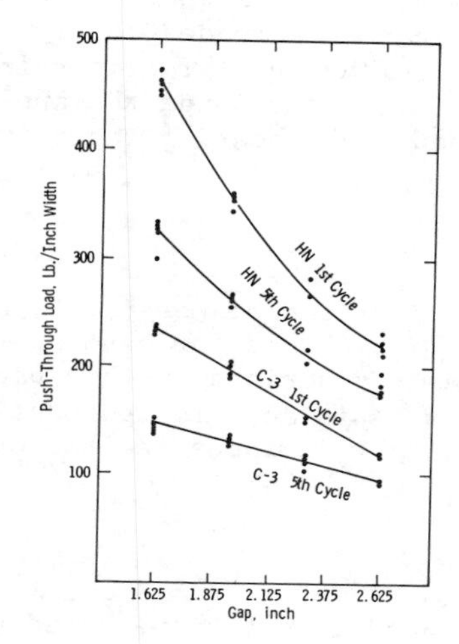

Figure 16. Push-through load vs. gap for tapered beam and shape B seal segments prepared from polyurethane C under controlled conditions

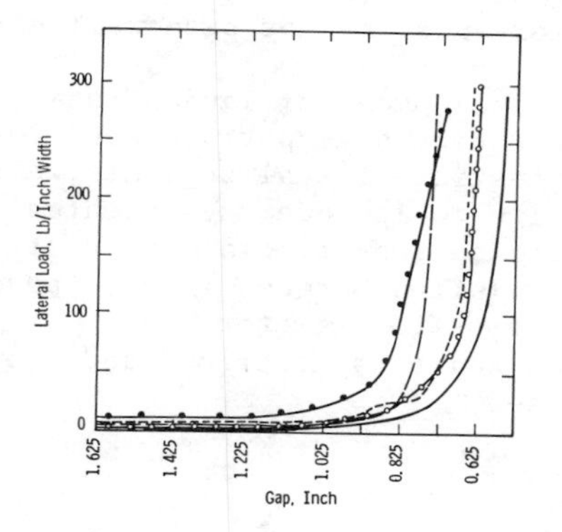

Figure 17. Seal segment lateral excursion tests for polyurethane C: (——) tapered beam shape A, (●) H–N shape B, (○) H–N shape C, (– – –) H–N shape D, (— —) H–N shape E; see Figure 2 for Seal A cross-section, see Figure 15 for Seals B–E cross-sections

shown in Fig. 9. Typical data are shown in Fig. 17 for a tapered beam seal segment and H-N seals of various shapes (see Fig. 15 for the seal cross-sections) cast from polyurethane C. Obviously, C is the most desirable H-N shape from a bottoming consideration.

Future Applications

Utilizing both theory and experimental techniques described in this paper, launch seal technology can be extended to high launch pressures at larger gaps. The influence of fabric reinforcement on seal performance can be investigated and the application of this technology to rotating and reciprocating shafts may be a possibility.

Acknowledgment

The authors acknowledge the contributions of Messrs. H. A. Steffey (retired March 1, 1977) and J. F. Chance for casting the seal segments and two-ft. diameter seals described in this paper. We also acknowledge Mr. J. O. Bowden's assistance for conducting all push-through testing and maintaining voluminous records over a four-five year period.

Abstract

Lip-type launch seals have been used in submarine based missile systems since the 1960's. The seals, rings of neoprene rubber that span the annular space between the missile and launch tube, are bonded to the launch tube. They have functioned adequately, but their mechanism of performance has not been well understood.

In this paper we summarize the development of launch tube mounted seals, outline a theory to describe seal performance, introduce the concept of push-through testing of seal segments, discuss material selection and seal design to include the patented humped-notched seal, present test data on two-ft. diameter cast polyurethane seals, and show the correlation between theory, push-through tests, and pressure tests on two-ft. diameter seals.

Preparation of full scale launch seals is described and possible future applications are discussed.

Literature Cited

1. Mendelsohn, M.A.; Black, R.G.; Runk, R.H.; Minter, H.F. J. Appl. Polym. Sci. 1965, 9, 2715,

2. Mendelsohn, M.A.: Black, R.G.; Runk, R.H.; Minter, H.F. J. Appl. Polym. Sci. 1966, 10, 443.

3. Mendelsohn, M.A., unpublished results.

4. Meier, J.F.; Rudd, G.E. J. Cell. Plas. 1975, 50, 40.

5. Rudd, G.E.; Meier, J.F. J. Cell. Plas. 1975, 164, 143.

6. Mendelsohn, M.A.; Meier, J.F.; Rudd, G.E.; Rosenblatt, G.B. J. Appl. Polym. Sci. 1979, 23, 325.

7. Mendelsohn, M.A.; Meier, J.F.; Rudd, G.E.; Rosenblatt, G.B. J. Appl. Polym. Sci. 1979, 23, 333.

8. Mendelsohn, M.A.; Meier, J.F.; Rudd, G.E.; Rosenblatt, G.B. J. Appl. Polym. Sci. 1979, 23, 341.

9. Meier, J.F.; Minter, H.F.; Connors, H.J. J. Appl. Polym. Sci. 1971, 15, 619.

10. Meier, J.F.; Rudd, G.E.; Rosenblatt, G.B. J. Appl. Polym. Sci. 1972, 16, 559.

11. Rudd, G.E.; Meier, J.F.; Rosenblatt, G.B. ASTM STP 515, ASTM, 1972, 180.

12. Meier, J.F.; Rosenblatt, G.B.; Sterling, R.F. Elastomerics 1978, 25.

13. Connors, H.J.; Mendelsohn, M.A.; Runk, R.H.; Rosenblatt, G.B. J. Environ. Sci. 1968, 27.

14. Mendelsohn, M.A.; Runk, R.H.; Connors, H.J.; Rosenblatt, G.B. Ind. Eng. Chem, Prod. Res. Dev. 1971, 10, No. 1.

15. Mendelsohn, M.A.; Rudd, G.E.; Rosenblatt, G.B. Ind. Eng. Chem., Prod. Res. Dev. 1975, 14, No. 3.

16. Meier, J.F.; Mendelsohn, M.A.; Rosenblatt, G.B.; Sterling, R.F. Elastomerics 1978, 21.

17. Meier, J.F.; Rudd, G.E.; Molnar, A.J.; Jerson, D.D.; Mendelsohn, M.A.; Rosenblatt, G.B., Chapter 15 in this volume.

18. Meier, J.F.; Rudd, G.E.; Molnar, A.J.; Jerson, D.D.; Mendelsohn, M.A.; Weir, D.F., paper presented at the ACS/CSJ Chemical Congress, Honolulu, HI, April 1-6, 1979.

19. Molnar, A.J.; Rudd, G.E.; Meier, J.F., (to Westinghouse Electric Corp.) U.S. Patent 4,033,593, 1977.

20. Meier, J.F.; Rudd, G.E.; Weir, D.F., Rubber Chem. and Technol. 1979, 52, No. 1.

APPENDIX 1

ANALYSIS OF MISSILE SEAL DURING PRESSURIZATION AND LAUNCH

From Fig. A the force equilibrium at T is

$$P_b \sin\alpha + p\frac{\ell}{2}\cos\alpha + N = 0 \quad \text{(Horizontal)} \tag{1}$$

$$P_b \cos\alpha - p\frac{\ell}{2}\sin\alpha - \mu N = 0 \quad \text{(Vertical)} \tag{2}$$

then

$$N = -\left(P_b \sin\alpha + p\frac{\ell}{2}\cos\alpha\right) \tag{3}$$

and

$$P_b \cos\alpha - p\frac{\ell}{2}\sin\alpha + \mu\left(P_b \sin\alpha + p\frac{\ell}{2}\cos\alpha\right) = 0 \tag{4}$$

and

$$P_b - p\frac{\ell}{2}\tan\alpha + \mu\left(P_b \tan\alpha + p\frac{\ell}{2}\right) = 0 \tag{5}$$

or

$$P_b = \frac{p\frac{\ell}{2}(\tan\alpha - \mu)}{(1 + \mu\tan\alpha)} \quad \text{(Stationary Missile)} \tag{6}$$

and

$$P_b = \frac{p\frac{\ell}{2}(\tan\alpha + \mu)}{(1 + \mu\tan\alpha)} \quad \text{(Moving Missile)} \tag{7}$$

Let M_A = moment to the left of point A (x,y) of the beam

then

$$M_A = \left(-p\frac{\ell}{2}x + p\frac{x^2}{2}\right)(1\text{ in.}) - P_b\,y \tag{8}$$

From elastic beam theory

$$M = \frac{EI}{R} \tag{9}$$

where R, the radius of curvature, is

$$R = \frac{\left[1 + \frac{dy}{dx}^2\right]^{3/2}}{\frac{d^2y}{dx^2}} \tag{10}$$

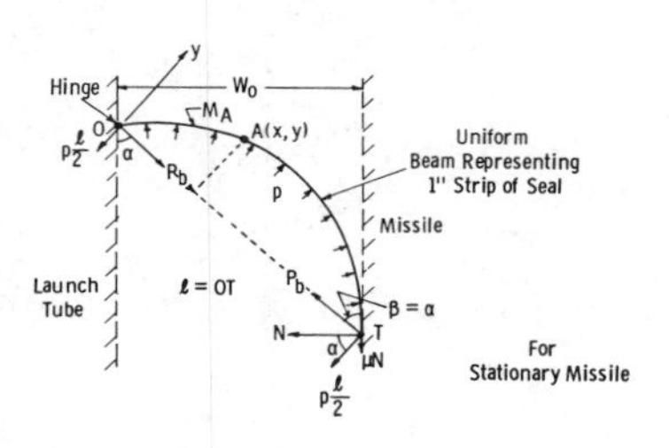

Nomenclature:

- E = Young's modulus (psi)
- I = Area moment of inertia (in^4)
- ℓ = Beam length OT (in)
- p = Pressure (psi)
- (x, y) = Coordinate axis for defining uniform beam equations
- μ = Coefficient of friction
- P_b = Axial load on beam (lbs)
- N = Normal force on missile at tangent point, T (lbs)
- μN = Friction force (lbs)
- M_A = Moment to the left of point A(x, y)

Figure A. Nomenclature and equilibrium forces used in the analysis of the missile seal for the pressurized and launch condition

then

$$\frac{d^2y}{dx^2} = \frac{[1 + \frac{dy}{dx}^2]^{3/2} M}{EI} \quad (11)$$

Sub. of Eq. (8) in Eq. (11) the beam equation is

$$\frac{d^2y}{dx^2} = \frac{[1 + \frac{dy}{dx}^2]^{3/2}}{EI} \left(- p \frac{\ell}{2} x + p \frac{x^2}{2} - P_b y\right) \text{ (Large Defl.)} \quad (12)$$

For small deflections $\frac{dy}{dx} \simeq 0$ and Eq. (12) becomes

$$\frac{d^2y}{dx^2} + \frac{P_b}{EI} y = \frac{1}{EI} \left(- p \frac{\ell}{2} x \; p \frac{x^2}{2}\right) \quad \text{(Small Defl.)} \quad (13)$$

The closed form solution of Eq. (13) can be found and the peak deflection at $x = \ell/2$ is

$$y_{(x=\ell/2)} = \frac{p\ell^4}{32 \; EI} \left(\frac{2 \sec u - 2 - u^2}{u^4}\right) \quad (14)$$

where

$$u = \frac{\ell}{2} \sqrt{\frac{P_b}{EI}} \quad (15)$$

also the slope at x =   is

$$\frac{dy}{dx}_{x=\ell} = \tan \beta = \frac{p\ell^3}{8EI} \left(\frac{\tan u - u}{u^3}\right) \quad (16)$$

Since a portion of the seal lies against the missile during higher launching pressures, the following constraint equation must be satisfied at the tangent point T between the missile and seal:

$$\tan \beta = \tan \alpha \quad (17)$$

where

$$\tan \alpha = \frac{Wo}{\sqrt{\ell^2 - Wo^2}} \tag{18}$$

or defining a function

$$g(\ell) = \tan \beta - \tan \alpha = 0 \tag{19}$$

when $g(\ell) = 0$ the equilibrium at T, the uniform beam equation and the slope constraint at point T are simultaneously satisfied.

A summary of the actual equations as they appear in the computer analysis using Eq. (13) are:

$$\tan \alpha = \frac{Wo}{\sqrt{\ell^2 - Wo^2}} \tag{20}$$

$$u = \sqrt{\frac{p\ell^3}{8EI} \frac{\tan \alpha + u}{1-u \tan \alpha}} \tag{21}$$

$$\tan \beta = \frac{p\ell^3}{8EI} \left(\frac{\tan u - u}{u^3}\right) \quad \text{at } x = \ell \tag{22}$$

$$g(\ell) = \tan \beta - \tan \alpha = 0 \tag{23}$$

The maximum displacement at the mid-point and the axial load P_b are also calculated in the computer program.

RECEIVED May 20, 1980.

15

Development of Missile-Mounted Polyurethane Seals for Missile Launch Systems

JOSEPH F. MEIER, GEORGE E. RUDD, ALBERT J. MOLNAR, DONALD D. JERSON, and MORRIS A. MENDELSOHN

Westinghouse Research and Development Center, Pittsburgh, PA 15235

DAVID F. WEIR

Marine Division, Westinghouse Electric Corporation, Sunnyvale, CA 94088

In the previous paper (Meier et al., 1980) the subject of launch seals for submarine based missiles was discussed and the advantages of the humped-notched (H-N) seal shape were enumerated. In this paper we describe the development of missile mounted H-N seals for a land-based missile system. A missile mounted seal is shown schematically in Figure 1.

Seal requirements for the two seal mounting geometries are considerably different as summarized below:

Launch Tube Mounted Seals	Missile Mounted Seals
Friction promotes seal inversion	Friction inhibits seal inversion.
Cold launch (~300°F for <1 sec.)	Hot launch (~4,800°F for <1 sec).
Seal inside diameter stretched during missile onloading.	Seal outside diameter reduced during missile onloading.
Near vertical launch.	Launch at angles <45° from vertical.
Operates in <5 pphm ozone for ~10 years.	Operates in 25 pphm ozone for at least 15 years.
Multiple launch requirement (seals retained in launcher).	One launch life (seal exits canister with the missile).

These parameters required that a new seal be developed, but the basic humped-notched (H-N) seal (Molnar et al., 1977) was directly transferrable from previous work.

0-8412-0567-1/80/47-132-177$06.00/0

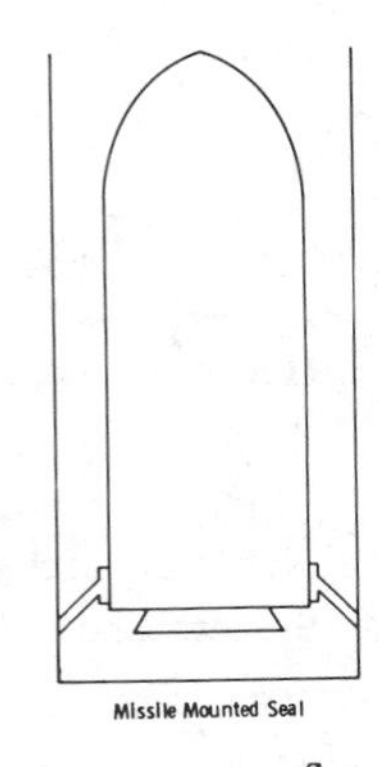

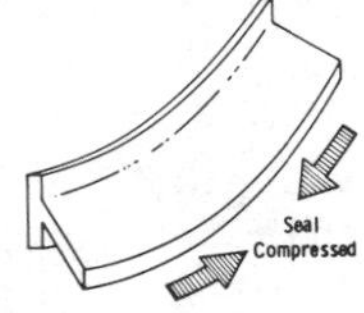

Figure 1. Missile-mounted seal

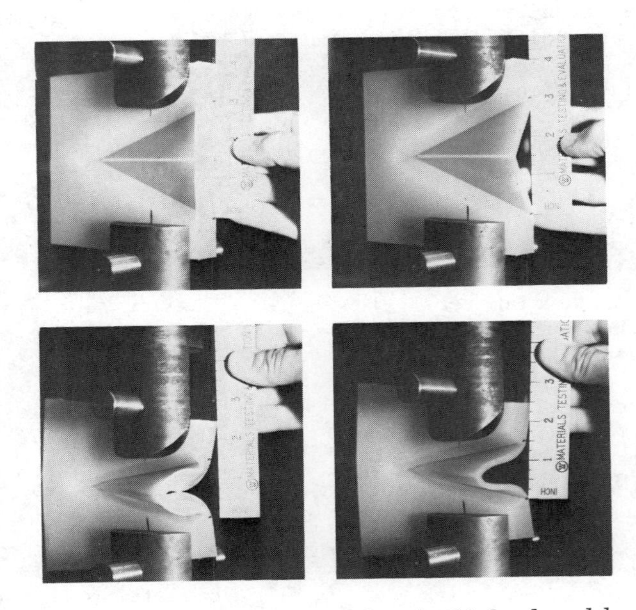

Figure 2. Flexing of cast-polyurethane buckled web models

The new seal development concentrated on the following areas:

- Selecting a seal elastomer and geometry to meet launch pressure and lateral excursion or bottoming requirements.
- Designing a seal to allow outside diameter reduction during missile onloading.
- Evaluating the effect of 4,800°F flame (to simulate launch) for periods up to 1 sec.
- Casting two-ft. diameter model polyurethane seals.
- Modifying the two-ft. diameter seals in the modified pressure test apparatus.
- Casting 54-in. diameter seals for larger-scale missile launch tests.
- Evaluation of the 54-in. diameter seals in missile launch tests.

Experimental Procedures

Buckled Web Section Model. The first task of this development was to investigate methods of accommodating the reduction in seal outside diameter for onloading. Flat buckled web sections were cast and flexed as shown in Fig. 2. The design seemed adequate and gave us confidence that similar sections could be designed and integrally cast into seals to take up the compressive hoop strain. Alternate slitting or wedge removal patterns were also envisaged as shown in Fig. 3.

Preparing Model Seals for Push-Through and Lateral Excursion Tests and Two-Ft. Diameter Seals for Pressure Testing. Processing the DuPont Adiprene L-167/MOCA was a well understood technique by the time we started on this new development. Initial samples for push-through testing and lateral excursion measurements were made by sawing the desired shape from uniform thickness seal sections similar to those shown in Fig. 4. Two-ft. diameter seals were cast in a mold made for this project and shown in Fig. 5. Two seals cast in this mold, one with eight buckled web sections (mold inserts were used), and one with eight flaps, are shown in Fig. 6. The flaps were band-sawed into the seal using a wood fixture for aligning and guiding the saw cuts.

Polymer Preparation - Adiprene L-167. Adiprene L-167 (Athey, 1967) is a liquid polyether urethane prepolymer which can be cured to a strong, rubbery solid by reaction of the terminal isocyanate groups with polyamine or polyol compounds. It is a honey colored viscous liquid (4,700-6,500 cps at 30°C) with an available isocyanate content of 6.35% (6.15-6.55%). This prepolymer is based on polyoxytetramethylene and tolylene diisocyanate and has an equivalent weight of about 670.

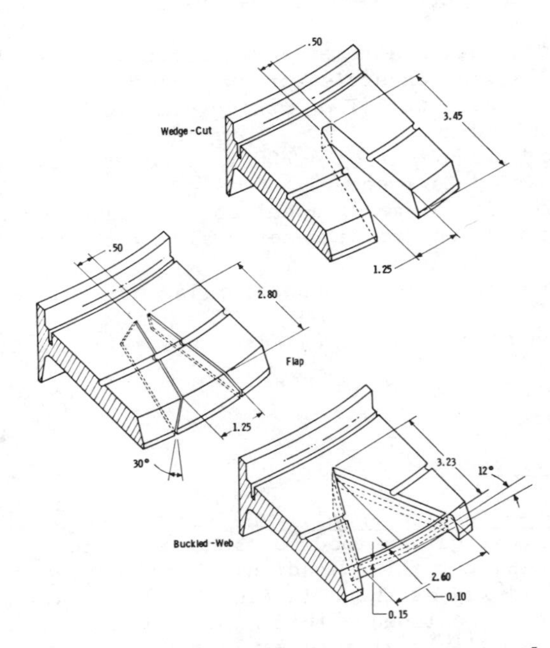

Figure 3. Three concepts evaluated to permit reduction of outside seal diameter

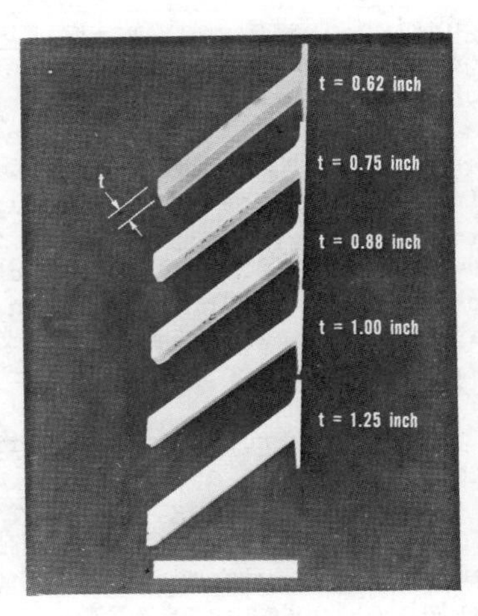

Figure 4. Uniform thickness seal sections

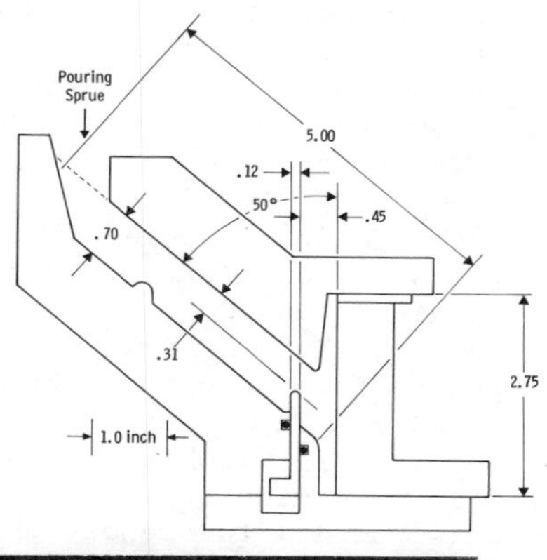

Figure 5. Two-foot-diameter H–N seal mold

Figure 6. Buckled web and flap-type seals cast from L-167/MOCA polyurethane

MOCA. MOCA 4,4'-methylene-bis(2-chloroaniline) , a diamine curing agent for isocyanate-containing polymers, has a molecular weight of 267. It is a light tan solid, in pellet form, having a specific gravity of 1.44 and a melting range of 100°-109°C. MOCA is considered to be a Class I carcinogen and has mild toxicity. Before using this material the vendor should be consulted for the latest safe handling procedures.

Seal Preparation. Bulk quantities of Adiprene L-167 were preheated to 40°-45°C. Two batches of Adiprene L-167, 3,000 g each, were weighed into two-gallon containers and degassed under vacuum until the vigorous bubbling stopped. Two 1/2 gal. cans, each containing 600 g of MOCA were opened, covered with aluminum foil, and heated to 120°-160°C in a small oven. One can of hot MOCA was poured into a container of degassed L-167 which had been cooling for an hour or more and mixed for two minutes with an air stirrer at low speed. The NCO/NH_2 ratio was 1:1 throughout this work. Mixed resin was gently poured into the preheated (70°C) two-ft. diameter seal mold while the second batch was being mixed. Then the second batch was poured into the mold. The filled mold was placed in the oven six to eight minutes after the first MOCA was blended into the Adiprene L-167, cured in the mold for two hours at 60°-65°C, demolded and allowed to remain at room temperature for at least three days before use.

Push-Through and Lateral Excursion Tests. Push-through tests on seal segments were used to predict the pressure performance of lip type seals. The test is conducted in an Instron machine at 5 in./min. crosshead speed while an autographic record of load versus displacement is made. Figure 7 shows a sequence of photos taken during the push-through test on several seal segments.

Lateral excursion tests are also conducted on seal segments in the Instron using a different fixture (Meier et al., 1979). These tests permit measuring the excursion of a particular design to see if it will meet requirements for inclined launch or shock. A basic performance guide for any seal design is that the shock isolation pads must bottom out before the seals do.

Oxy-Acetylene Torch Tests. During launch, the underside of the seal is briefly exposed to a 4,800°F gas generator flame. Naturally, it is required that the seal not degrade to the point where it can no longer retain pressure throughout the launch. A simple bench test was devised using an oxy-acetylene torch to simulate this exposure. A tungsten-rhenium thermocouple was used to measure the exposed surface temperature and a chromel-alumel thermocouple to measure either the temperature 0.02-in. below the surface when a bare coupon was tested or the temperature at the polyurethane/protective coating interface when a protected specimen was tested. Photographs of this setup are shown in Fig. 8.

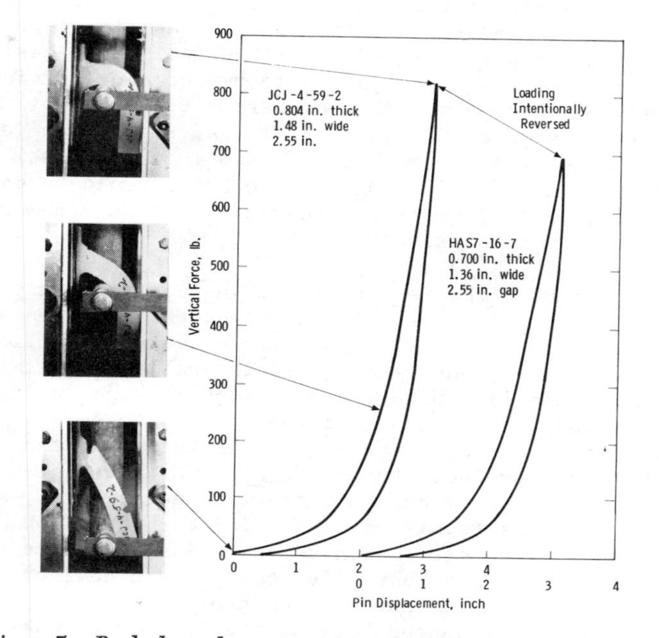

Figure 7. Push-through test on L-167/MOCA H–N seal segments

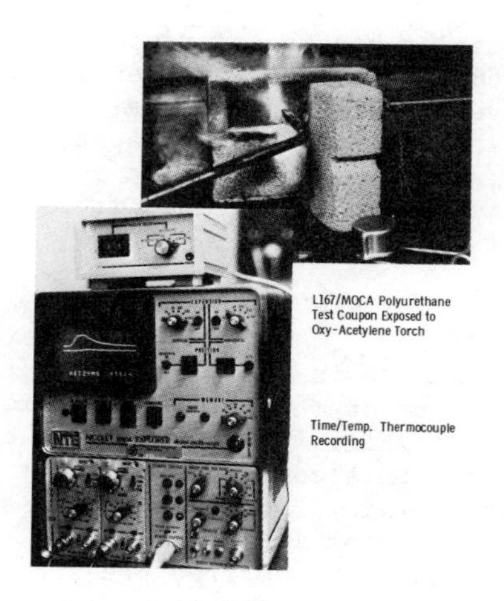

Figure 8. Test apparatus for flame testing polyurethane coupons and recording temperatures

Seal Outside Diameter Reduction With an Instrumented Clamp. The outside diameter of the seal must be reduced from 26.7-in. to 24.0-in. for installation into the test canister. To accomplish this a clamp was designed to fit into a groove cast into the seal (see Fig. 3). The clamp assembly consists of four modified commercial clamps joined together by 0.125-in. diameter stainless steel aircraft cable. Strain gages were applied to one of the clamps and the assembly calibrated in a universal testing machine so that the tensile load in the cable could be measured as the seal outside diameter was reduced. Figure 9 shows several views of the clamps in use.

Modification to the Seal Pressure Tester to Measure Gas Flow Rate and Valve Opening Time. The buckled webs or flaps that are designed into the seal to allow outside diameter reduction are also gas leak paths. Some leakage is tolerable, but limits were not defined initially. However, measuring leakage during pressure testing permits comparison of the two design concepts and provides performance data that can be used in computer-simulated launches. The pressure test produces transient flow for a useful duration of 0.5 second or less since the potential energy (i.e., accumulator size and pressure) is somewhat limited. This obviates use of conventional flow meters such as venturi and orifice meters and turbines which are essentially steady state devices. An initial thought was to measure accumulator and gas weight during the test. This was rejected because the empty accumulator weighs over 300 lb. and only 4-6 lb. of nitrogen was discharged during a test. Thus a load cell of sufficient capacity would have marginal resolution to accurately measure the weight of nitrogen expelled. It would also be difficult to structurally isolate the accumulator for weighing because of rigid pipe connections. A third difficulty would be separating inertial loading effects during the explosive-like release of gas.

The method selected involves some approximation because it assumes that the mass of nitrogen in the accumulator at any instant during blowdown can be determined from the equation of state for a perfect gas,

$$pV = mRT$$

Since the blowdown process is transient, we decided to measure pressure, p, and temperature, T, at two locations within the accumulator as shown in Fig. 10 to see if the calculated mass, m, varied with position. Data collected, but not presented in detail, show agreement within 5% between the mass calculations based on the two measurement locations.

After about one-half the desired pressure testing had been conducted, some slightly unusual seal pressure versus time oscillograms were obtained. They were unusual in that the rise time to peak seal pressure was longer than obtained previously. The

Figure 9. Clamp used to reduce o.d. of 2-ft-diameter seal for installation into the canister

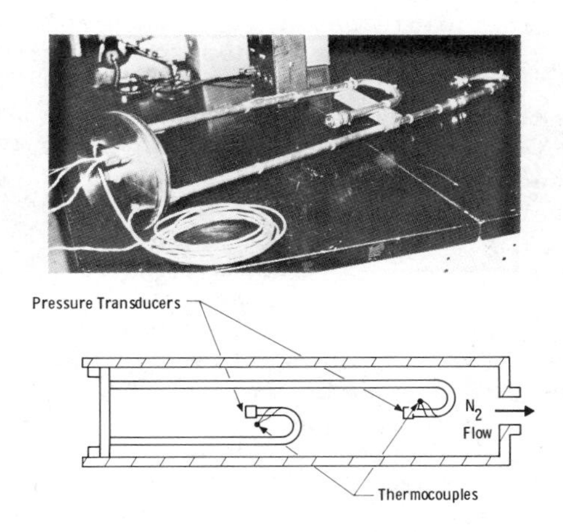

Figure 10. Pressure and temperature transducers mounted in accumulator A1 to measure nitrogen flow

remedy was to increase the pressure in accumulator A2 used to open valve C2 which dumps the main accumulator A1. These components are identified in Fig. 11. To establish that the valve opening time was indeed the undesirable variable, an LVDT was added so valve spool travel could be recorded during the pressure test and correlated with the pressure record.

Pressure Testing. The test fixture originally developed to test launch tube (canister) mounted seals was modified to test missile mounted seals. This required an inner cylinder of the proper diameter to simulate the missile to which the seal could be bonded. Figure 11 schematically shows the test apparatus with a seal installed and also outlines in the operation of the apparatus. Figure 12 is a photograph of the tester.

Experimental Results and Discussion

"Weak" Sections to Allow Seal Outside Diameter Reduction. The inside diameter of a launch tube or canister mounted lip-type seal is smaller than the outside diameter of the missile and is therefore stretched when the missile is onloaded. The seal can accommodate the tensile hoop strain quite readily. However, when the seal is mounted on the missile or a separable base ring attached to the missile, the seal outside diameter must be reduced for onloading. The latter condition is shown schematically in Fig. 1. To rectify the unstable buckling which would occur at an indeterminate location on the seal periphery and only after considerable compressive hoop stress was generated, it was decided that "weak" sections would be designed into the seal at evenly spaced intervals. These weak sections could be merely wedge-shaped openings (Fig. 3) cut into the seal lip. Indeed this approach was tried but leakage was excessive when pressure tested. Even though it is possible that this concept could be improved, it was not pursued further since it was felt to have inherently poor sealing characteristics that would only get worse when tested at eccentric gap conditions (i.e., the missile not centered in the launch tube).

The buckled web section approach was initially thought to be the most leak resistant since the seal was not cut through but only reduced in thickness. Since the flat model described earlier (Fig. 2) appeared to deform in a manner that could be expected to provide a reasonable seal, inserts were machined to produce similar buckled webs in the conical surface of the 2-ft. diameter scale seal (Figs. 3 and 6). This seal sustained pressure beyond that required and also exhibited a reduced leak rate.

The flapped seal design (Figs. 3 and 6) was also evaluated. This seal performed well during pressure testing and calculated leak rates were similar to the buckled web concept.

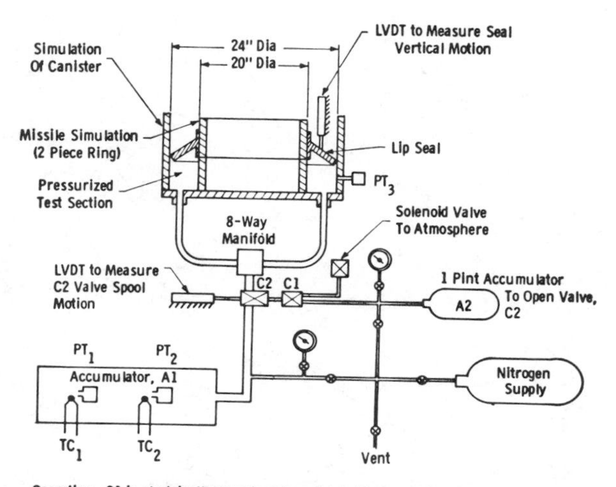

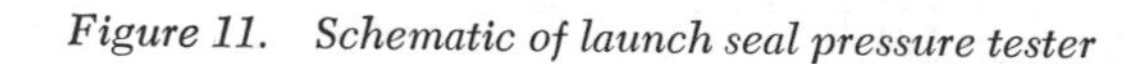
Figure 11. Schematic of launch seal pressure tester

Figure 12. Two-foot-diameter launch seal test apparatus

TABLE I

SUMMARY OF PRESSURE TEST DATA ON TWO-FOOT DIAMETER CAST POLYURETHANE LIP SEALS

Seal No.	Test No.	Test Age of Seal (Days)	Missile Dia. (in.)	Annular Gap (in.)	Push-in Load (kips)	Al Press. (psig)	Max. Seal Press. (psig)	Rise Time to Peak Press. (sec.)	Dwell Time (sec.)				Remarks
JCJ-2-76[a]	1	0	20.75	1.62	3.7	300	--	--	--				Test Malfunction
	2	0			--	300	157	0.64	several mins.				No Flip
	3	0			--	300	229	0.40	0.28				Flip
	4	2			2.7	450	--	--	--				Test Malfunction
	5	15			--	450	186	0.28	0				Test Malfunction
	6	52			3.7	450	244	0.43	0.06				Flip
	7	55			2.4	450	240	0.36	0.05				Flip
	8	57			2.0	400	202	0.32	0				Flip
	9	78			1.6	400	212	0.36	0				Flip
	10	85			1.9	400	200	0.33	0				Flip
DJ3-46B[b]	1	0	18	3.00	10.7	450	232	0.42	0.12				Test Malfunction
									Seal Pressure (psi) vs. Time (sec.)				
									0.5 sec.	1.0 sec.	2.0 sec.	5.0 sec.	
	2	7			--	440	219	0.40	186	168	143	115	No Flip
	3	8			--	500	237	0.41	196	171	143	113	No Flip
	4	37			8.4	600	292	0.33	0.50				Bond Failed
	5	42			7.7	600	282	0.39	0.29				Seal Flipped
	6	45			6.1	550	273	0.31	0.07				Seal Flipped
	7	46			6.1	550	259	0.32	0.06				Seal Flipped
	8	49			6.2	550	272	0.30	0				Seal Flipped
	9	52			5.5	550	268	0.30	0				Seal Flipped
	10	59			5.7	550	242	0.27	0				Seal Flipped

(a) Adiprene L-167/MOCA tapered beam seal.

(b) Humped-notched seal with 0.75-in. thick beam. Adiprene L-167/MOCA.

<u>Initial Considerations Regarding the Pressure Capabilities of the New Seal Design</u>. Initially, a tapered beam seal cast from polyurethane was considered since we had developed considerable understanding of this combination. The maximum gap, g, for the new design was 2.0-in., therefore,

$$W_o = g - h - t/2 = 2.8 - 0.5 - 0.8/2 = 1.9\text{-in.}$$

for a seal of assumed thickness, t - 0.8-in.,

$$p_f = \frac{Ct^{2.3}}{W_o^{2.05}} = \frac{760 \times 0.8^{2.3}}{1.9^{2.05}} = 122 \text{ lb./in.}^2$$

This equation, based on unpublished experimental results yields about one-half the required 250 lb./in.^{2} pressure, neglecting the fact that friction inhibits seal inversion for a missile mounted seal. Obviously p_f can be doubled by increasing C, a modulus related term, or t or a combination thereof. These remedies are undesirable because the force required to reduce the seal outside diameter would increase and the lateral excursion would decrease. In addition, doubling C would be difficult as polyurethane B is already a fairly high modulus material.

One logical alternative was to apply the humped-notched (H-N) seal design since it had been proven to possess higher pressure capabilities than the tapered beam design for a given material and thickness. Fortunately, an initial H-N design was already available from our previous work (Meier et al., 1979). Test results for this design are shown in Table I where values as high as 292 lb./in.^{2} at a gap of 3.0-in. are given. However, two potentially deleterious factors had not been dealt with - reduced capability because of launch flame and the "weak" sections for buckling.

<u>Push-Through Test Results</u>. To substantiate the design described above, a push-through sample was prepared to investigate the influence of the groove required for the outside diameter-reducing clamp. Test data showed that the groove is in a region of the seal lip that is fully against the canister wall when the seal is pressurized and thus essentially inactive in bending. Concern over the effect of the clamp groove on reducing the seal capability was therefore eliminated.

Some push-through tests were also conducted on seal samples that were exposed to actual gas generator flame. The H-N seal segments suffered some mechanical abuse during exposure causing a 0.2-in. tear at the intersection of the seal lip lower surface and the bonding surface. Tapered beam seal segments that were also exposed were not mechanically damaged. In spite of losing 0.045 to 0.070-in. of material by ablation, the seals withstood the flame quite well. Table II summarizes the results of these exposures and push-through tests.

TABLE II

PUSH-THROUGH DATA ON VIRGIN SEAL SEGMENTS[a] & SEGMENTS EXPOSED TO GAS GENERATOR TESTS

	Width, in.	Thickness, in.	Push-through cycle, load in lb. at W_o = 1.0					
Tapered Seal Segment			1	2	3	4	5	
Virgin	1.0	0.485	240	240	190	121	121	
#1 exposed to gas generator	1.0	0.470	275	222	185	175	165	Seal material darkened but
#2 exposed to gas generator	1.0	0.472	250	205	155	150	145	not cut or torn.
Segments From Two-ft. Diam. H-N Seal			Push-through cycle, load in lb. at W_o = 1.75					
Virgin	1.5	0.693-0.701	490[b]	--				
Exposed to gas generator	1.5	0.648-0.701	362	325[c]				
Exposed to gas generator	1.5	0.648-0.630	378	340[c]				

(a) L-167/MOCA NCO/NH_2 = 1.0, no Teflon on seal segments.

(b) Bond failed at steel plate.

(c) Seal segment had a 3/16-1/4 in. crack at the base of the hinge and it fractured on the 2nd cycle.

Oxy-Acetylene Torch Test Results. The bench test shown in Fig. 8 was used to expose polyurethane coupons to a flame to approximate a hot launch. Bare samples were first exposed to provide base line data. For these tests the tungsten-rhenium thermocouple was located on the surface for direct exposure to the torch flame. A chromel-alumel thermocouple was located 0.02-in. below the surface in a hole drilled partially through from the back side of the coupon. The remaining test coupons were covered with one of the protective coatings and the chromel-alumel thermocouple was sandwiched between the coupon and the covering. The results of these tests, listed in Table III, show that even the bare sample provides a considerable temperature drop through the 0.02-in. polyurethane. We judged that attenuation of this magnitude was adequate and that no additional protection would be required. However, the 3M ceramic fiber belting or the DuPont Armalon could be added if subsequent launch testing indicates that additional protection is required. The intumescent coating supplied by Avco showed a tendency to be swept away by the flame and was judged to be less efficient than the other coverings. Photographs of the exposed samples are given in Fig. 13.

Use of the Seal Outside Diameter-Reducing Clamp. Figure 9 shows the clamp that was assembled to reduce the seal outside diameter for installation into the two-ft. diameter test canister. Four commercial clamps were modified by replacing the jaw blocks with pulley-like wheels to position the cable well into the groove in the seal. This is important to insure that the tensile force in the cable does not create a moment which would cause the cable to pull out of the groove. The clamp functioned quite well but we did discover that the catalog rated load of 1,500 lb. meant that 1,500 lb. force could be applied to the jaws by external means without failure - it did not mean that 1,500 lb. could be generated with the 3/8-16 thread and T-handle. For our use, loads exceeding 600 lb. caused considerable galling in the threads and thrust end of the threaded shank. Conversation with the manufacturer revealed that some of their larger model clamps use Acme threads and alloy steel shanks to improve the clamping capacity.

In spite of the limited load capacity, the clamps performed quite well. Figure 14 shows plots of clamp force versus seal outside diameter for the buckled web and flap-type seal designs. The flap-type seal required more force to reduce the outside diameter to 24-in. than did the buckled web seal. However, the flaps were cut into the seal with a bandsaw and thus the slots were not smooth and did not slide very well. To alleviate this situation, 0.03-in. Teflon was inserted in the slots during clamping. Perhaps clamping forces would be reduced if the slots were cast rather than saw-cut.

Two-Ft. Diameter Seal Pressure Test Results. The first seal that was pressure tested was a L-167/MOCA seal with 0.7-in. thick seal lip and eight wedge cuts as shown in Fig. 3. An accumulator

TABLE III

LAUNCH FLAME EXPOSURE TEST DATA

Specimen[a]		Time in Sec./Temp. in °F[b]							Max. Temp.
		0.1	0.2	0.3	0.5	1.0	2.0	3.0	
Bare-1	Exposed surface	210	375	600	2800	1250			3190[c]
	Sub-surface	80	80	83	92	135			145
Bare-3	Exposed surface	1860	3100	3800	4050	3950	1150	550	4118
	Sub-surface	80	90	105	320	780	500	390	831
AB312-2	Exposed surface	1810	3215	3720	3970	3975	2450	1550	3975
	Sub-surface	82	89	113	170	350	800	825	827
AB312-6	Exposed surface	1400	3000	3140	3050	3730	1800	1300	3730
	Sub-surface	80	88	94	116	250	465	530	540
Armalon-4	Exposed surface	645	1200	2100	3350	3825	1100		3825
	Sub-surface	84	95	125	165	490	845	760	850
Armalon-7	Exposed surface	1200	2200	3200	4050	>4200	2350	1450	>4200[d]
	Sub-surface	80	80	82	95	240	536	560	563
Avco-5	Exposed surface	900	2210	2100	2100	2600	1450	650	3210
	Sub-surface			100	2500	1400	750	400	2500[e]

(a) All were prepared from 0.16 in. x 1.0 in. x 3.0 in. coupons of L-167/MOCA.

AB312 A ceramic fiber yarn woven into a 2 lb./yd.^{2} material (0.06 in. thick) intended for use as long-life conveyor belting in furnaces up to 2000°F, 3M, St. Paul, MN. Bonded to polyurethane coupon with Chemlok 305.

Armalon Trade name for a family of fabrics, laminates, tapes and felts coated with TFE resin. We evaluated Product No. TG-1408, a fiber glass substrate, overall thickness of 0.008 in., weighing 9.5 oz./yd.^{2}, Dupont, Wilmington, DE. Bonded to polyurethane coupon with Chemlok 305.

Avco Flamarest 1400S, brush applied 0.040 mils thick. Avco Systems, Lowell, MA.

(b) 0.010 diam. W 3% Re/W 25% Re and 0.005 in. diam. Cr/Al (Type K) thermocouples.

(c) Exposure too brief and flame not as close as in other tests.

(d) Exceeded calibrated range of thermocouple.

(e) Flame burned through coating directly exposing the sub-surface thermocouple, test error in duplicate sample - no data.

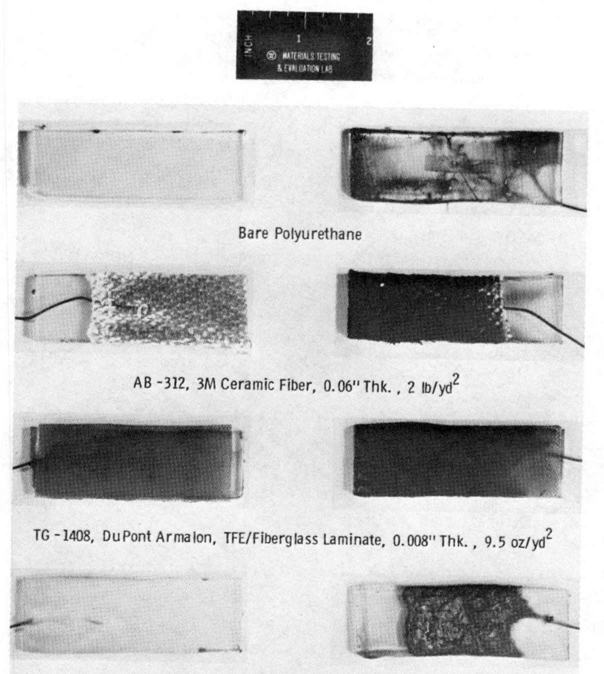

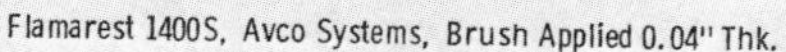

Figure 13. Before and after photos of flame-exposed polyurethane coupons

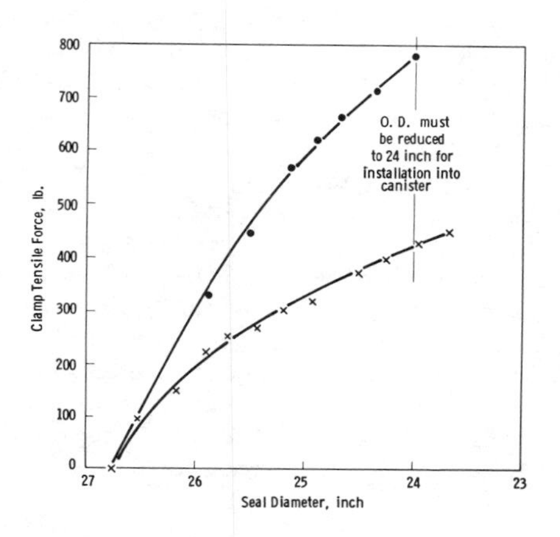

Figure 14. Clamp force vs. seal diameter measurements: (●) flap-type seal, (×) buckled web seal

pressure (A1) of 515 lb./in.^{2} (abs) was used but the resulting maximum seal pressure was only 31 lb./in.^{2} (ga). Subsequent testing of the flapped and buckled web seals showed that A1 pressures of 800-1,000 lb./in.^{2} (abs) might be required but even so the extremely low seal pressure caused us to reject the wedge-cut design and proceed with the other two concepts. Pressure units are inconsistent because absolute pressure is required to calculate N_2 mass values but gage pressure is more appropriate for seal pressure performance.

Figure 15 summarizes the pressure-time characteristics of the buckled web and flapped seals. Specific test results were selected to compare the two designs under similar test conditions. Since performance of the blow-down valve, C2, was not monitored during all testing, a complete set of comparative data are not shown.

In addition to pressure capability, leak rate can also be used to grade seal performance. As stated earlier, nitrogen mass flowing out of accumulator A1 was calculated using the equation of state for a perfect gas. There are some restrictions, however, regarding the conditions under which this flow can be interpreted as leakage. These are that the seal pressure and displacement must be constant - actually the gas temperature in the test chamber should also be constant but was not measured. If the seal did not leak, flow would be zero when these parameters were constant, therefore flow which does occur under these conditions is interpreted as leakage.

The test oscillograms (not presented) and Fig. 15 show that at maximum seal pressure there is about a 0.1 second period where the pressure is nearly constant. Seal displacement is also nearly constant at this time and for some time thereafter. Test data plotted in Fig. 15 consider nitrogen flow values that occur during this time interval so that leakage comparisons can be made under similar test conditions. These data show that there is no significant difference in the leak performance of the two seal designs.

A source of minor error in calculating nitrogen mass is the response time of thermocouples TC_1 and TC_2. These thermocouple junctions are welded beads of the 0.005-in. diameter wires. The Omega Temperature Measuring Handbook (Omega Engineering, Inc., a manufacturer of temperature measuring equipment), indicates that the response time for this configuration is about 0.1 second (time to reach 63.2% of an instantaneous temperature change). By comparison, the galvanometers used in the oscillograph have a response time of 10^{-3} sec, or 0.01 times the thermocouple response. However, shifting the temperature data by 0.1 second and calculating new mass and mass flow values yields no significant difference in the leakage. Therefore, the effects of the thermocouple response time have been neglected.

Performance objectives for the launch seal entailed achieving a pressure capability of 220 lb./in.^{2} (ga) @ a nominal gap of 2.0-in. At maximum missile eccentricity (gap = 3.0-in.) the seal

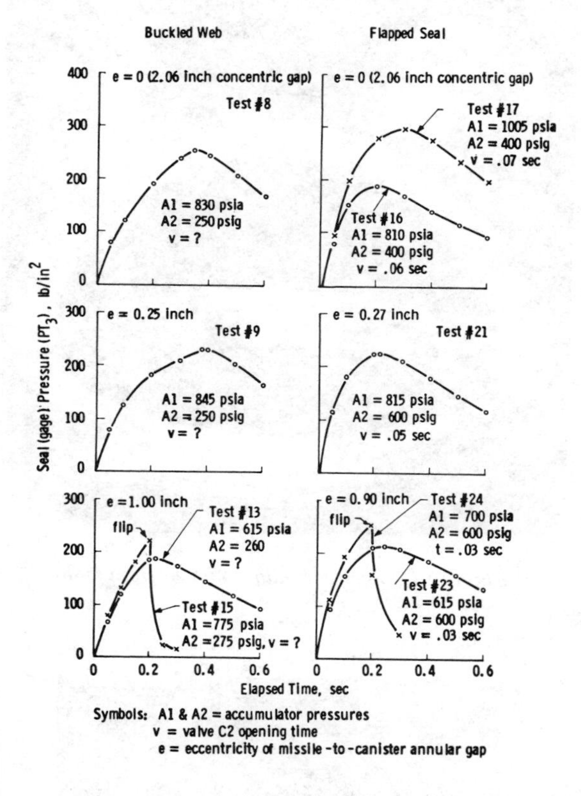

Figure 15. Pressure–time test curves for the buckled web and flapped seals at various gap eccentricities; symbols: A1, A2, *accumulator pressures;* v, *valve* C2 *opening;* e, *eccentricity of missile-to-canister annular gap*

Figure 16. Seal partially removed from mold

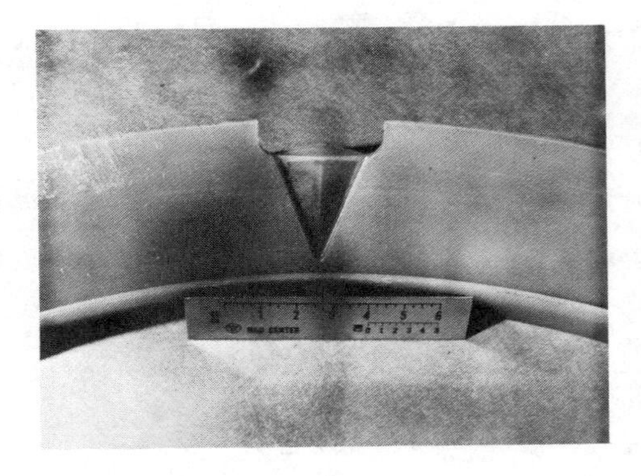

Figure 17. Buckled web segment of a 54-in. i.d. launch seal

pressure requirement reduced to 105 lb./in.^{2} (ga). Both the buckled web and flapped seals exceeded these objectives by achieving an invert pressure of 220 lb./in. @ the maximum annulus eccentricity.

Fabrication of 54-In. Diameter Seals. Following a successful two-ft. diameter seal test program, we scaled up to 54-in. diameter seals. An aluminum seal mold was fabricated and seals were cast from Adiprene L-167/MOCA. A partially demolded and untrimmed seal is shown in Fig. 16 and a close-up of the buckled web design is shown in Fig. 17. Sixteen evenly spaced buckled web sections were molded into the seal.

These seals were subsequently evaluated in "cold launch" experiments using compressed gas as the eject medium (Fig. 18). Also, "hot launches" using an actual gas generator were successful. In Fig. 19 the seal, mounted at the base of the missile, is obscured by the combustion products of the gas generator. In both cases, using a 36,000 lb. simulated missile @ near vertical and 45° inclined launch angles, the launch simulation tests were successful. Measured maximum breech pressures were approximately 70 lb./in.^{2} (ga) for the "hot launch" tests. These pressures were somewhat below seal development objectives since launch pressurization characteristics for this test program were dictated by missile launch kinematics rather than launch seal requirements.

Segments cut from the tested seals, shown in Fig. 20, exhibit the effect of abrasion against the launch tube as well as the influence of heat in the case of the hot launch.

Although only designed for a single launch, several seals were tested in multiple launches and retained sealing integrity - a testimony to the durability of the cast polyurethane seal material.

Summary

Based on laboratory experiments with two-ft. diameter cast polyurethane, missile mounted seals it was possible to develop test data which met missile onloading, launch pressure, and leakage requirements. As a result of these tests, 54-in. diameter seals cast from Adiprene L-167/MOCA were mounted on a 36,000 lb. missile and tested successfully. Under both "hot" and "cold" launch conditions, the seals met leakage requirements and withstood launch pressures up to 90 lb./in.^{2} (ga) and annular gaps as large as 2.3 in.

Acknowledgment

The authors acknowledge the contribution of Mr. J. F. Chance for casting the seal segments, two-ft. diameter seals, and 54-in. diameter seals. We also acknowledge Mr. J. O. Bowden's assistance for conducting all push-through tests on two-ft. diameter seals.

Figure 18. "Cold Launch" seal test using compressed gas for propellant

Figure 19. "Hot Launch" seal test using a gas generator for launch propellant

Figure 20. Segment of seals showing abrasion and ablation caused by test launches

Many missile systems employ a tube to guide the missile during launch. There are at least two basic sealing techniques to minimize launch pressure blow-by during this period - one is to attach several seals along the tube inside diameter which are successively "uncorked" as the missile travels out the tube, a second is to attach one seal to the base of the missile so that the seal exits the tube with the missile. The reasons for selecting one scheme over the other are many and varied and may include such factors as type of propulsion, mechanical shock environment during stowage, and the ability of the missile to withstand pressure and seal forces, etc. It is not our purpose here to discuss how these factors are analyzed to determine which manner of sealing is to be employed - it is sufficient to note that for a given set of design conditions, one system or the other will be favored.

Literature Cited

Athey, R. J., DuPont Bulletin No. 12 (1967).
Meier, J. F., Rudd, G. E., Molnar, A. J., Jerson, D. D., Mendelsohn, M. A. and Rosenblatt, G. B., Chapter 14 in this volume.
Molnar, A. J., Rudd, G. E. and Meier, J. F. (to Westinghouse Electric Corp.), U.S. Patent 4,033,593 (1977).

RECEIVED May 20, 1980.

REINFORCED PLASTICS

16

Molecular Composites

Rodlike Polymer Reinforcing an Amorphous Polymer Matrix

G. HUSMAN, T. HELMINIAK, and W. ADAMS

Air Force Materials Laboratory, Nonmetallic Materials Division, Wright–Patterson Air Force Base, OH 45433

D. WIFF and C. BENNER

University of Dayton Research Institute, Dayton, OH 45469

Recent developments in the synthesis of rodlike aromatic heterocyclic polymers have generated a great amount of interest in the development of these polymers as structural materials. A large effort is currently being expended to characterize these polymers and to develop them into useful product forms, such as fibers, films or sheets. The Air Force Materials Laboratory and the Air Force Office of Scientific Research are engaged in a research and development program directed toward the preparation and processing of very high strength, environmentally resistant polymers for use as structural materials in aerospace vehicles. The objective is the attainment of mechanical properties for a structural material comparable with those currently obtained with fiber reinforced composites, but with significantly higher environmental resistance and without the use of a fiber reinforcement. The materials chosen for this effort are the rigid rod, extended chain, aromatic-heterocyclic polymers whose physical and chemical properties show promise for achievement of the program objectives. However, these materials present special processing problems because of the extended chain, rigid rod structural character of molecules. Present processing requires strong mineral or organic acid solvents and there is little opportunity to influence the polymer morphology once the material is in the solid state. One potential concept for the utilization of the rodlike polymers is molecular composites. This concept consists of blending a rodlike aromatic heterocyclic polymer with a coil-like aromatic heterocyclic polymer. The intent is to reinforce the coil-like or amorphous polymer with the rodlike polymer, thus forming a composite on the molecular level analogous to chopped fiber reinforced composites. The subject of this paper is a study to demonstrate the feasibility of this concept.

Materials and Processing

A variety of rodlike and amorphous polymers were studied in this investigation. The chemical structures of the various

TABLE I

POLYMERS STUDIED

CHEMICAL STRUCTURE ACRONYM

Rod-Like Polymers

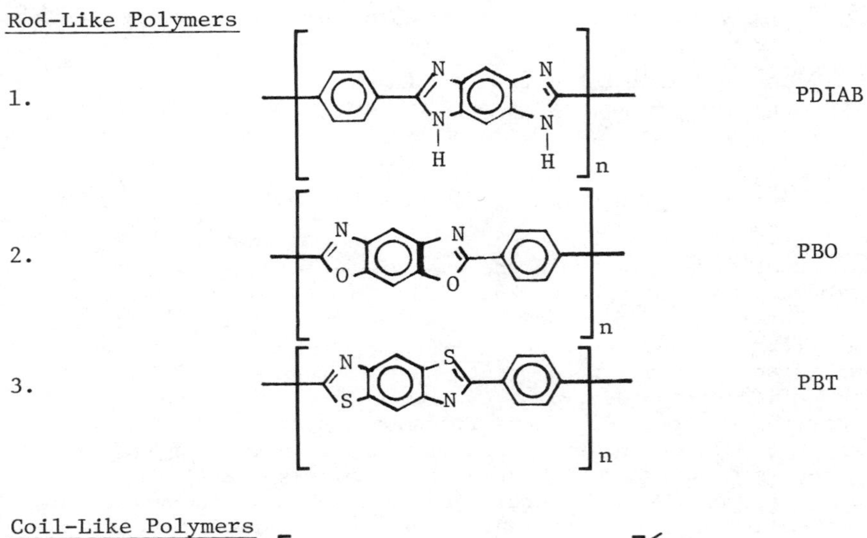

Coil-Like Polymers

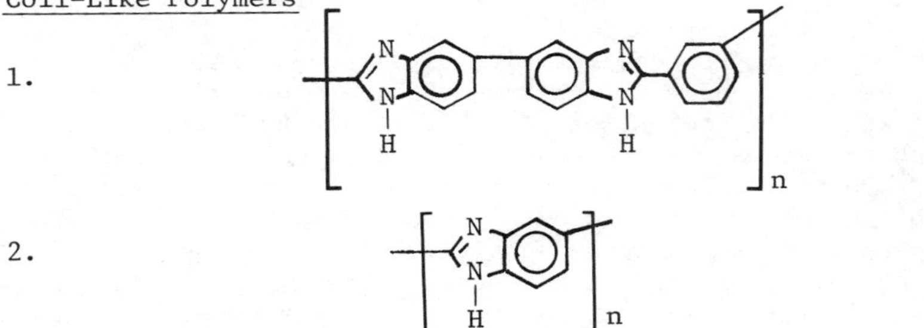

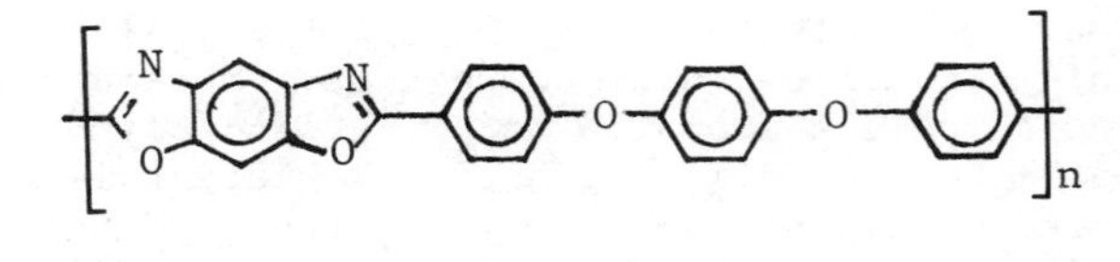

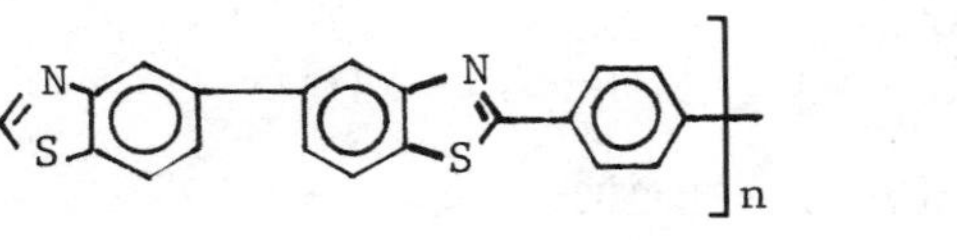

polymers are shown in Table I. The polymer blends studied and their weight percents are listed in Table II.

The polymer blends studied were processed as thin films by vacuum casting from dilute solutions. The general procedure followed was to prepare a 1-2% polymer solution in methane sulfonic acid and put the solution in a specially fabricated circular flat bottomed casting dish. The dish was then placed and leveled in the bottom of a sublimator. The cold finger of the sublimator was maintained at 25°C and the sublimator was continuously evacuated and heated to 60°C to facilitate the removal of the methane sulfonic acid. After the films were formed and removed from the casting dish, they were generally dried at 100°C in a vacuum oven for 24 to 48 hours. The films produced were approximately 5 cm in diameter and varied from 1.3×10^{-3} to 16.5×10^{-3} cm in thickness. Most of the films retained approximately 20 to 30% residual solvent.

Specimen Preparation and Testing

The films were cut with a razor blade into .635 cm strips. Strips at least 2.54 cm in length were used for testing, while shorter pieces were used for as cast morphological studies. Tests were performed on an Instron universal test machine at a crosshead speed of .02 inches per minute. After initial specimen breaks occurred, remaining pieces were retested until the length became too short to reasonably grip and test (approximately 1.5 cm). This provided not only as-cast data, but also mechanically stretched data. In addition, some specimens were plasticized with methanol to permit larger amounts of stretching. After these specimens were dried of the methanol, they were mechanically tested to determine the effect of the stretching.

The results of the testing of the AB-PBI/PDIAB blends are presented in Table III. These blends were studied in the most detail because they appeared to be the most compatible polymer blends and gave the most interesting results. Several general observations can be made about the data. In the as-cast (no stretching) polymer blends, the rod-like polymer appears to act more as a filler than a reinforcement. However, stretching (both mechanical and solvent) appears to provide some orientation and demonstrates a real reinforcing effect (strength and modulus increases). The reinforcing effect, however, does not follow a rule-of-mixtures behavior, the 10 percent blend being proportionally better than the 20 percent or 30 percent blends. The 57 percent and 75 percent blends could not be stretched because of their low strain-to-failure and no significant reinforcing was observed. Similar trends were observed with other polymer blends studied; however, the data presented represents the most significant results.

TABLE II

POLYMER BLENDS STUDIED

	Matrix	Reinforcement	Weight Percents of Reinforcements
1.	M-PBI	PDIAB	0, 20, 50, 75
2.	AB-PBI	PDIAB	0, 10, 20, 30, 57, 75
3.	PPBT	PBT	0, 25, 50, 75
4.	PEPBO	PBO	0, 25, 50, 60, 75
5.	AB-PBI	PBO	0, 10, 20, 30

TABLE III

MECHANICAL PROPERTIES - AB-PBI/PDIAB BLENDS

% ROD POLYMER	STRETCH/ % AREA REDUCTION	MODULUS (G Pa)	STRENGTH (M Pa)	STRAIN (%)
0	None	1.03	79.92	98
0	Mech./20	2.00	134.36	43
0	Solvent/60	3.37	105.42	12
10	None	2.00	70.28	46
10	Mech./37	4.60	161.92	28
10	Solvent/57	6.86	315.56	14
20	None	1.58	44.10	26
20	Mech./37	2.38	82.68	15
20	Solvent/70	7.17	253.55	9
30	None	1.25	36.52	14
30	Mech./20	2.16	71.66	22
30	Solvent/60	8.96	189.48	4
57	None	1.34	28.73	5
75	None	1.51	22.32	4

Morphological Studies

The morphology of the films has been studied using scanning electron microscopy and x-ray diffraction. A careful study of SEM photographs explains many of the observed test results. Figures 1-4 show SEM photographs of the surfaces of the 0%, 10%, 20%, and 30% rod films respectively. The morphological changes are obvious. Figures 5-8 show edge views of liquid nitrogen fractures of the same four films. Several observations have been made from the study of the SEM photographs. A second phase or conglomerate is present in the blends. These conglomerates in the as-cast (not stretched) films appear to be symmetric and saucer shaped with an aspect ratio (length/thickness) of 2-3. The absolute size of the conglomerates increases with increasing rod content. The volume content of the conglomerates in the film is greater than the volume content of rod-like polymer, indicating that the conglomerates contain both rod-like polymer and amorphous polymer. Stretching the films changes the shape of the conglomerates, increasing the length and decreasing the width and thickness. This can be seen in Figures 9 and 10 which show edges perpendicular to and parallel to the stretch direction of a 30% rod, solvent stretched film.

A tabulation of some of the morphological phenomena is presented in Table IV. Although the measurements made were relatively crude, the magnitudes of the measurements and phenomenological trends can be derived from this data. One important observation is that volume calculations indicate that the percent rod-like polymer in the conglomerates is constant, approximately 57%. This assumes that all of the rod is in the second phase. If this is true, it indicates that 57% may be an equilibrium mix condition for the two polymers. To verify this, a 57% PDIAB/43% AB-PBI film was prepared. As can be seen in Figure 11, no second phase was observed. Results of the x-ray diffraction studies are still being analyzed; however, definite signs of ordering and orientation have been observed.

Analysis

In order to develop a better understanding of the results obtained, analysis procedures developed for composite materials were utilized for determining the effective moduli of the conglomerates. The Halpin-Tsai equations (1) used are shown below:

Halpin-Tsai Equations

$$\frac{\bar{E}}{E_M} = \frac{(1 + \zeta \; \eta \; v_f)}{(1 - \eta \; v_f)}$$

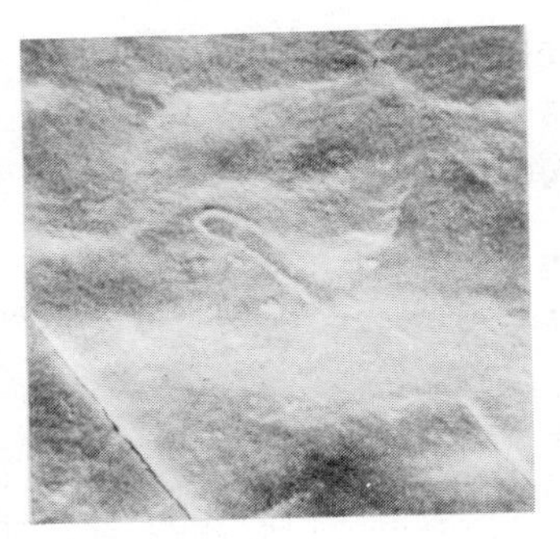

Figure 1. 100% AB–PBI surface (1500×)

Figure 2. 90% AB–PBI/10% PDIAB surface (1500×)

Figure 3. 80% AB–PBI/20% PDIAB surface (1500×)

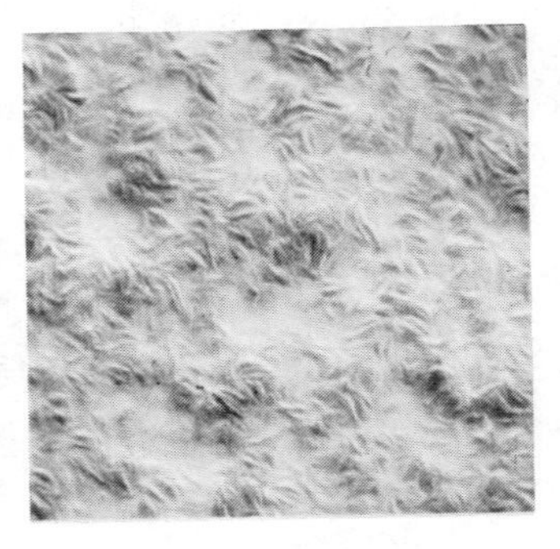

Figure 4. 70% AB–PBI/30% PDIAB surface (1500×)

Figure 5. 100% AB–PBI edge (2000×)

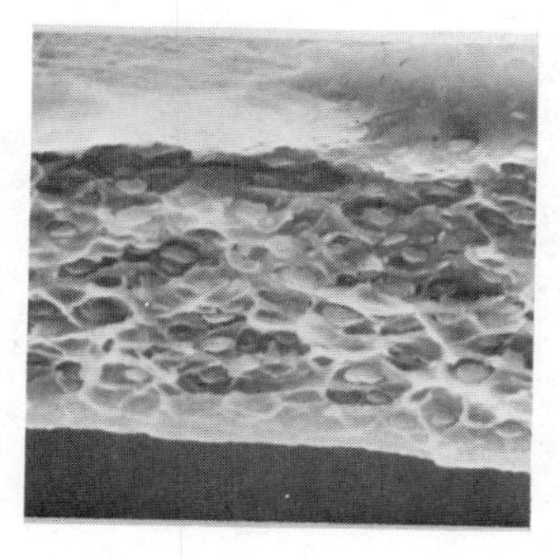

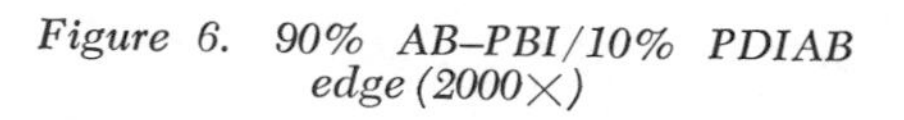

Figure 6. 90% AB–PBI/10% PDIAB edge (2000×)

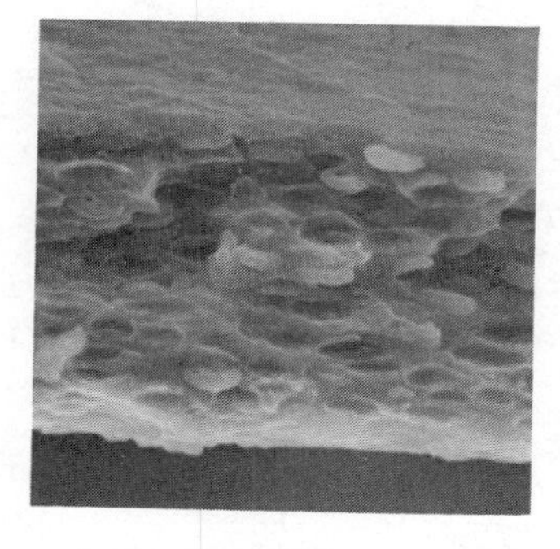

Figure 7. 80% AB–PBI/20% PDIAB edge (2000×)

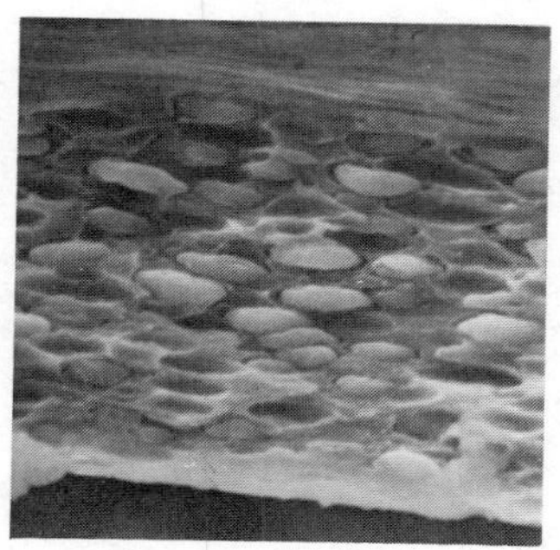

Figure 8. 70% AB–PBI/30% PDIAB edge (2000×)

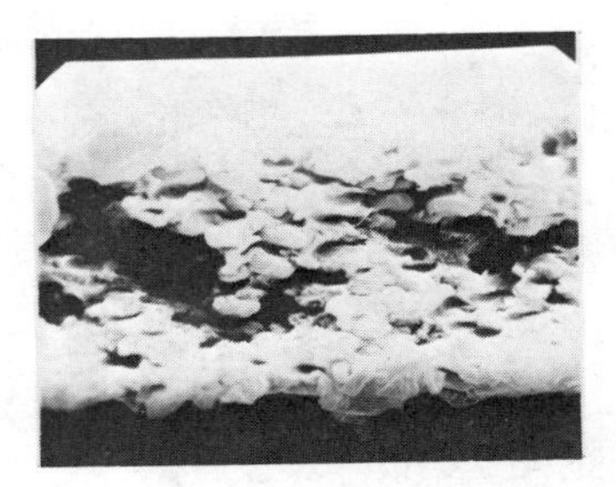

Figure 9. 70% AB–PBI/30% PDIAB solvent-stretched edge ⊥ to stretch (2500×)

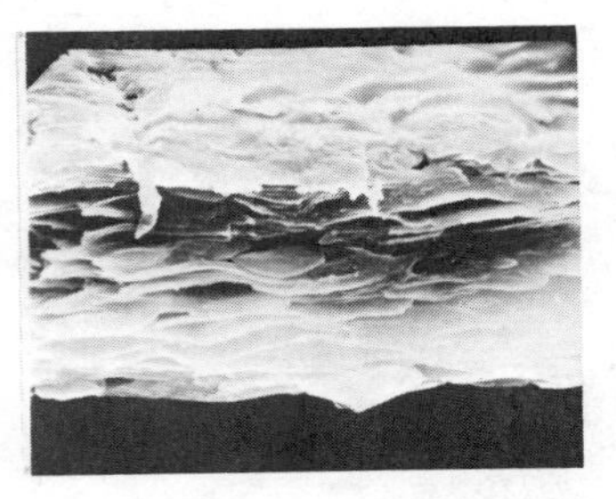

Figure 10. 70% AB–PBI/30% PDIAB solvent-stretched edge || to stretch (2500×)

TABLE IV

MORPHOLOGY OF AB-PBI/PDIAB BLENDS

Conglomerate	90/10	80/20	70/30
Length (cm x 10^{-4})			
Initial	1.78	3.81	5.84
Mechanically Stretched	2.54	5.08	7.62
Solvent Stretched	3.81	5.08	6.35
Thickness (cm x 10^{-4})			
Initial	.635	1.27	1.91
Mechanical Stretched	.508	.889	1.27
Solvent Stretched	.254	.508	.762
Width (cm x 10^{-4})			
Initial	1.78	3.81	5.84
Mechanically Stretched	1.52	3.30	5.08
Solvent Stretched	1.02	1.14	1.27
Aspect Ratio (L/T)			
Initial	2.8	3.0	3.1
Mechanically Stretched	5.0	5.7	6.0
Solvent Stretched	15	10	8
Volume % in Film	18%	35%	53%
Volume % in Conglomerate	56%	57%	57%

$$\eta = \frac{\left(\frac{\overline{E}_f}{E_M} - 1\right)}{\left(\frac{\overline{E}_f}{E_M} + \zeta\right)}$$

where,

$\overline{E}$ = Composite or film modulus.

E_M = Corresponding matrix modulus.

v_f = Volume fraction of reinforcement.

$\overline{E}_f$ = Corresponding effective reinforcement modulus.

ζ = Measure of reinforcement dependent on boundary conditions (for these calculations, taken as $\zeta = 2\ (a/_b)$.

The results of the analysis are shown in Table V. Several observations can be made from these calculations. The higher effective modulus of the conglomerates in the stretched 10% rod film indicates a higher degree of orientation in these conglomerates. This increased orientation and corresponding higher aspect ratio of these conglomerates accounts for the higher relative modulus observed in the 10% rod films. The magnitudes of the conglomerate moduli are high indicating relatively good translation of rod properties. Even in the most highly stretched 10% rod film, the conglomerate aspect ratio and degree of orientation are not nearly sufficient to obtain desired properties. Therefore, the conglomeration is considered undesirable and complete dispersion of the rods plus the ability to orient them in the film is the desired goal.

Processing Studies

In an attempt to achieve dispersion of the rods, various processing techniques were studied. Instead of vacuum casting the films, a technique of precipitating from dilute solution in a high humidity environment was developed. A 10% PDIAB/90% AB-PBI film made by this process is shown in Figure 12. As can be seen, no visible second phase is present. Mechanical properties of this film are given in Table VI. The properties are much better than those obtained from the vacuum cast films indicating excellent translation of rod properties.

Figure 11. 43% AB–PBI/57% PDIAB edge (1500×)

TABLE V

CALCULATED CONGLOMERATE MODULUS - AB-PBI/PDIAB BLENDS

90/10	AB-PBI/PDIAB (Initial)	$\overline{E}_f$ = 16.15 G Pa
90/10	AB-PBI/PDIAB (Mech. Stretched)	$\overline{E}_f$ = 33.35 G Pa
90/10	AB-PBI/PDIAB (Solvent Stretched)	$\overline{E}_f$ = 26.23 G Pa
80/20	AB-PBI/PDIAB (Solvent Stretched)	$\overline{E}_f$ = 15.43 G Pa
70/30	AB-PBI/PDIAB (Solvent Stretched)	$\overline{E}_f$ = 14.94 G Pa

Figure 12. 90 % AB–PBI/10% PDIAB water-precipitated edge (1500×)

TABLE VI

MECHANICAL PROPERTIES - 90% AB-PBI/10% PDIAB (PRECIPITATED)

STRETCH % AREA REDUCTION	MODULUS (G Pa)	STRENGTH (M Pa)	STRAIN (%)
None	3.08	92.39	15
Mech./5	4.00	122.09	13
Solvent/55	9.65	243.95	3

Summary

The concept of molecular composites has been demonstrated. Although the studies to date are very preliminary, the concept appears to be very promising. Future work will be oriented toward characterizing the solution behavior of the polymer blends and developing processing techniques to achieve better morphology control as well as orientation control.

Literature Cited

1. Halpin, J. C., Tsai, S. W., "Environmental Factors in Composite Materials Design," AFML-TR-67-423.

RECEIVED February 6, 1980.

17

Analytical Techniques Applied to the Optimization of LARC-160 Composite Lamination

A. WERETA, JR. and D. K. HADAD

Lockheed Missiles & Space Company, Incorporated, Sunnyvale, CA 94086

LARC-160 is a commercially available polyimide resin system developed by NASA Langley which starts out as a mixture of monomeric reactants in an ethanol solution (1). The constituents of LARC-160, shown in Figure 1, are the diethylester of benzophenone tetracarboxylic acid (BTDE), the ethyl ester of norborene dicarboxylic acid (NE), and Jeffamine AP-22, a mixture of ditri- and tetrafunctional isomers, as well as higher molecular weight species of methylene dianiline. These constituents are mixed in molar ratios of 0.335:0.610:0.539 (BTDE:NE:Jeffamine AP-22) which give an average molecular weight of 1600.

Ideally, the monomers undergo a condensation reaction to form imidized oligomers end capped with norborene rings. Tri- and tetrafunctional amine isomers provide some crosslinking via condensation; however, the bulk of crosslinking is attributed to an addition reaction as the norborene rings open at higher temperatures.

Lockheed Missiles & Space Company, Inc. (LMSC) is currently evaluating LARC-160 prepregged on graphite fabric in a combined materials, design, analysis, and manufacturing program aimed at developing a fiber-reinforced organic matrix composite for use at 317°C (600°F). Autoclave cure cycles are of primary interest because they are required for large part fabrication. A workable autoclave cycle, Figure 2, was empirically developed at LMSC and demonstrated on an integrally stiffened cylinder measuring 30 in. in diameter by 12 in. high (2). Since the results show promise, an optimized process is now desired.

Processing LARC-160 prepreg into well-consolidated composites of high quality requires removal of both prepregging and condensation volatiles and timely application of consolidation pressure. Premature consolidation results in excessive flow and/or trapping volatiles which can lead to voids, blistering, delamination, or reduced performance at elevated temperatures (plasticization). Delayed pressurization results in poor flow because of incipient

0-8412-0567-1/80/47-132-215$05.00/0

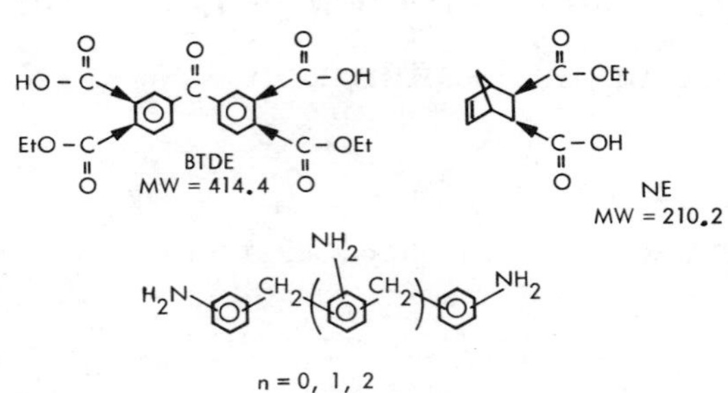

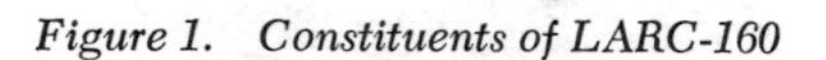

Figure 1. Constituents of LARC-160

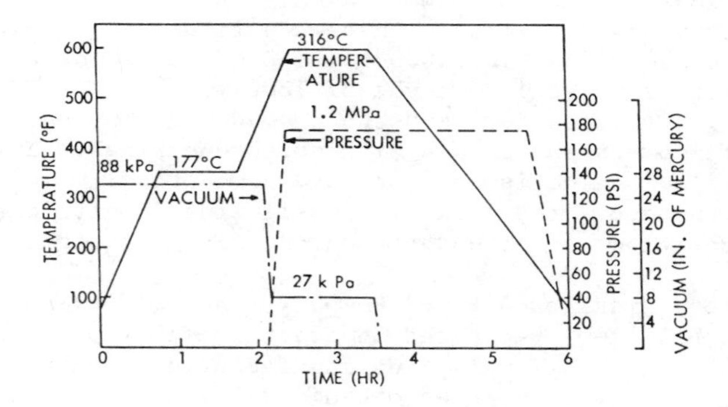

Figure 2. Empirically developed autoclave cycle for LARC-160 lamination

network formation. Optimum processing requires definition of a processing window between these extremes which is subject to the constraints of the particular manufacturing process, be it autoclave, hydroclave, or matched-metal molding. Process variables of interest include heating rate(s) and hold temperature(s), degree of vacuum and both the degree and rate of pressurization. In addition, a pidicous choice of ancillary materials such as bleeder plies and resin barriers must be made on the basis of the resin's flow characteristics.

The chosen path to optimization centers around defining the rheological changes which occur during processing and relating them to volatile evaluation and chemical kinetics. The initial phase of this work entailed evaluation of analytical methods which could follow chemical and physical changes during staging. Infrared spectroscopy (IR), differential scanning calorimetry (DSC), and liquid chromatography (LC) were used to follow chemical changes. Volatile evolution was traced by mass spectroscopy (MS) coupled to a heated cell and physical changes were visually observed on a programmed hot plate. Rheological changes, as determined by dielectric spectroscopy and dynamic viscosity measurements, were compared. All results will ultimately be correlated with dielectric spectra because they are amenable to real time process monitoring and control of production hardware (3,4).

Experimental

The material used in this investigation was a commercially produced LARC-160 prepreg. The graphite fabric used for reinforcement was a 24 x 24 8 harness satin weave of Thornel 300 fiber tows containing 3,000 filaments per tow. The fibers were sized with a C 309 epoxy coating at a concentration of 2% or less. Fabric was prepregged by dipping into a LARC-160/ethanol solution and tower drying to obtain acceptable levels of tack and drape. Resin solids, volatile content and flow were determined to be 41.8, 13.9, and 37.4%, respectively.

Samples for IR, DSC and LC experiments were all previously staged in an Audrey II dielectric cell heated at 4°C/min to a selected temperature. This assured good correlation with the dielectric data and allowed a look at imidization under process-like conditions where diffusion of trapped volatiles can influence the reaction rate. Kinetic studies often ignore the influence of by-product diffusion on reaction rates (5) but such considerations are of prime engineering concern.

Infrared spectra were obtained on a Perkin-Elmer Model 180 Grating Spectrophotometer. Staged samples were ground up and formed into KBr pellets whereas resin from the control prepreg was smeared onto a salt plate.

DSC thermograms were obtained with a DuPont Model 990 Thermal Analyzer. Samples of staged resin were first ground up and then weighed to 7.0 ± 0.03 mg in presorted aluminum pans of equal weight. Sample pans were covered but not crimped before scanning at a heating rate of 10°C/min.

Liquid chromatograms were obtained on each of the staged samples as well as the control prepreg and Jeffamine AP-22 using a Waters Associates Model 244 Liquid Chromatograph. The separation was obtained by gel permeation chromatography, a size exclusion technique using STYRAGEL columns (500 + 10^3A) and dimethylformamide as the mobile phase.

A Hitachi Perkin Elmer Model RMU-6 mass spectrometer of the single focusing magnetic variety with a heated inlet was used to detect volatiles released from a prepreg sample as it was heated at 4°C/min. It was necessary to outgas the chamber for several minutes at approximately 1 microtorr before heating so that the initial prepreg concentrations were altered.

Visual observations of bubbling and hardening were performed between glass slides on a hot plate heated at 4°C/min. A fiberglass cloth blanket was used to minimize heat losses. A low magnification microscope aided observation and relative deformation of the glass slides provided a means for determining the resin's hardness at any given time.

The dielectric dissipation factor was measured with a Tetrahedron Audrey II dielectrometer at a frequency of 1 kHz. One layer of cloth prepreg was placed in an aluminum foil dish and covered with successive layers of 181 weave glass fabric and 5 mil thick Kapton film before inserting into an Audrey sample cell. The glass fabric extracts resin from the conductive graphite fabric so that the dielectric properties of the matrix resin can be monitored during cure. The Kapton film is generally considered stable and therefore without influence on a curing epoxy resin, but when polyimides or other high temperature resins are being cured the presence of the Kapton must be considered.

A Rheometrics mechanical spectrometer was used to measure relative viscosity changes during cure of LARC-160 prepreg. Four layers of stacked prepreg were sandwiched between 50mm dia. parallel plates separated by a 1 mm gap. The bottom plate remains stationary while the top plate is sinusoidally oscillated, in this case at a selected frequency (1 Hz). The sample cell was heated in a stepwise fashion which closely approximated a ramp function of 4°C/min. Similar experiments were attempted with unreinforced LARC-160 at heating rates of 2 and 4°C/min.

Results and Discussion

Advances in cure chemistry are discussed relative to data obtained from DSC, IR, and LC experiments before proceeding to chemorheological results obtained from dielectric and dynamic viscosity measurements.

A series of DSC thermograms for LARC-160 resin which had been staged at 120° and 150°C are shown in Figure 3. The thermograms are vertically shifted for ease of comparison. In order to aid in

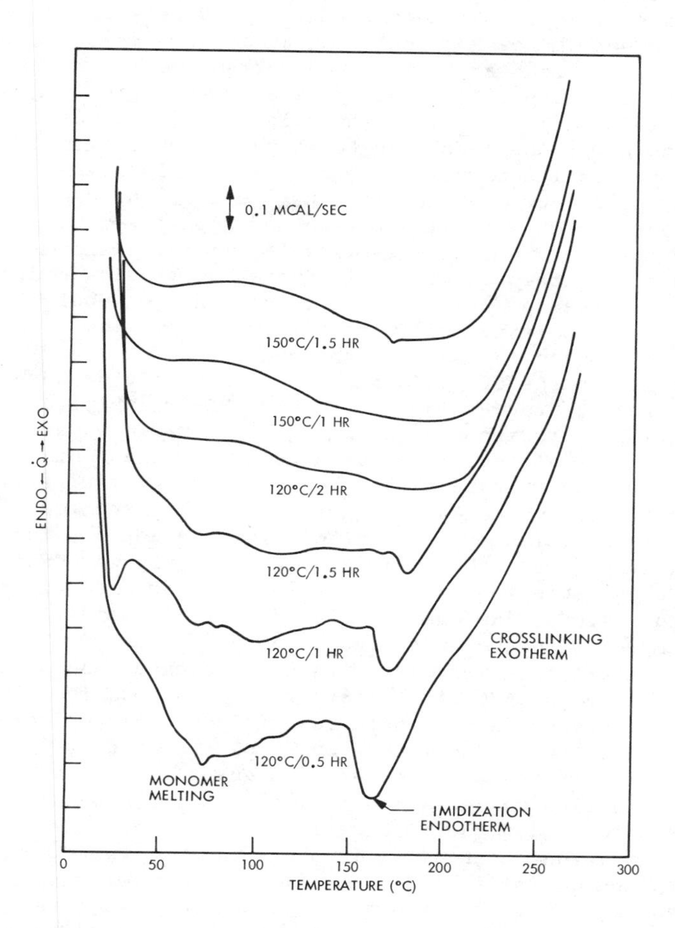

Figure 3. Thermograms for LARC-160 prepreg samples staged at 120° and 150°C

the discussion, several samples were first scanned to 320°C, cooled, and then rescanned as shown in Figure 4. While the reheated samples are certainly not fully cured, they indicate the behavior of highly imidized resin. Several comments can be made. First, the thermograms are rather noisy below 150-175°C for samples staged less than 2 hours at 120°C, no doubt because of the retention of volatiles. Second, there is a persistent, although small endothermic peak around 75°C in these samples which can be associated with monomer melting and perhaps salt formation (6). This peak is no longer evident for the more severe staging conditions in this series. Third, there is a strong endothermic peak around 160°C for the least staged sample which shifts to higher temperatures as its magnitude decreases with increased staging until it is no longer apparent for the 120°C/2 hour sample. This peak is associated with imidization and appears in the same region reported for the chemically similar PMR-15 formulation (6). A shift to higher temperatures with increased staging could be the result of decreased mobility as imidization progresses. Additional evidence supporting this possibility will follow. Finally, there is an upward tail on all the thermograms which was initially associated with the crosslinking reaction involving norborene rings. However, Lauver (6) shows that PMR-15 exhibits another end of them in this region when scanning at 10°C/min under 60 PSI of nitrogen and that the crosslinking exotherm does not occur until 325°C. LARC-160 thermograms in the vicinity of the rise were rather erratic as indicated in Figure 4 for the reheated samples, and in fact satisfactory peak resolution was not obtained when scanning as high as 460°C. It appears that high pressure DSC may be a more valuable tool and its utility is now under investigation in our laboratory. No quantitative measure of imidization rate was attempted with the data presented in Figure 4.

IR spectra for a sample of LARC-160 prepreg and a sample of LARC-160 staged for 30 minutes at 120°C are shown in Figure 5. Some of the key bands have been identified and labelled on the spectra. Spectra were obtained on samples staged at 120, 150, 177, and 204°C for times up to 2 hours. A quantitative measure of imidization kinetics was obtained from this data. A band at $1380cm^{-1}$ was chosen to follow the imidization reaction because it was shown elsewhere (5) that Beer's law is obeyed, i.e., there is no change in absorption coefficient during the reaction. This is not the case for a band at $1780cm^{-1}$. A band at $1510cm^{-1}$ was chosen as an internal standard since each point shown in Figure 10 represents an individually staged sample. The band at $1510cm^{-1}$, associated with a substituted aromatic ring, showed no apparent changes during imidization. A straight baseline was drawn between 1900 and 1030 cm^{-1} and relative absorbance ratios of the $1380cm^{-1}$ band to the internal standard band at $1510cm^{-1}$ were calculated from measured values of transmittance. The degree of imidization during cure was determined by comparing the above ratio for each staged sample (time=t min.) with that of the baseline prepreg (t=o min.) and a

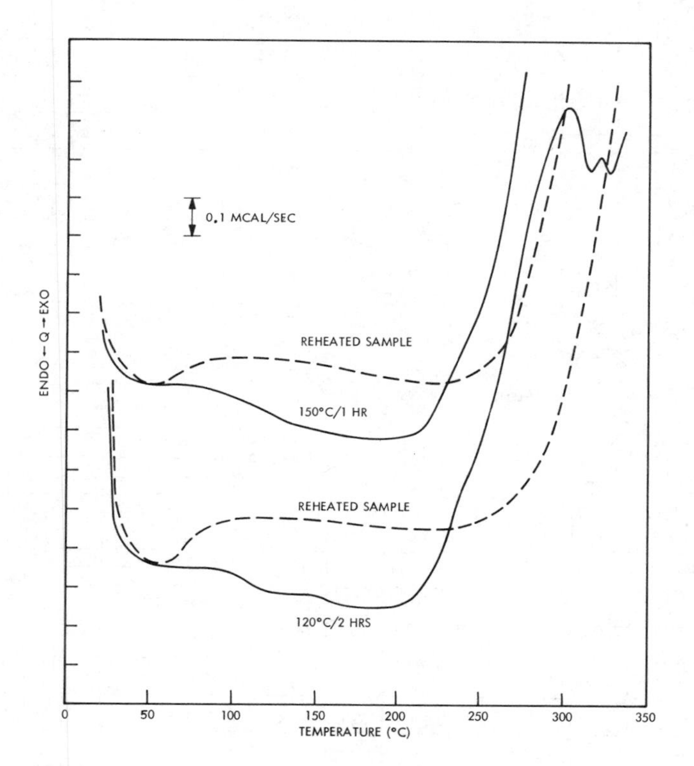

Figure 4. Comparison of original and reheated thermograms for staged LARC-160 prepreg

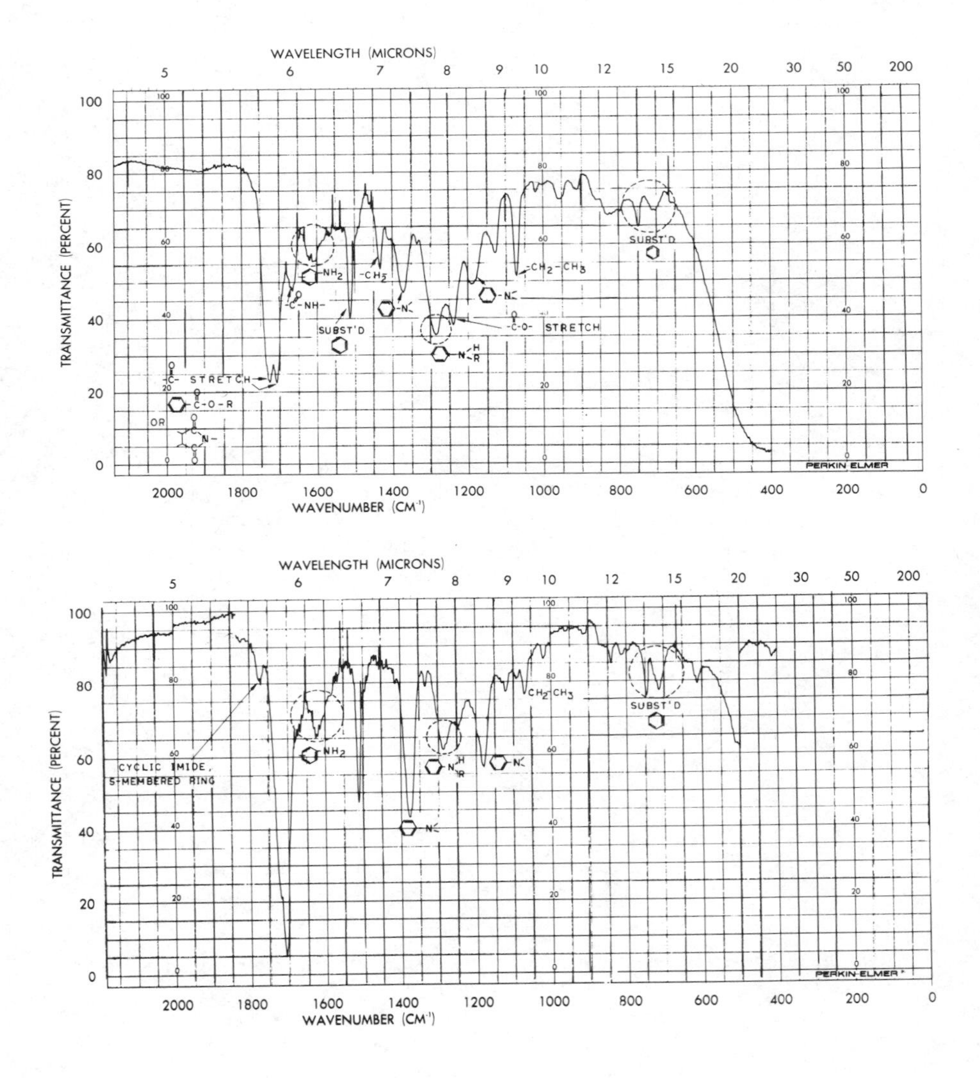

Figure 5. IR spectra for (a) LARC-160 prepreg control, and (b) LARC-160 prepreg that was staged for 30 min at 120°C

fully imidized sample ($t=\infty$). A completely imidized material was somewhat arbitrarily chosen as that obtained by heating a staged sample to 316°C and holding for one hour. An isothermal expression for the fraction imidized at any time is given by

$$\text{Fraction Imidized} = \frac{\frac{A\ 1380}{A\ 1510}(t) - \frac{A\ 1380}{A\ 1510}\ (\text{prepreg})}{\frac{A\ 1380}{A\ 1510}\ (1\ \text{hr} \ @\ 316°\text{C}) - \frac{A\ 1380}{A\ 1510}\ (\text{prepreg})}$$

where A 1380 and A 1510 represent the absorbences corresponding to the respective band lengths. The results are shown in Figure 6. Data for the 120 and 150°C staging temperatures appear well behaved, showing smooth, monotonic increases in the fractions imidized as a function of time. While data at the 177°C staged temperature displays the same initial behavior, there is a sharp drop in the fraction imidized for the 1 1/2 hr. sample. Since it represents a single sample, its significance is unknown. An unexpected result occurs for the 204°C staging temperature where it appears that less imidization occurs than at 177°C and in fact not much more than that found at 150°C. This behavior may be explained by the dielectric results which are discussed later.

The chromatograms of Figure 7 show definite effects of staging as indicated by the appearance and growth of peaks associated with unidentified molecular species. Judging from the similarities in evolution times, it appears that the species may be associated with Jeffamine reaction products. The present separation techniques have not been optimized, but the method appears quite sensitive to chemical advances caused by staging. One must realize, however, that the analysis is limited to soluble species, and that its utility is therefore limited to the early stages of cure.

A very simple, yet quite useful experiment shedding insight into the rheology of LARC-160 during cure was conducted with a programmed hot plate. By following the cure cycle illustrated in Figure 8 and noting the behavioral changes indicated thereon, a basic understanding of the processing behavior of LARC-160 was obtained. These results were correlated with those from mass spec analysis in order to find out relative volatile evolution rates and with the dielectric dissipation factor in order to aid in its interpretation. Visual observation showed that the prepreg began softening at about 85°C. Bubbling began at 95°C, became violent near 115°C, and subsided near 150°C as the resin thickened. Within 15 minutes after reaching 177°C, the prepreg had become quite hard and could not be deformed by relative motion of the glass slides.

The mass spec shows that water and ethanol are the major volatiles emerging during cure. Initial concentrations of ethanol and water may be associated with residual prepregging solvent and absorbed atmospheric moisture, but between 100-120°C the concentration profiles change. This may be due to the decreased resin viscosity or the onset of imidization or both. Both concentration

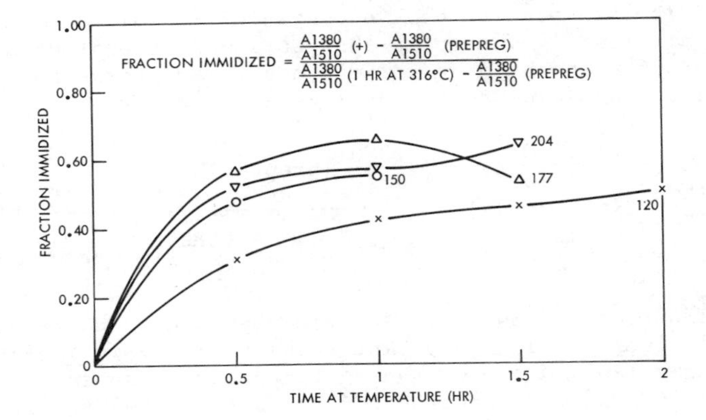

Figure 6. Fraction of LARC-160 prepreg imidized after heating at 4°C/min and holding at the indicated temperatures

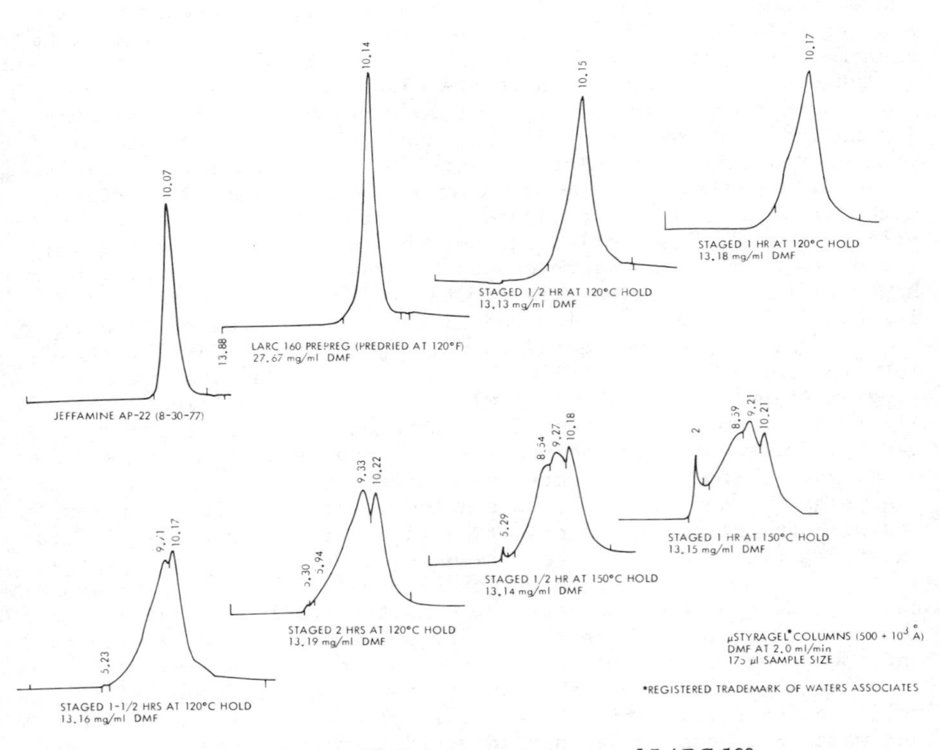

Figure 7. Liquid chromatograms for staged LARC-160

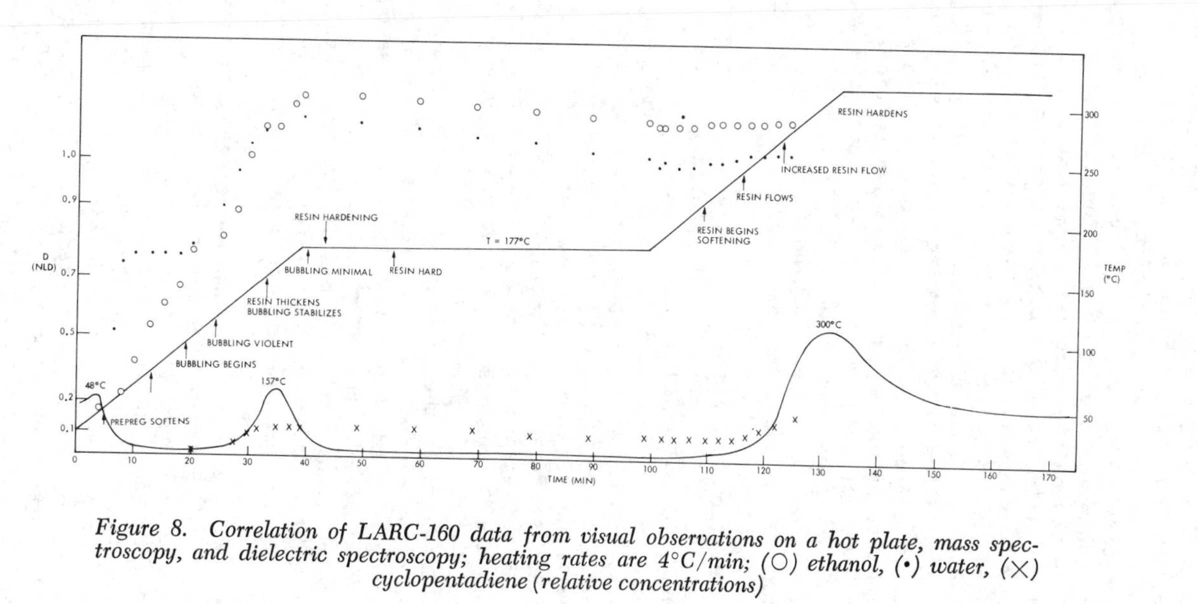

Figure 8. Correlation of LARC-160 data from visual observations on a hot plate, mass spectroscopy, and dielectric spectroscopy; heating rates are 4°C/min; (○) ethanol, (•) water, (×) cyclopentadiene (relative concentrations)

levels remain high during the temperature hold and change very little, even when the resin softens during resumption of heating. It would appear that the evolution of these volatiles is not hindered by the nearly glass state of the resin, but one must remember that the mass spec experiment was conducted under high vacuum and the prepreg was only a single ply. In addition, it is not known if the evolution rate reflects the imidization rate or the diffusion rate to the surface. In any case it is clear that a large concentration of volatiles remain in the resin when the prepreg is heated to the cure temperature and that they could get trapped within a composite as increased crosslinking gives rise to a glossy state. A free standing postcure therefore appears desirable for composite hardware with thick cross sections. The only other volatile measured in reasonable concentrations was cyclopentadiene. It first appeared at about 100°C and rose to a shallow maximum near the onset of the 177°C hold. Its concentration began rising again once the imidized resin softened during reheating (260-300°C). Cyclopentadiene is known to arise from a reverse Diels-Alder reaction associated with polymerization through the norborene ring (1,7). However, its appearance prior to the final cure reaction is unexplained at this time.

The Audrey dielectric dissipation factor shows an initial peak (48°C) associated with the prepreg softening and perhaps flow of the resin into the adjacent bleeder ply. A second peak (157°C) appears when the resin hardens due to imidization and therefore increased chain stiffness. A third peak (300°C) appears when the imidized resin softens and rehardens as cure continues. Dissipation peaks appear in regions of limited mobility which are bounded by regions of high or low relative viscosities. This interpretation is consistent with the visual observations described above and with the dynamic viscosity measurements shown in Figure 9 . Upon rescanning a cured sample both the first and second peaks are gone, but one similar to the third remains. Although the third peak seems to be meaningful in terms of LARC-160 rheology, it appears to be influenced by the Kapton film used in the dielectric cell. Note that the dissipation factor is plotted on a nonlinear scale.

A series of dielectric dissipation curves are shown in Figures 10 and 11. Heat up to staging temperatures of 93, 105, 121, 149, 177, and 204°C along with the corresponding dielectric responses are shown in Figure 10. Resumption of heating from each of these staging temperatures up to a 316°C cure temperature and their corresponding dielectric responses are shown in Figure 11. Referring first to the staging temperatures one finds no change in dipole mobility after 40 minutes at 93°C. Temperatures at or below 93°C may be useful for débulking thick parts of complex geometry if such a step is deemed necessary. By increasing the staging temperature to 105°C, one finds a very broad, drawn-out peak extending out to 310 minutes indicative of a slow reaction rate. The rate increases considerably by raising the hold temperature to

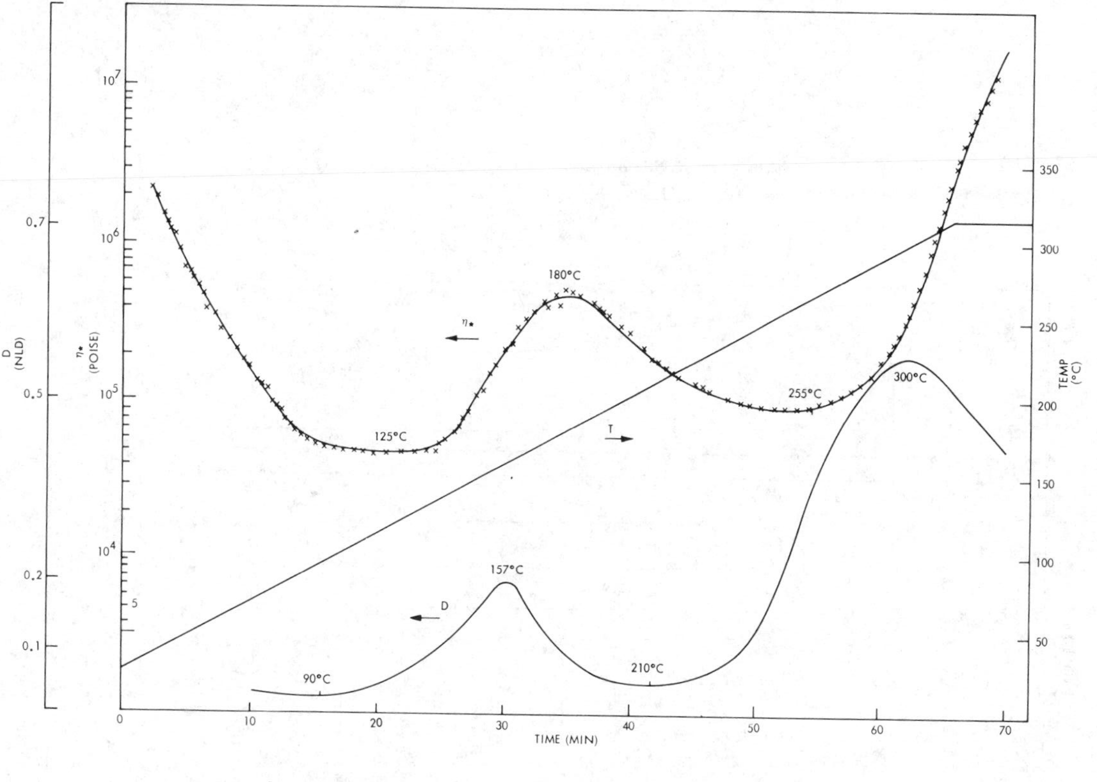

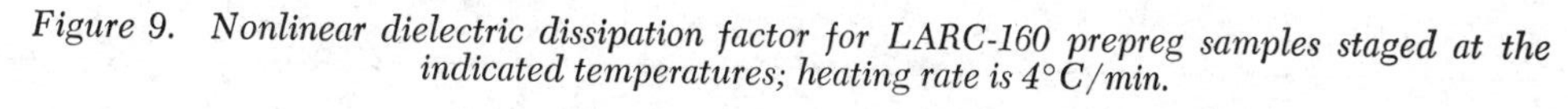

Figure 9. *Nonlinear dielectric dissipation factor for LARC-160 prepreg samples staged at the indicated temperatures; heating rate is 4°C/min.*

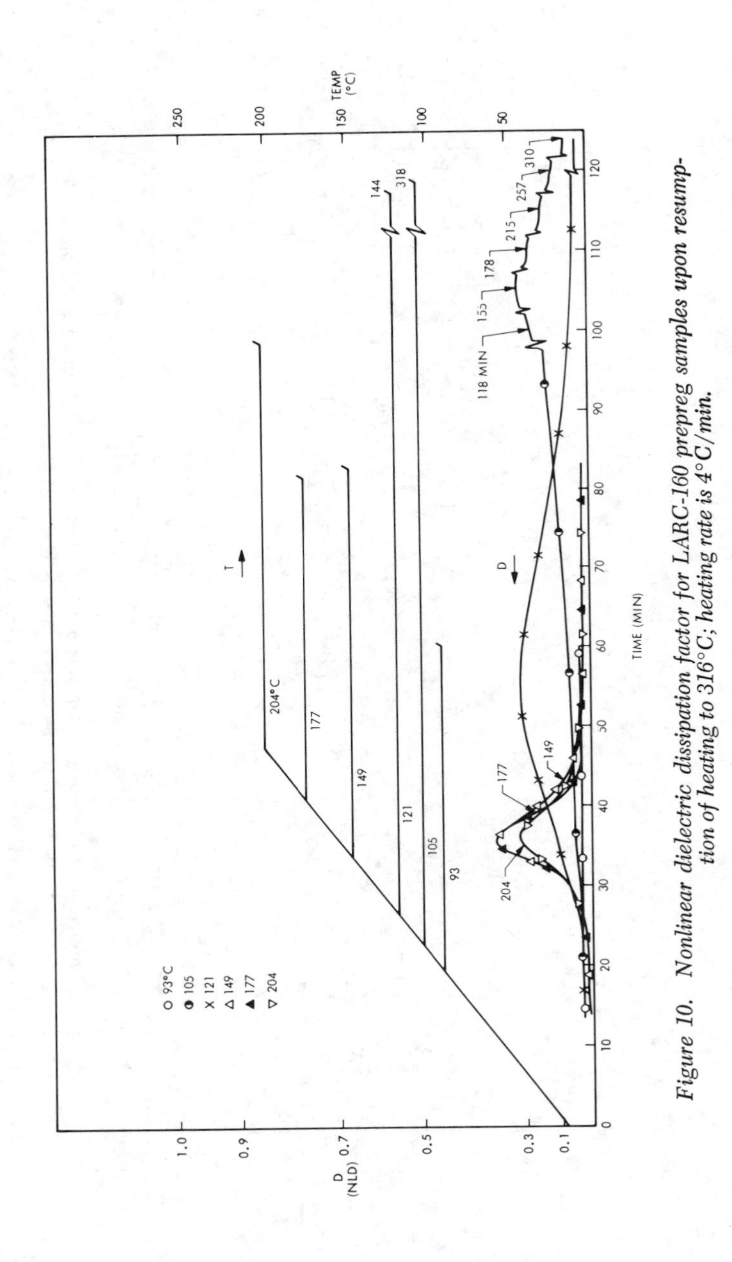

Figure 10. Nonlinear dielectric dissipation factor for LARC-160 prepreg samples upon resumption of heating to 316°C; heating rate is 4°C/min.

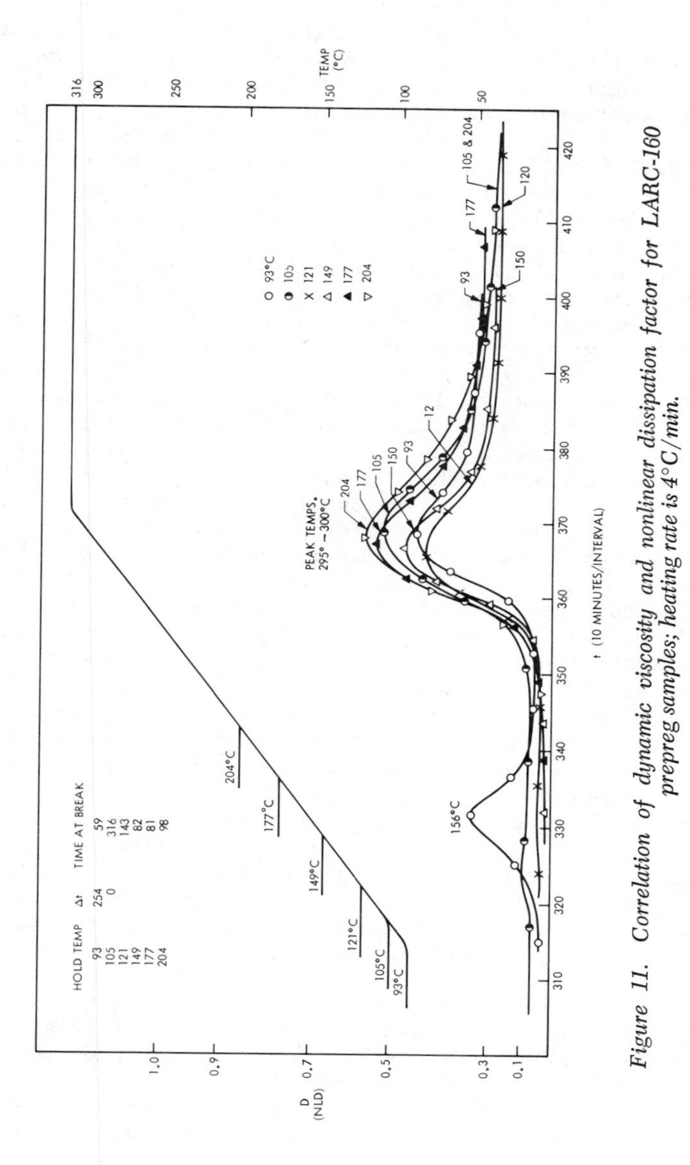

Figure 11. Correlation of dynamic viscosity and nonlinear dissipation factor for LARC-160 prepreg samples; heating rate is 4°C/min.

121°C where dipole mobility continues for about 140 minutes. A staging temperature of 149°C shows a dissipation peak appearing immediately after reaching temperature and the cessation of dipole mobility after a total time of 55 minutes. For staging temperatures of 177 and 204°C, the position of the dissipation peak is virtually the same as that for the 149°C staging temperature. However, the peak occurs during heat up for the two higher staging temperatures. It should be emphasized that the peak reflects a change in modulus and not actually extent of chemical reaction as was exemplified by the programmed hot plate experiment. The dielectric experiments lend credance to the IR findings which indicate less imidization occurs at 204°C than at 177°C. Since the resin hardens during heating, a significantly lower diffusion rate in the solid state could account for the diminished reaction rate. The dielectric staging experiments also suggest that heating rate is a very important process parameter which warrants investigation for LARC-160.

Figure 11 shows the dielectric dissipation during resumption of heating from a series of staging temperatures. In the case of a 93°C staging temperature, the imidization peak appears at nearly the same temperature at which it appeared during a straight heat up with no staging. One may conclude that 40 minutes at 93°C does not significantly affect the imidization rate of LARC-160. There seems to be a remnant of the imidization peak for a 105°C staging temperature although it appears at a lower temperature than anticipated. The cure peak occurs at one of two temperatures in the vicinity of 300°C. The lower temperature peak corresponds to staging temperatures of 93 and 105°C and the higher one to staging temperatures of 121°C and higher. A lower peak temperature indicates a lesser extent of reaction for the molder staging conditions. The relative heights of the cure peak show no smooth trend with staging temperature and the significance of this observation is without explanation.

When comparing the dielectric dissipation factor to the dynamic viscosity data as shown in Figure 9 , it becomes clear that although the shapes of both curves are similar, the dielectric peaks precede the viscosity peaks. This is explained by the fact that a more localized motion is satisfactory for dipole oscillation while cooperative segmental chain motion is necessary for flow.

The dynamic viscosity data of Figure 9 gives a relative measure of η^* during cure of LARC-160 prepreg. Only relative values of η^* on the ordinate are significant because of the composite nature of the sample. Regions of high and low viscosity are delineated with respect to temperature when LARC-160 is cured at a heating rate of 4°C/min, about the maximum available with an autoclave. When the LARC-160 formulation is first heated, the viscosity decreases as temperature increases until imidization occurs. The viscosity then levels off and then increases until the partially imidized resin softens with continued heating.

Further imidization and crosslinking account for the final rise in viscosity. Measurements were terminated at this point in order to facilitate sample removal.

Absolute viscosity measurements are not readily obtained on unreinforced LARC-160 because of extensive volatile evolution during heating. Bubbling was visually observed in the region of low viscosity for both neat resin and prepreg samples of LARC-160. As a result, the signal became somewhat erratic in this region. Bubbling of unreinforced resin at a heating rate of 4°C/min causes excessive resin loss from the sample cavity and the experiments had to be terminated. By reducing the heating rate to 2°C/min, the problem was reduced, but further improvement is required before meaningful data can be obtained. A pressurized sample cell is now being developed for further studies.

The use of prepreg rather than unreinforced resin is viewed as an expedient way of obtaining relative viscosities for cure cycle development. The problem of volatile evolution is much less pronounced with prepreg because the fabric helps maintain resin in the sample cell.

Summary

A number of experimental techniques were evaluated for optimizing LARC-160 composite lamination. Processing characteristics were determined by a combination of chemical, thermal, and rheological techniques. A good quantitative measure of imidization kinetics was obtained from infrared spectroscopy. Combined with results from dielectric spectroscopy and visual observations on a programmed hot plate, the infrared findings suggest that too high a staging temperature reached at a reasonable process heating rate can lead to a diffusion-limited imidization rate. At 4°C/min the imidization rate at 204°C actually lagged behind that at 177°C. Slower heating rates would presumably lower the staging temperature at which the same phenomenon could be observed. Thermal analysis by DSC gave marginal results because of residual voltiles contained in staged samples. Liquid chromatography was not developed to its potential, but the preliminary findings showed distinct differences between samples staged to varying degrees. Mass spectroscopy confirmed that water and ethanol are the main outgassing constituents during LARC-160's cure, while cyclopentadiene appears around 100°C and maintains a low concentration until the addition reaction begins and then its concentration rises. Dielectric spectroscopy showed that well defined differences in dissipation factor can be attributed to the influence of staging temperature. Three peaks were found and identified. One was associated with prepreg softening, another with hardening due to imidization, and a third with softening of the imidized resin prior to crosslinking. Dynamic viscosity measurements aided greatly in interpretation of the dielectric data. Relative viscosity measurements aided greatly in interpretation of the dielectric data. Relative viscosity measurements were obtained for prepreg samples, but the high percentage of volatiles liberated by the condensation reactions (about 10%) prevented absolute viscosity measurements for the neat resin.

As a result of these conclusions, the cure cycle shown in Figure 2 was modified. The vacuum level was lowered to prevent violent bubbling at the onset of the cure cycle from filling the bleeder plies with low viscosity resin. Second, the staging temperature was lowered to 120°C while the dwell was increased to 90 minutes. Third, a hold was added at 250°C so that full autoclave pressure could be applied during minimum increase in resin viscosity. Several 6 ply, 50 cm x 50 cm panels autoclaved with this modified cure cycle showed excellent compaction (circa. 65% F.V.) with near zero void content. Flexural strengths of 15.2 MPa at RT and 6.4 MPa at 600°F determined for these panels are quite satisfactory for woven cloth composites.

Acknowledgments

This work was supported by LMSC Independent Development funds. The authors wish to thank Amando G. Aquino and Cheryl Bostwick of LMSC for their experimental assistance, John Newner of Hescel for his helpful comments on the infrared data, and Rheometrics, Inc. for the use of their rheological equipment.

References

1. St. Clair, T. L., Jewell, R. A., National SAMPE Technical Conference Series, 1976, 8, 82.

2. Bailie, J. A., Mace, W. C., Wereta, A., Jr., Menke, G. D., Procedures of the Fourth Conference on Fibrous Composites in Structural Design, San Diego, CA, Nov. 1978.

3. Wereta, A., Jr., Fritzen, J. S., May, C. A., unpublished results, 1977.

4. May, C. A., National SAMPE Symposium and Exhibition, 1975, 20, 108.

5. Androva, N. A., Bessonov, M. I., Laius, L. A., Rudakov, A. P., "Polyimides - A New Class of Thermally Stable Polymers", Progress in Material Science...Volume VII, Technomic Pub. Co. Stamford, CT, 1970; 34.

6. Lauver, R. W., Journal of Polymer Science, Part A-1, Polymer Chemistry, 1978.

7. Kaplan, S. L., Helf, D., Hirsch, S. S., 27th Annual Technical Conference Proceedings, Reinforced Plastics/Composites Institute, The Society of the Plastics Industry, 1972, 2A-1

RECEIVED February 15, 1980.

18

Structure–Property Relations of Composite Matrices

ROGER J. MORGAN and ELENO T. MONES

Lawrence Livermore Laboratory, University of California, Livermore, CA 94550

The need to conserve energy has provoked increased interest in the use and development of high-performance, light-weight, fibrous composites for the transportation industry, e.g., for use in aircraft and automobiles, and in energy-storage systems such as flywheels (1). The matrices utilized in these high-performance composites, which can be exposed to extreme environments, have generally been epoxies. The question of the durability of the epoxy matrix and of the overall composite in such extreme environments is a cause for concern (2,3). However, such durabilities cannot be predicted accurately without a knowledge of the structure, modes of deformation and failure, mechanical response relationships of the epoxy matrices, and the possible modification of such relationships by fabrication procedures and the service environment. The structure-property relations of epoxies, however, have received little attention compared to other commonly utilized polymer glasses. The epoxy glasses, because of variation of their chemical and physical structure with fabrication conditions and because of their infusible, insoluble nature, are more difficult to study than noncrosslinked polymeric glasses.

In this paper, we will review our structure-property studies of epoxies that are commonly utilized as composite matrices. The two systems primarily studied were: (1) diethylene triamine (Eastman)-cured bisphenol-A-diglycidyl ether (Dow, DER 332) epoxy (DGEBA-DETA); and (2) diaminodiphenyl sulfone (Ciba-Geigy, Eporal)-cured tetraglycidyl 4,4'diaminodiphenyl methane (Ciba-Geigy, MY 720) epoxy (TGDDM-DDS).

Physical Structure

The major physical and structural parameters that control the modes of deformation and failure as well as the mechanical responses of epoxies are their crosslinked network structures and microvoid characteristics (4-9).

0-8412-0567-1/80/47-132-233$05.00/0

The cure process and final network structure of epoxies have been deduced from the chemistry of the system, providing the curing reactions are known and can be assumed to go to completion, and by experimental techniques such as infrared and carbon-13 nuclear magnetic resonance spectroscopy and swelling, ultrasonic, dynamic mechanical, thermal conductivity, and differential scanning calorimetry measurements (see Ref. 8). However, in many epoxy systems the chemical reactions are diffusion controlled and incomplete, and there is a heterogeneous distribution in the crosslink density.

High crosslink-density regions from 6 to 10^4 nm in diameter have been observed in epoxies (8). The most recent studies on the network morphologies of epoxies were carried out by Racich and Koutsky (10) and Mason and co-workers (11). These workers studied both etched and non-etched surfaces of epoxies by scanning electron microscopy as well as by transmission electron microscopy of carbon-platinum surface replicas. In our studies, we also strained films directly in the electron microscope (6). For example, in DGEBA-DETA epoxies, we observed that 6- to 9-nm-diam particles remain intact and flow past one another during the flow processes. We suggested that the 6- to 9-nm-diam particles were molecular domains that were intramolecularly crosslinked and that formed during the initial stages of polymerization. These domains were interconnected by regions of low crosslink density, which allowed flow to occur in this network. The ductile mechanical response of the DGEBA-DETA epoxies and the strain-rate and thermal-history dependence of this response are consistent with the premise that regions of low crosslink density control the flow processes (12). Hence, details of the network morphology are needed for a basic understanding of the structural parameters that control the deformation processes and the mechanical response of epoxies.

More recently, we have observed under polarized light a network of larger 1-mm-sized nodules in BF_3 catalyzed TGDDM-DDS systems (Fiberite 934). These birefringent networks, one of which is illustrated in the transmission optical micrograph in Figure 1, break up under combined stress and temperatures of >175°C. The birefringence originates from the preferential alignment of the macromolecules within the birefringent network. This molecular alignment only occurs in BF_3-catalyzed TGDDM-DDS epoxies. These networks may result from an inhomogeneous distribution of the catalyst within the TGDDM-DDS system, which would result in the high catalyst concentration regions polymerizing at a faster rate than their surroundings. The stresses caused within the resultant heterogeneous system during gelation and glass formation could produce the birefringent network illustrated in Figure 1.

The microvoid characteristics of the epoxy are also important in controlling the mechanical response of the glass. Microvoids can have a deleterious effect on the mechanical properties of epoxies by acting as stress concentrators and by serving as a sink for the accumulation of sorbed moisture. Microvoids can result from trapping of air in the system during cure and from trapping low-molecular-weight material, which is subsequently eliminated during postcure, in the glass. This low-molecular-weight material results either from inhomogeneous mixing of epoxide and curing agent or from the inability of the constituents to react, with the resultant aggregation of these constituents. In polyamide-cured DGEBA epoxies, crystals of DGEBA epoxide monomer trapped in the partially cured resin at room-temperature can produce microvoids by melting and volatilizing during certain postcure conditions (7). Also, in TGDDM-DDS epoxies, thermal anneal, moisture sorption, and mechanical property studies indicate that the melting and volatilization of unreacted DDS crystallites during cure produces microvoids in these resins (9). In Figure 2, voids and their travel paths during cure are illustrated in a sheet of a commercial TGDDM-DDS epoxy (Narmco 5208). Such voids only appear when the cure temperature exceeds the crystalline melting point (162°C) of the DDS.

Chemical Structure

The chemical structure of epoxies can be complex. The structure will depend on specific cure conditions, because more than one reaction can occur and the kinetics of each reaction exhibits different temperature dependencies. In addition, the structure is affected by factors such as steric and diffusional restrictions of the reactants during cure (8,9,13,14,15,16), the presence of impurities that can act as catalysts (17), the reactivity of the epoxide and curing agent (18), isomerization of epoxide groups (18,19,20), inhomogeneous mixing of the reactants (7,9), and cyclic polymerization of the growing chains (18). These factors can lead to network structures that are physically and chemically heterogeneous. In amine-cured epoxides, networks are generally assumed to result from addition reactions of epoxide groups with primary and secondary amines (18), as illustrated in Figure 3. Epoxides and amines with functionalities >3 can form highly crosslinked network structures. However, considerable evidence suggests that amine cured-epoxies, such as DGEBA-DETA and TGDDM-DDS systems, are not highly crosslinked. Such evidence includes their high temperature ductility (4,5,7,8,9,12), the effects of thermal history and strain rate on their mechanical response (12), and their microscopic deformation and failure processes (4-9).

Such observations suggest either that few epoxide-secondary amine reactions occur, thus limiting the number of crosslinks, or that polyether chains are formed by trans-etherification through a ring-opening homopolymerization of the epoxide (17,18,19,21), as illustrated in Figure 4. The epoxide-amine reactions are controlled by the presence of H-bond donors, such as OH groups, which are necessary to open the epoxide rings (14,15,17). The trans-

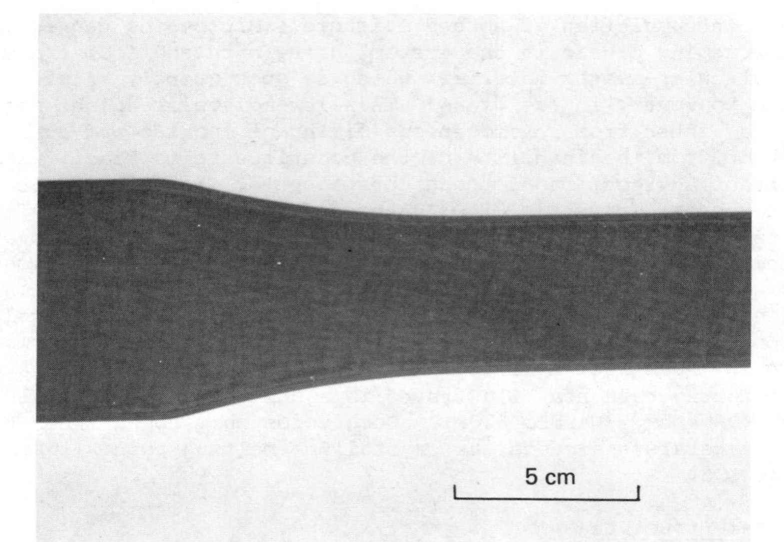

Figure 1. Transmission optical micrograph (under polarized light) of a network of aligned molecules in a BF_3-catalyzed TGDDM–DDS epoxy

Figure 2. Transmission optical micrograph of voids and their travel paths caused by the melting and volatilization of unreacted DDS crystallites during cure in a TGDDM–DDS epoxy

etherification reaction requires a tertiary amine as a catalyst and a H-bond donor as a cocatalyst (17,19,21). Hence, the final chemical structure of the epoxy system can be complex, because it will depend on such parameters as: (1) the relative rates of the chemical reactions at room temperature, (2) the final postcure temperature (3) the concentrations of catalysts such as sorbed moisture in the system, and (4) steric restrictions that inhibit reactions at secondary-amine sites.

In the case of the TGDDM-DDS epoxy systems, there is growing evidence that such networks do not form exclusively from epoxide-amine addition reactions. In Figure 5, the percentage of epoxide groups consumed during cure, as determined by the disappearance of the epoxide band at 910 cm^{-1} in the infrared spectra, are plotted versus cure conditions for both a BF_3-catalyzed TGDDM-DDS epoxy (Fiberite 934) and a non-catalyzed TGDDM-DDS epoxy (Narmco 5208). For a standard 177°C cure for 2.5 h, all the epoxide groups are consumed, within experimental error, for the BF_3-catalyzed system despite the weight percent of DDS in this system being well below the stoichiometric quantity necessary to consume all the epoxide groups by normal epoxide-amine addition reactions. In the case of the non-catalyzed system, ~35% of the epoxide groups remain unreacted after curing at 177°C for 2.5 h, and all such groups only react after exposure to 300°C for 1 h. Fourier-transform infrared-spectroscopy studies utilizing spectral stripping, which reveals differences in the spectra recorded at different stages of cure, should reveal information on the chemical reactions occurring that form these TGDDM-DDS epoxy networks.

Deformation and Failure Processes

Localized plastic flow has been reported to occur during the deformation and failure processes of epoxies; in a number of cases, the fracture energies were two to three times greater than the expected theoretical estimate for purely brittle fracture (see Ref. 8). However, no systematic studies have been made to elucidate the microscopic flow processes occurring during the deformation of epoxies and to determine the relation of such flow processes to the network structure.

Our recent investigations reveal that both DGEBA-DETA and TGDDM-DDS epoxies deform and fail by a crazing process (4-9). Crazes were observed in films either strained directly in the electron microscope or strained on a metal substrate. The fracture topographies of these epoxies, fractured as a function of temperature and strain rate, are interpreted in terms of a crazing process. The TGDDM-DDS epoxies also deform to a limited extent by shear banding, as indicated by multiple, unique right-angle steps in the fracture-topography initiation region, as illustrated in Figure 6. Shear-band propagation in these partially crosslinked glasses produces structurally weak planes because of bond cleavage

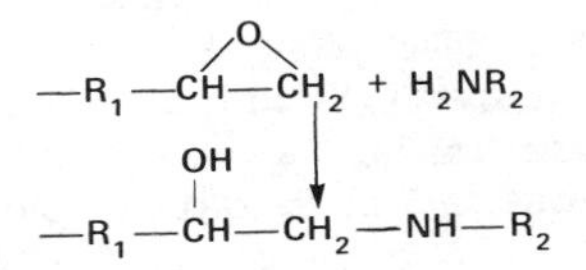

Figure 3. Epoxide–amine addition reaction

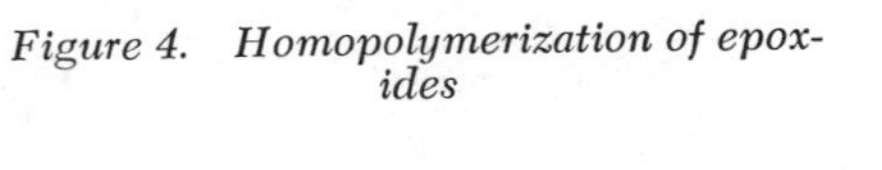

Figure 4. Homopolymerization of epoxides

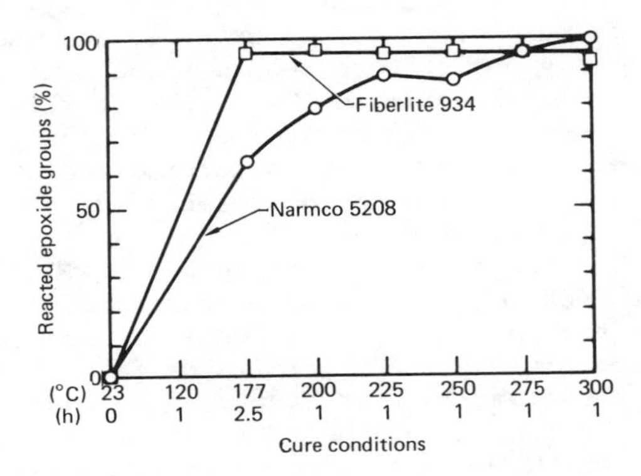

Figure 5. Percentage of reacted epoxide groups as a function of cure conditions for (TGDDM–DDS)-based epoxies

caused during molecular flow. Hull (22) and Mills (23) have both noted that the intersection of shear bands, which occurs at right-angles, causes a stress concentration. This stress concentration is sufficient to cause a crack to propagate through the structurally weak planes caused by shear band propagation. These phenomena will produce the multiple right-angle steps in the fracture topography. Mixed modes of deformation that involve both crazing and shear banding were also observed in the fracture topography of TGDDM-DDS epoxies. The type of microscopic deformation that occurs in epoxy matrices can play a direct role in the environmental sensitivity and mechanical response of the composite.

Durability

The durability of epoxy matrices depends on many complex interacting phenomena. The factors that control the critical path to ultimate failure or unacceptable damage depend specifically on the particular prevailing environmental conditions. These environmental factors include service stresses, humidity, temperature, and solar radiation. The combined effects of thermal history, moisture exposure and stress have a deleterious effect on the physical and mechanical integrity of epoxies.

We have recently studied the effect of specific combinations of moisture, heat, and stress on the physical structure, failure modes, and tensile mechanical properties of TGDDM-DDS epoxies. Our main findings from these studies were as follows.

Sorbed moisture plasticizes TGDDM-DDS epoxies and lowers their tensile strengths, ultimate elongations, and moduli. The fracture topographies of the initiation cavity and mirror regions of these epoxies indicate that sorbed moisture enhances the craze initiation and propagation processes. The crazing process is more susceptible to sorbed moisture than to the glass transition temperature (T_g), which can be explained in terms of local moisture concentrations enhancing the local cavitation and flow processes. Hence, modification of T_g by sorbed moisture cannot be utilized alone as a sensitive guide to predict deterioration in the mechanical response and, hence, the durability of epoxies.

We also investigated the effect of increasing stress levels on the subsequent moisture sorption characteristics of initially dry TGDDM-DDS epoxies. In Figure 7, the equilibrium moisture sorption levels after ~40 days exposure to 100% relative humidity (RH) at room temperature are plotted versus the stress levels that were applied to the epoxies prior to moisture sorption. All data points fall within the shaded areas. Stresses in the 0- to 38-MPa range had no detectable influence on the subsequent moisture sorption levels. However, moisture sorption levels increase sharply by up to ~11% in the 38- to 43-MPa stress range. At higher stress levels in the 43- to 65-MPa range, in which a few specimens actually broke, there is only a slight trend towards higher moisture sorption levels with increasing stress.

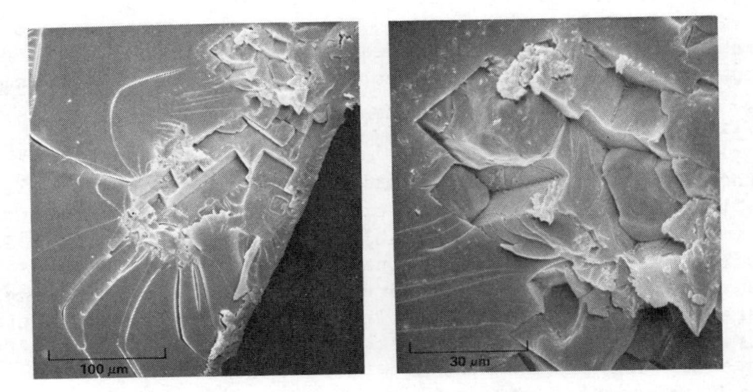

Figure 6. Scanning electron micrographs illustrating multiple right-angle steps in the fracture-topography initiation region of a TGDDM–DDS epoxy

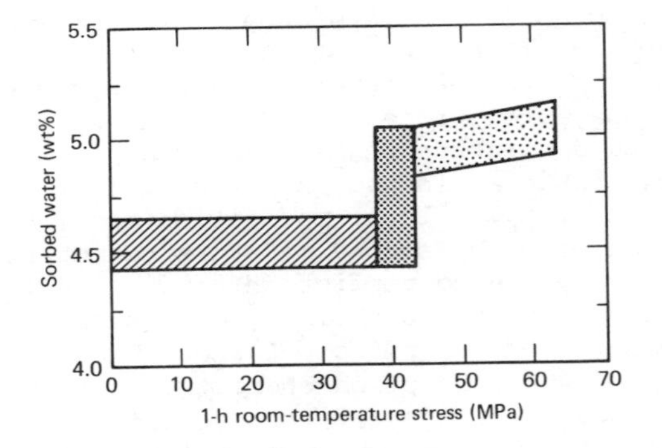

Figure 7. Equilibrium weight percent moisture sorbed by a TGDDM–DDS epoxy at relative humidity of 100% and at 23°C vs. 1-hr constant-stress levels that were applied prior to moisture exposure

The data in Figure 7 indicate that the initial stages of craze-crack growth enhance the accessibility of moisture to sorption sites to a greater extent than the later stages of growth. (The primary sorption sites within the TGDDM-DDS epoxy are the hydroxyl, sulfonyl, and primary and secondary amine groups, all of which are capable of forming hydrogen bonds with water molecules.) The TGDDM-DDS epoxy specimens that fractured under constant load were found to exhibit similar fracture topographies as previously studied specimens that fractured in shorter times in the 10^{-2} to 10^{1} min^{-1} strain-rate region (9). Such topographies have been interpreted in terms of a craze-crack growth process (9) with crazing, followed by crack propagation, predominating in the initial stages of failure and crack propagation alone predominating during the later stages of failure. The dilatational changes produced in the epoxy glass by the crazing process will enhance the accessibility of moisture to sorption sites within the epoxy to a greater extent than will crack propagation alone. Hence, the initial stages of failure in TGDDM-DDS epoxies will enhance the accessibility of moisture to sorption sites to a greater extent than in the later stages of failure.

One of the more extreme environmental conditions experienced by an epoxy composite matrix on a fighter aircraft occurs during a supersonic dash. The aircraft dives from high altitudes (outer surface temperature -20 to -55°C) into a supersonic, low-altitude run during which the surface temperature rises in minutes to 100 to 150°C as a result of aerodynamic heating. On reduction of speed, the outer surface temperature drops extremely rapidly at rates up to ~500°C/min, thus exposing the epoxy composite to a thermal spike. Simulation of such thermal spikes has been shown to increase the amount of moisture sorbed by the epoxy or epoxy composite (24-30). However, after a certain number of consecutive thermal spikes, the amount of moisture sorbed ceases to increase. Browning (26,29) has suggested such increases result from microcracks caused by the moisture and temperature gradients present during the thermal spike. McKague (28) has recently noted that damage does not occur unless the thermal-spike maximum temperature exceeds the T_g of the moist epoxy.

We observed that the amount of moisture sorbed by TGDDM-DDS epoxies is enhanced by ~1.6 wt% after exposure to a 150°C thermal spike. The surfaces of the thermally-spiked epoxies were examined by scanning electron microscopy for the presence of surface microcracks. No significant areas of microcracking were observed in any of the specimens when examined under magnifications of up to 30,000×. Hence, the additional moisture sorbed by the epoxies after exposure to thermal spikes is not primarily caused by microcracking.

The primary mechanism by which thermally-spiked epoxies sorb additional amounts of moisture can be explained in terms of moisture-induced free-volume changes. The molecular mobility of the epoxy is enhanced as the T_g of the epoxy-moisture system is approached at the high temperatures experienced during the thermal spike. This molecular mobility is sufficient to enhance the dissociation of H-bonds between the water molecules and active sites within the epoxy. Although the ruptured H-bonds can reform at active sites, there is an overall decrease in the amount of hydrogen bonding and a corresponding increase in the mobility of the water molecules. The more mobile water molecules with fewer H-bonds require a greater free volume because hydrogen bonding generally causes a volume decrease. The molecular mobility of the epoxy-moisture system during a thermal spike is sufficient to allow configurational changes to occur within the epoxy network that accommodates the greater free volume required by both the more mobile water molecules and the normal moisture-induced swelling stresses imposed on the epoxy. Such free volume increases, which involve permanent rotation-isomeric population changes within the epoxy network, are frozen into the epoxy glass during the rapid cool-down portion of the thermal spike. The additional free volume allows water molecules access to previously inaccessible active sites within the epoxy.

To a lesser extent, the rupture of crosslinks, crazing and/or cracking, and the loss of unreacted material can also contribute to enhanced moisture sorption after thermal-spike exposure. Thermal-spike exposure can cause surface crazing and/or cracking of epoxies if the moisture-induced swelling stresses, together with those stresses that result from temperature gradients and relaxation of fabrication stresses, exceed the craze-initiation stress at the maximum thermal-spike temperature. Thicker epoxy specimens are more susceptible to the growth of permanent damage regions during thermal-spike exposure, because they are exposed to larger temperature gradients and shrinkage stresses during cure, which in turn produce larger fabrication stresses and strains.

Conclusions

The structure-property relations of amine-cured epoxies and the modification of such relations by fabrication and environmental factors have been reviewed and our primary findings are as follows:

(1) The physical structural parameters that control the mechanical response of the epoxies are the network topography and microvoid characteristics. The network structures can be heterogeneous because of variations in the crosslink density in the 5- to 10-nm range and, also, from the alignment of macromolecules in regions ~1 mm in size. Microvoids can be produced in these epoxies as a result of the elimination of unreacted material.

(2) The chemical reactions that produce the amine-cured epoxies can be complex and do not exclusively occur by epoxide-amine addition reactions.

(3) Microscopic deformation occurs in amine-cured epoxies by either crazing and/or shear banding.

(4) Sorbed moisture enhances the craze cavitation and propagation processes in epoxies and deteriorates their mechanical properties.

(5) The initial stages of failure, which involve both dilatational craze- and subsequent crack-propagation, enhance the accessibility of moisture to sorption sites within the epoxy to a greater extent than in the latter stages of failure, which involve crack propagation alone.

(6) The amount of moisture sorbed by epoxies is enhanced after exposure to a thermal spike as a result of moisture-induced free-volume increases that involve rotational-isomeric population changes.

Acknowledgment

This work was performed under the auspices of the U.S. Department of Energy by Lawrence Livermore Laboratory under contract No. W-7405-Eng-48. Reference to a company or product name does not imply approval or recommendation of the product by the University of California or the Department of Energy to the exclusion of others that may be suitable.

Abstract

The structure, deformation and failure processes, and mechanical property relations of composite matrices, and the modification of such relations by fabrication and environmental factors are presented. The primary composite matrices considered are epoxies. The physical structure of epoxies is discussed in terms of the network topography and microvoid characteristics. Such parameters directly control the mechanical response of these glasses. The chemical reactions that produce epoxy networks can be complex and do not exclusively occur by epoxide-amine addition reactions. The complex nature of these reactions and the factors that control them are discussed. Epoxies deform microscopically and fail by either crazing and/or shear banding. The durability of these resins is discussed in terms of the effect of specific combinations of moisture, heat, and stress on their physical and mechanical integrity.

Literature Cited

1. "Proceedings of 1977 Flywheel Technology Symposium"; Conf.-771053, U.S. Dept. of Energy: Washington, D.C., March 1978.
2. "U.S. Air Force Durability Workshop"; Battelle-Columbus: Columbus, Ohio, September 1975.
3. "U.S. Air Force Conference on the Effects of Relative Humidity and Temperature on Composite Structures", University of Delaware; AFOSR-TR-77-0030, U.S. Air Force: Washington, D.C., 1977.
4. Morgan, R. J.; O'Neal, J. E. in "Toughness and Brittleness of Plastics, Advances in Chemistry", Series 154, American Chemical Society: Washington, D.C., 1976; Chap. 2.
5. Morgan, R. J.; O'Neal, J. E. in "Chemistry and Properties of Crosslinked Polymers", S. S. Labana, Ed., Academic Press: New York, 1977; p. 289.
6. Morgan, R. J.; O'Neal, J. E. J. Mater. Sci., 1977, 12, 1966.
7. Morgan, R. J.; O'Neal, J. E. J. Macromol. Sci. Phys., 1978, B15(1), 139.
8. Morgan, R. J.; O'Neal, J. E. Polym. Plast. Technol. Eng. 1978, 10(1), 49.
9. Morgan, R. J.; O'Neal, J. E.; Miller, D. B. J. Mater. Sci., 1979, 14, 109.
10. Racich, J. L.; Koutsky, J. A. J. Appl. Polym. Sci., 1976, 20, 2111.
11. Manson, J. A.; Sperling, L. H.; Kim, S. L. "Influence of Crosslinking on the Mechanical Properties of High T_g Polymers", AFML-TR-77-109, U.S. Air Force Materials Lab.: Dayton, Ohio, 1977.
12. Morgan, R. J. J. Appl. Polym. Sci., 1979, 23, 2711.
13. French, D. M.; Strecker, R. A. H.; Tompa, A. S. J. Appl. Polym. Sci, 1970, 14, 599.
14. Horie, K.; Hiura, H.; Sawada, M.; Mita, I.; Kambe, H. J. Polym. Sci, 1970, A-1(9), 1357.
15. Acitelli, M. A.; Prime, R. B.; Sacher, E. Polymer, 1971, 12, 335.
16. Prime, R. B.; Sacher, E. Polymer, 1972, 13, 455.
17. Whiting, D. A.; Kline, D. E. J. Appl. Polym. Sci., 1974, 18, 1043.
18. Lee, H.; Neville, K. "Handbook of Epoxy Resins"; McGraw-Hill: New York, 1967.
19. Sidyakin, P. V. Vysokomol. Soyed, 1972, A14, 979.
20. Bell, J. P.; McCavill, W. T. J. Appl. Polym. Sci., 1974, 18, 2243.
21. Narracott, E. Brit. Plast., 1953, 26, 120.
22. Hull, D. Acta. Met., 1960, 8, 11.
23. Mills, N. J. J. Mater. Sci., 1976, 11, 363.
24. Verette, R. M. "Temperature/Humidity Effects on the Strength of Graphite/Epoxy Laminates"; AIAA Paper No. 75-1011, 1975.

25. McKague, E. L., Jr.; Halkias, J. E.; Reynolds, J. D. J. Compos. Mater., 1975, 9, 2.
26. Browning, C. E., Ph.D. Thesis, University of Dayton: Dayton, Ohio, 1976.
27. Hedrick, I. G.; Whiteside, J. B. Effects of Environment on Advanced Composite Structures, in "AIAA Conference on Aircraft Composites: The Emerging Methodology of Structural Assurance", San Diego, California, Paper No. 77-463, 1977.
28. McKague, E. L. in "Proceedings of Conference Environmental Degradation of Engineering Materials", Louthan, M. R. and McNitt, R. P., Eds., Virginia Polytechnic Inst. Printing Dept.: Blacksburg, Virginia, 1977; Chap. V, p. 353.
29. Browning, C. E. The Mechanism of Elevated Temperature Property Losses in High Performance Structural Epoxy Resin Matrix Materials After Exposure to High Humidity Environments, 22nd National SAMPE Symposium and Exhibition, San Diego, CA, 1977, 22, 365.
30. "Advanced Composite Materials -- Environmental Effects", ASTM STP 658, Vinson, J. P., Ed. American Society for Testing and Materials, 1978.

RECEIVED May 23, 1980.

Viscoelastic Properties of Fiber-Reinforced Plastics

HARUO YOSHIDA

Osaka Municipal Technical Research Institute, 1-1, Ogimachi-2, Kitaku, Osaka 530, Japan

The most typical form of the composite materials is the fiber reinforced plastics composed of glass or carbon fibers and resins. The reinforcements such as glass or carbon fibers may be elastic, but resins must be regarded as viscoelastic materials(1). When these composite materials are subjected to various loads, they show viscoelastic complicated behavior caused by the mechanical properties of the fiber and resin, specially by the internal viscosity of resin and the friction at the interface (2) of fiber and resin.

Then the strain of these composite materials is not proportional to the stress, and the stress-strain diagram becomes the curved line affected by the time scale. This phenomenon is the characteristic property of the reinforced plastics and becomes the special merit in practical uses.

Under the repeated loads, the stress-strain diagram draws hysteresis loops. Corresponding to the area of the loop, the strain energy is consumed in the material at each alternating stress. This energy loss affects the fatigue of the material and the other hand is effective to damping the vibration or screening the acoustic emission.

In this research, the dynamic viscoelastic properties of the fiber reinforced plastics were studied by measuring the complex moduli of elasticity of the materials with the technics of the vibrating reed method (3,4,5).

Experiments and Discussions

The measuring arrangements consist of the complex modulus apparatus, the oscillator, the amplifier and the level recorder. The test sample is clamped at one end and the other end is free. Two little iron discs are bonded on the sample bar opposite to the transducers to be responsive to the magnetic force. The free end is excited at the frequency from 20 to 20000 Hz by the exciter transducer, then the bending vibration of the sample bar occurs and the pick-up transducer detects the deflection amplitude

0-8412-0567-1/80/47-132-247$05.00/0

of the sample bar.

The samples tested are as follows:

Sample(a); Polyester resin laminates reinforced with glass fiber roving cloths. Table I.

Sample(b); Epoxy resin laminates reinforced with carbon fibers.

Sample(c); Hybrid laminated composites. One of the reinforcements is glass mat(M), another is combined woven cloth with glass fibers and carbon fibers in warps(C). Resin is epoxy.

The complex moduli of elasticity of the materials mentioned above were measured.

Table I. Constitutions and dimensions of polyester resin laminates reinforced with glass fiber roving cloths.

Sample's Mark	PE	PE8G	PE8GV	PE8GH	PE12G
Plies	0	8	8(*1)	8(*2)	12
Specific gravity	1.18	1.60	1.14	1.39	1.83
Glass content (volume %)	0	31.5	17.5	16.2	48.6

(*1) Warps only, (2*) Woofs only
Specimen: Length 210, Width 22, Thickness 5.5 mm
Reinforcement: Glass roving cloth (EWR55)
Resin: Polyester (EPOLAC N317)

The complex moduli of elasticity of the samples(a) at the various temperature from -10 to 110 degree C are shown in Figure 1. The composites which have the gross content of glass fibers such as the sample PE12G and PE8G obtained the higher values of the dynamic modulus of elasticity and kept the elasticity at the elevated temperature. The sample PE has not the fiber and is resin only, then the dynamic modulus of elasticity decreased at the temperature of 110 degree C. The loss factors of the same samples are shown in Figure 2. The loss factors of the samples PE and PE8GH increased remarkably in accordance with the rise of temperature. The sample PE8GH has not the warps. As the results, it could be found that the materials which contained much volume of glass fibers were heat resistant.

Various kinds of composite materials were laminated and some hybrid composites were constructed as shown in Figure 3. The complex modulus of elasticity of this hybrid composite material E^* is theoretically expressed in the equation(1) as the function of the individual complex modulus of elasticity E^*_K of each layer.

$$E^*I=\sum_{K=1}^{n} E^*_K I_K \qquad (1)$$

I and I_K are respectively the moment of inertia of the cross

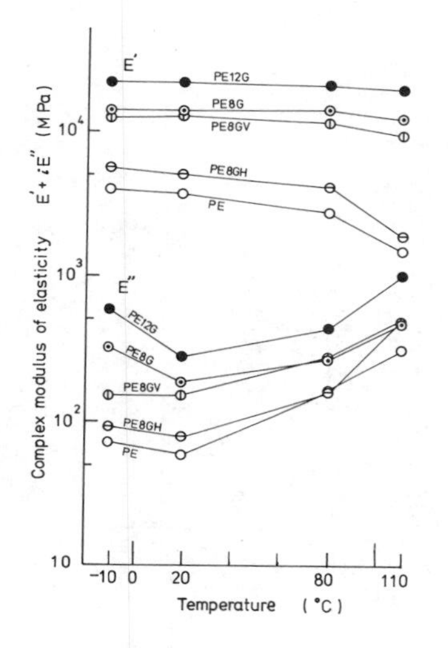

Figure 1. Complex moduli of elasticity of reinforced polyester resin laminates

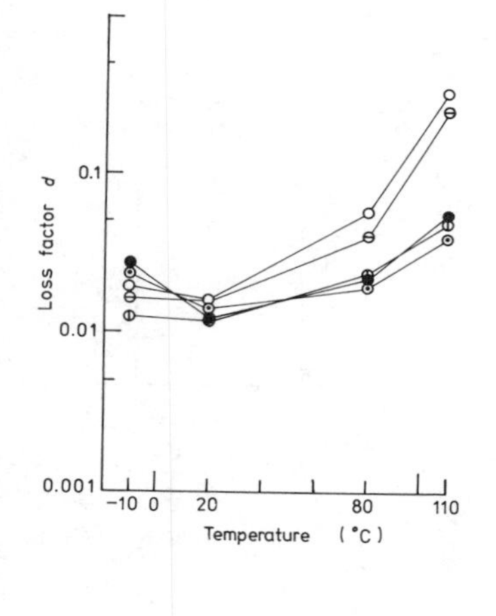

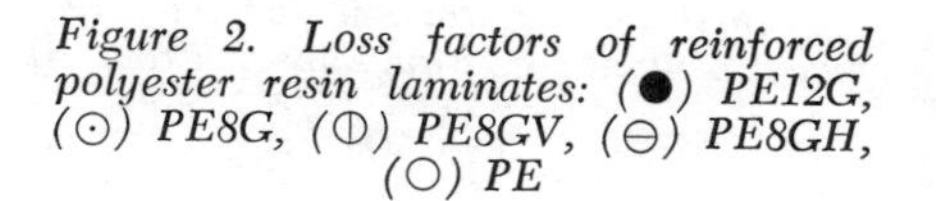

Figure 2. Loss factors of reinforced polyester resin laminates: (●) PE12G, (⊙) PE8G, (⦶) PE8GV, (⊖) PE8GH, (○) PE

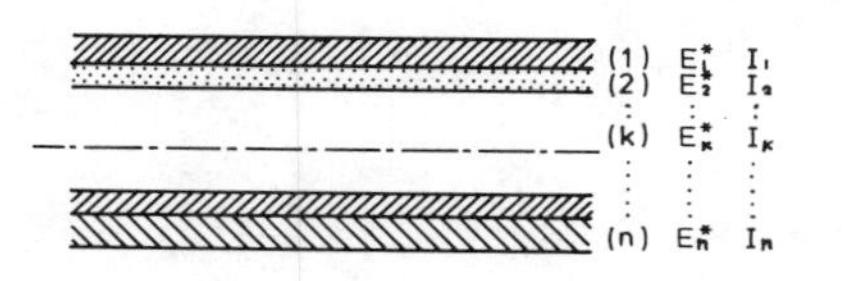

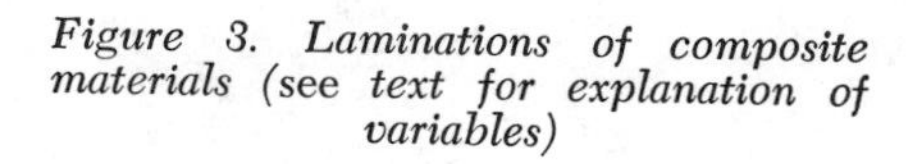

Figure 3. Laminations of composite materials (see text for explanation of variables)

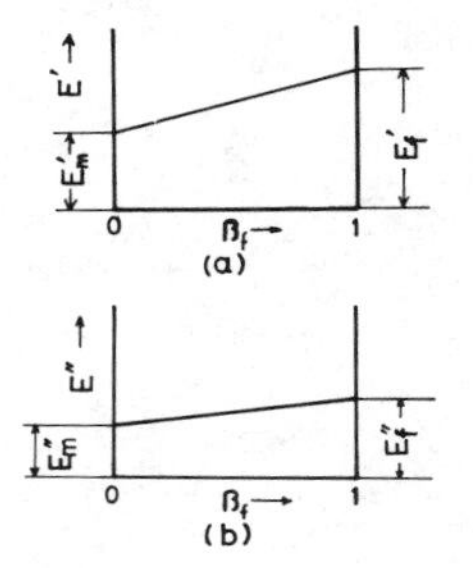

Figure 4. Theoretical relationship between the modulus of elasticity and the reinforcement content (see *Equations 6, 7, and 8*)

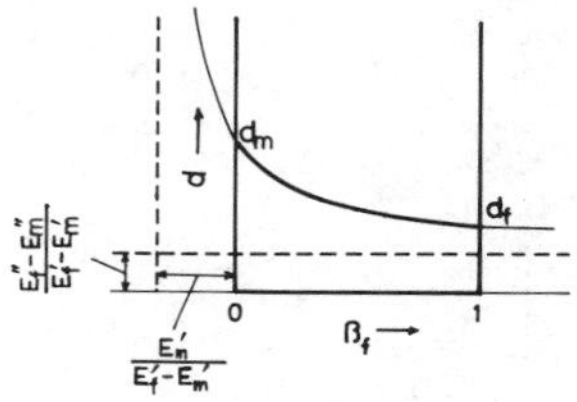

Figure 5. Theoretical relationship between the loss factor and the reinforcement content (see *Equations 6, 7, and 8*)

section concerning the neutral axis of the hybrid composite. The symbols with suffix k mean that they belong to the k'th layer. Generally $E^* = E' + iE''$, $(i = \sqrt{-1})$, E^* is the complex modulus of elasticity, E' is the dynamic modulus of elasticity, E'' is the loss modulus of elasticity. From the equation(1) the following equations are introduced. d is the loss factor.

$$E' = \sum_{k=1}^{n} \frac{I_k}{I} E'_k \tag{2}$$

$$E'' = \sum_{k=1}^{n} \frac{I_k}{I} E''_k \tag{3}$$

$$d = \frac{E''}{E'} \tag{4}$$

If the material of reinforcement is only one kind and uniformly distributed in the composite, the fraction of volume content β of the reinforcement is expressed as follow.

$$\beta = \frac{I_k}{I} \tag{5}$$

Then the equations(2,3,4) are rewrote as follows.

$$E' = \beta E'_f + (1-\beta) E'_m \tag{6}$$

$$E'' = \beta E''_f + (1-\beta) E''_m \tag{7}$$

$$d = \frac{\beta E'_f d_f + (1-\beta) E'_m d_m}{\beta E'_f + (1-\beta) E'_m} \tag{8}$$

The symbols with suffix f and m mean that they belong to the reinforcement(fiber) and to the matrix(resin) respectively.

The theoretical relations of the dynamic and the loss modulus of elasticity and the loss factor to the content of the reinforcement expressed in the equations(6,7,8) are displayed in Figures 4,5. The dynamic and the loss modulus of elasticity are respectively proportional to the content of fibers but the loss factor has the hyperbolic relationship to the content of fibers.

The experimental results concerning the relation between the complex modulus of elasticity of polyester resin laminates reinforced with glass roving cloths(Sample(a)) and the content of glass fibers are shown in Figures 6,7. The solid bars express the range of the values from the minimum to the maximum at the temperature from -10 to 110 degree C. In Figure 6, except PE8GH and PE8GV, the end points (in the case of E', the upper end points are the values at -10 degree C, the lower end points are the values at 110 degree C, in the case of E", the uppers are at 110,

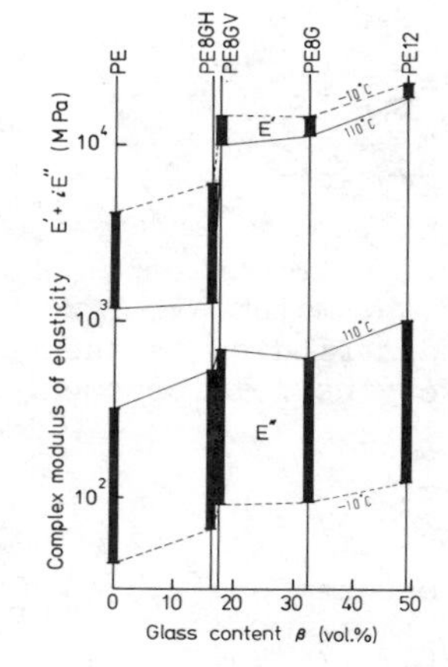

Figure 6. Experimental relationship between the complex modulus of elasticity of reinforced polyester resin laminates and the glass fiber content

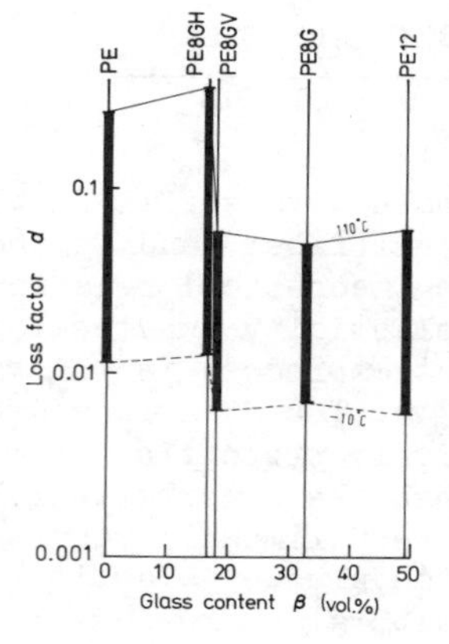

Figure 7. Experimental relationship between the loss factor of reinforced polyester resin laminates and the glass fiber content

the lowers are at -10 degree C) are arranged approximately in a straight line respectively. This fact proves the reasonability of the proportionality law expressed in the equations(6,7).

The sample PE8GV has not woofs and has warps only, then the glass content is smaller than that of the sample PE8G, but the values of the complex modulus of elasticity is nearly the same as that of PE8G. If the consideration is restricted within the warps only, the content of glass fibers is the same at PE8G and PE8GV. This means that only warps which lie to the same direction as the direction of the stress are effective to improve the complex modulus of elasticity. The sample PE8GH has woofs only and has not warps, then the value of the complex modulus of elasticity is the same as that of the resin sample PE.

In Figure 7, the end points of the solid bars are arranged on the hyperbolic curves except PE8GV and PE8GH. The values of the loss factor of PE8GV and PE8GH are respectively the same as those of PE8G and PE by the same theory concerning the dynamic and the loss modulus of elasticity explained above.

The samples of the epoxy resin laminates reinforced with carbon fibers(Sample(b)) are shown in Figure 8. The left plate is unidirectionally reinforced laminate (5 plies). The right plate is cross plied reinforced laminate (8 plies). The test samples were cut out as shown. The stress direction is the longitudinal direction of the sample bar. Then the direction of fibers of the sample CFRP-L agrees with the direction of the stress, the directions of fibers and stress of the sample CFRP-T are perpendicular to each other. The experimental results are shown in Figures 9,10. The dynamic moduli of elasticity of the samples CFRP-L and CFRP+L were pretty high and almost did not be affected by the temperature, but that of the sample CFRP-T evenly decreased in accordance with the rise of the temperature. The loss factors are shown in Figure 9. The value of the loss factor of the sample CFRP-T is higher than that of the sample CFRP-L. When the direction of fibers is 45 degree to the direction of the stress, for example the sample CFRP-45 has a pretty high value of the loss factor. It is deduced that this is caused by the shearing strain in the resin among fibers and by the friction at the interface of resin and fiber. The mechanical properties of the cross ply laminates such as the samples CFRP+L and CFRP+T are influenced by the direction of the fibers existing in the surface layers of the samples, because of the bending stress.

To improve the mechanical properties of the composite materials, two or more kinds of reinforcements are combined and the hybrid composites are made. In this research, two kinds of reinforcing materials (M) and (C) were used, the resin is epoxy(Sample(c)).

The sequence of laminating with these two kinds of materials was changed and five kinds of hybrid composites were set as shown in Figure 11. The theoretical values of the complex modulus of elasticity calculated by the equation(1) well agreed with the experimental results (6).

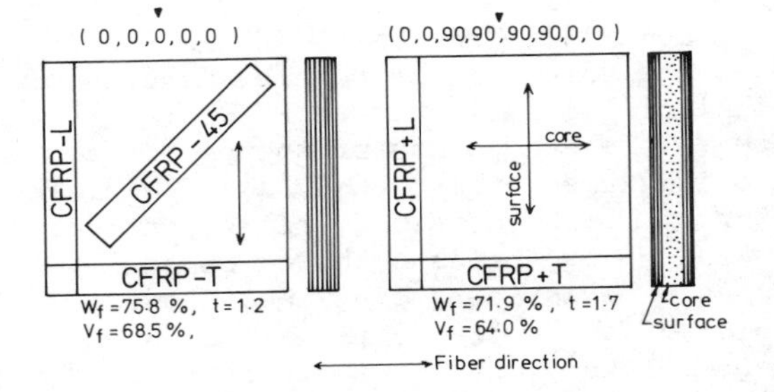

Figure 8. Epoxy resin laminates reinforced with carbon fibers: the left plate is 5-ply, unidirectionally reinforced; the right plate is 7-ply, cross-plied reinforced

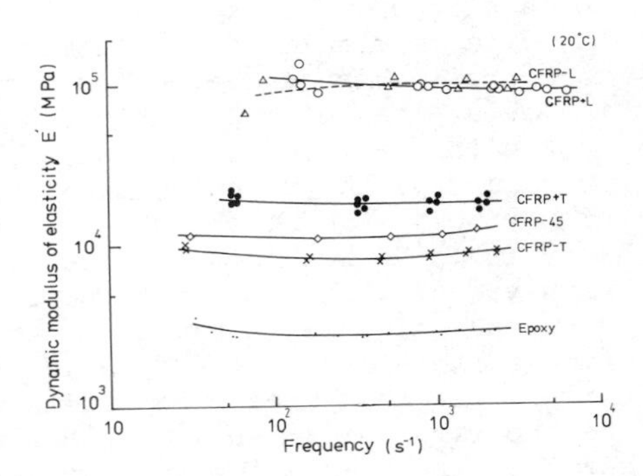

Figure 9. Dynamic moduli of elasticity of reinforced epoxy resin laminates

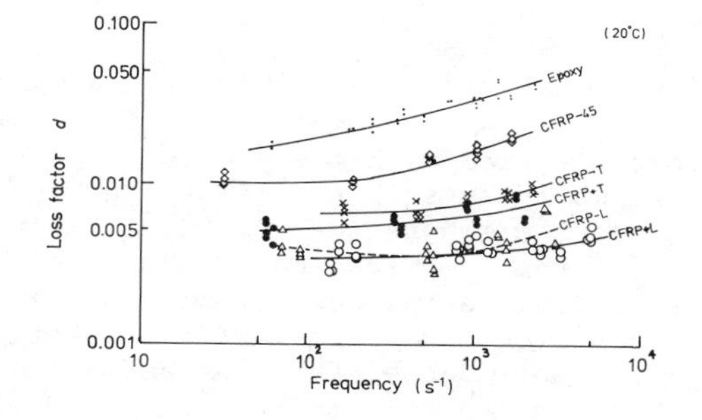

Figure 10. Loss factors of reinforced epoxy resin laminates

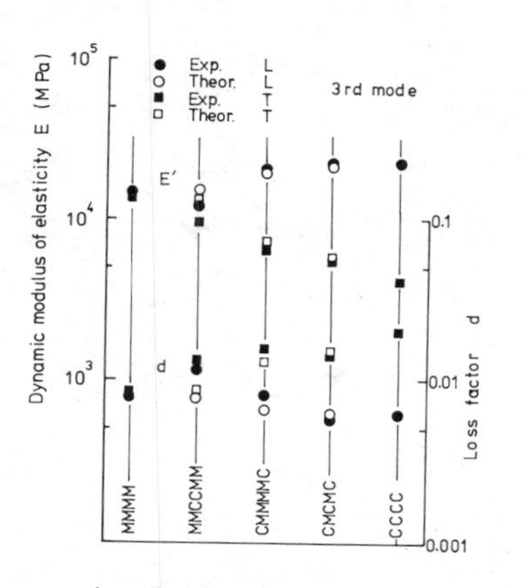

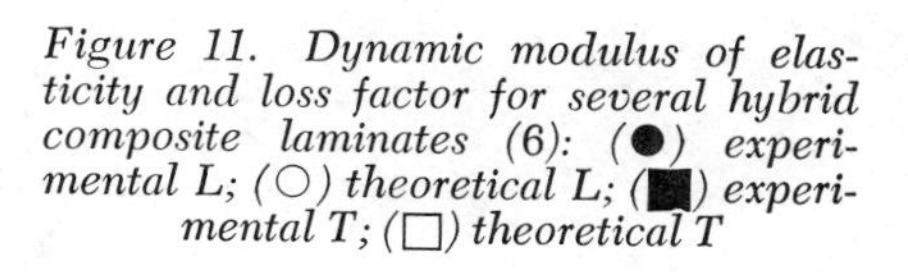

Figure 11. Dynamic modulus of elasticity and loss factor for several hybrid composite laminates (6): (●) experimental L; (○) theoretical L; (■) experimental T; (□) theoretical T

Conclusions

As the key to finding the viscoelastic properties of the composite materials, the complex modulus of elasticity of the materials were measured. The following results were obtained.

The values of the complex modulus of elasticity were not affected by the vibration mode at the resonance within the frequency range of this study.

The dynamic modulus and the loss modulus of elasticity were respectively proportional to the content of the reinforcement.

The relation between the loss factor and the content of the reinforcement was hyperbolic.

With an increase in temperature it was generally observed that the elastic response decreased, whereas the viscous response increased.

The reinforced plastics with the greatest content of the reinforcements were more resistible against the high temperature.

The direction of the fibers were influential in the viscoelastic properties of the composite materials.

The theory to calculate the complex modulus of elasticity of the hybrid composites from the viscoelastic properties of each constituent layer was introduced, then the mechanical properties of the hybrid composites are able to be controlled at the designing process.

Literature Cited

1. Flugge,W.,"Viscoelasticity",Blaisdell Pub.,London, 1967,p.3.
2. Endo,K.;Watanabe,M.,Proc.J.C.M.R.,1971,14,120.
3. Onogi,S.,"Kobunshikagaku",Maruzen,Tokyo,1957,p.308.
4. Nolle,A.W.,J.Appl.Phys.,1948,19,753.
5. Horio,M.;Onogi,S.,J.Appl.Phys.,1951,22,977.
6. Yoshida,H.,J.Soc.Mat.Sci.Japan,1976,25,442.

RECEIVED January 30, 1980.

20

Interphase Resin Modification in Graphite Composites

R. V. SUBRAMANIAN and JAMES J. JAKUBOWSKI[1]

Department of Materials Science and Engineering, Washington State University, Pullman, WA 99164

A comprehensive program of research is being conducted in our laboratories to investigate the applicability of electrochemcial processes for interphase resin modification in graphite composites (1-6). Research on asbestos and glass fiber composites has shown that the introduction of a polymer interlayer between the fiber and the polymer matrix leads to significant alterations in the mechanical properties of composites (7-11). Since carbon fiber is an electrically conducting material, it was coated with polymer by electropolymerization or by electrodeposition and the electrocoated fiber was used as reinforcement in an epoxy matrix to evaluate resulting improvements in composite shear and impact strengths. The results of this investigation of particular interest to aerospace applications are discussed here. Furthermore, the extension of this research to minimize the hazards of release of electrically conductive fiber fragments from composites is also included in this discussion. The concept of interphase modification has been expanded thus to deal with not only mechanical, but also other properties of composites.

Electropolymerization involves the polymerization of monomers in an electrolytic cell. Electrodeposition utilizes the migration of preformed polymers carrying ionized groups to the oppositely charged electrode under an applied voltage. The major advantage of utilizing these electrodic processes for carbon fiber coating is that a uniform layer of controlled thickness and variable polymer structure and properties can be expected to be formed.

It is useful, in this context, to recognize that the fiber and matrix are not bridged in the composite by a well defined interface but by an interphase polymer. The properties of the interphase polymer can be significantly different from the bulk matrix polymer properties. When a polymer coating is formed on the fiber by electrodeposition or electropolymerization,

[1] Current address: Dow Chemical Co., Midland, MI 48640

0-8412-0567-1/80/47-132-257$05.00/0

the properties of the interphase are further modified by the polymer coating. Such modification is manifested in corresponding changes in composite properties, mainly shear and toughness.

Prompted by anticipated Air Force requirements of a thermosetting polymer which could be fabricated into high strength fiber reinforced structural composites, acetylene terminated polyimides have been developed recently (12) as shown in the following example.

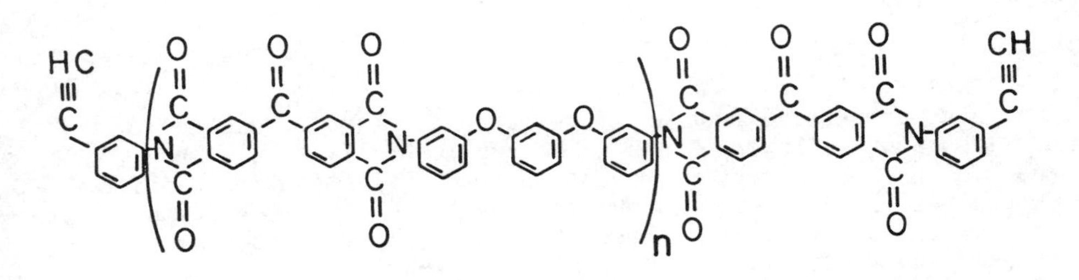

HR600

The major problem addressed in this program was the thermal cure, through an addition mechanism, without evolution of volatiles, of the polyimide intermediate. The electropolymerization of acetylene terminated polyimides and of a model compound, viz., phenylacetylene was therefore undertaken. It was necessary to establish that the polymerization of the polyimide intermediate on graphite fibers did in fact proceed through the terminal acetylenic groups. The results would have considerable significance in view of the development of a new family of polyphenylquinoxalines with terminal acetylene groups which offer great potential for use in high temperature composite and adhesive applications (13,14).

A parallel development in high temperature resistant matrix resins for composites is the synthesis of nitrile crosslinked polyphenylquinoxaline (15). The participation of nitrile groups in electroinitiated addition polymerization was therefore studied through the use of benzonitrile as monomer.

Experimental

Monomers of 99 percent or better purity were used as received from the suppliers. Obtained from Cardova Chemical N-(2-hydroxyethyl)ethyleneimine (HEEI) was used as received. Dimethyl formamide solvent, Mallinckrodt chemical, analytical grade was refluxed over calcium hydride for 24 hours before distillation at reduced pressure. Epoxy matrix resin, Epon 828 was obtained from Shell Chemical Co. HR600 was obtained from Gulf Chemical Co.

Graphite fibers in the form of continuous tow were specially obtained free of commercial treatments for use in electropolymerization. The usual commercial fibers were used as control for comparison with the experimental treatment. The fibers were all PAN based fibers, Hercules AS and AU from Hercules Inc., Fortafil 3T, 4T, 5T, CG3 and CG5 from Great Lakes Chemical Co., and Thornel 300 from Union Carbide.

Electropolymerization.

Generally, electropolymerizations were conducted in a three compartment cell in which the two end compartments were separated from the middle one by two fritted glass discs.

Graphite fiber electrodes were placed in the central compartment containing monomer and two platinum counter electrodes were placed in each of the two end compartments which contained only solvent and electrolyte. The use of fritted discs as a barrier minimized monomer and polymer migration into the end compartments.

Electropolymerizations were conducted at constant DC voltage provided by a Hewlett Packard Model 6438B power supply. Cell voltage and cell current, as monitored by measuring the voltage across a fixed 10 ohm, 100 watt, wire wound register placed in series with the cell, were recorded using a HP 7128A, two channel, strip chart recorder.

Composite Fabrication.

After polymerization, the graphite fibers were rinsed with water or acetone and then dried under vacuum to determine weight increase due to polymer coating and for further testing. When composite specimens were desired, Hercules AU fiber tow was wound around H-type frames and immersed in the electolytic cell. Prepregs were prepared by brushing an epoxy resin catalyzed by m-phenylenediamine and heating at 80°C for an hour. Strips were cut from the prepregs parallel to fiber·alignment and stacked in steel molds to obtain composite bars by compression molding at 150°C/30 min, 200 psi. Test specimens for impact strength and shear tests were cut from these pieces. ASTM standard methods were adopted to measure density, void content (less than 0.2%), and fiber content. Short beam shear tests were conducted in an Instron model TTCML testing machine at a crosshead speed of 1.0 mm/min, and span to thickness ratio of 4:1 (ASTM D2344-76). At least five specimens were tested for each composite; and only the values obtained for shear failure were used. A Tinius Olsen plastic impact tester was used to measure impact strengths by Method A, Izod of ASTM D256-73. Arithmetic averages of two or three specimens for each volume fraction were obtained. Measurements at nine different fiber volumes were conducted for each electrocoating treatment. A simple device was set up for continuous electrodeposition of polymer on graphite fiber tow from a spool drawn successively through an electrolytic cell and rinse bath before being rewound on a drum (2). Electrodepositions were carried out from a 2.5 percent solution of the selected polymer at 10 volts for a period of 1 minute. Epoxy prepregs and test specimens were prepared from the coated tow following procedures described above. A variety of maleic anhydride copolymers partially hydrolyzed were used for coating the fibers by electrodeposition.

Electrodepositions on a smaller scale were carried out as follows. Weighed lengths of carbon fiber tow, in the form of bundles 12.0± 0.5 cm long, tied at both ends, were placed in the center of a single compartment cell containing the electrodeposition solution. Cell dimensions were 8 x 7 x 12 cm. The carbon fiber bundle was immersed to a depth of 10.0± 0.2 cm. Platinum electrodes were placed on both sides of the bundle at a distance of 3.0 cm. Constant DC voltage was applied to the cell for a selected period of time, after which the fibers were removed, rinsed, and dried at 50°C for 18 hours in a vacuum oven. The increase in weight of the fiber bundle was then measured and the average weight increase of at least two specimens was recorded. Electrodeposition of polyamic acids was carried out by this procedure.

Pyre-ML (Du Pont) is a solution of polyamic acids formed by the reaction of aromatic diamines with aromatic dianhydrides. When Pyre-ML is baked, it is converted to an inert polyimide. Received as a 16.5 percent polymer solids solution in N-Methyl-2-pyrrolidone and aromatic hydrocarbons, a colloidal dispersion of Pyre-ML (#RC-5057) in acetone was prepared as follows (22). Twentyfive milliliters of Pyre-ML was mixed in 100 ml of dimethylsulfoxide. Five milliliters of triethylamine was added and the solution was heated to 40°C, with stirring, for 15 minutes. The solution was then slowly added to 500 ml of acetone in a Waring blender. Pyre-ML was electrodeposited from this solution on carbon fiber anodes. Upon completion of the deposition, the coated fibers were dried at 150°C for one hour and were then placed in a vacuum oven for 18 hours at 50°C. No rinsing procedure was employed since it was found experimentally that rinsing in water washed away almost all of the electrodeposited Pyre-ML on the fiber surface.

Thermal Analysis.

All thermogravimetric analyses (TGA) were performed with a Perkin Elmer TGS-1 Thermobalance. An atmosphere of flowing dry air was delivered from a cylinder of dry air at a rate of 25 ml per minute. Dynamic thermal analyses were conducted at a heating rate of 10°C per minute. The variation of weight with increasing temperature was recorded on a strip chart recorder. Data obtained as actual weights as a function of temperature were converted to show the percentage of residual weight and plotted as a function of temperature. Isothermal analysis was performed by raising the temperature of the sample to 500°C at a heating rate of 80°C per minute. The sample was allowed to decompose at constant temperature and the variation of weight was recorded as a function of time.

Cured resins and mixtures of epoxy (EPON 828-mPDA) resin and polyimide intermediate were ground into a fine powder in a mortar and pestle and screened through a #40 (0.417mm mesh) Tyler Standard screen and subjected to thermal analysis as described. Preparation of the cured resins prior to thermal analysis was as follows: Stoichiometric amounts of EPON 828 and meta-phenylenediamine (mPDA) were heated to 80°C, thoroughly mixed, and precured at 80°C for one hour. Final curing was done at 150°C for one hour. Thermid 600 (HR 600) or Pyre-ML as received, was placed in an aluminum cup and cured at 315°C for three hours, or two hours respectively. Fortafil 5U carbon fibers treated by electropolymerization and electrodeposition were subjected to thermal analysis as short lengths, cut from the treated fiber bundle. Untreated Fortafil 5U fibers were run as received in the same manner. All treated fibers were subjected to a 60 s, 24 VDC treatment prior to analysis.

Composites prepared from treated and untreated Fortafil 5U carbon fibers in an EPON 828-mPDA matrix were subjected to analysis in the form of a single solid chunk cut from the composite specimen. As before, all treated fibers were exposed to a 60 sec, 24 VDC treatment prior to incorporation into a composite. Composite specimens were prepared in the usual manner by compression molding as described earlier.

Results and Discussion

Detailed results of the screening of a variety of monomer-solvent-electrolyte systems in electropolymerization, and investigations of polymer grafting and stereoregularity, have been described elsewhere (1). Not only vinyl monomers but others containing a variety of cyclic function groups such as epoxy or aziridinyl groups were found to electropolymerize on graphite fibers which thus proved to be a good substrate for polymer coating by electroinitiation. The principal effect of electropolymerization of monomers on graphite fibers was expected to be in altering the interfacial bond strength of the coated fibers when incorporated in a composite. The short beam shear test is a popular compromise between purity of stress and ease of procedure and was employed here to measure interlaminar shear strength (ILS). The corresponding values for impact strength were also measured to follow the trend in toughness of the composite specimens. These extensive results (1) can be summarized as follows:

The ILS of composites prepared from Hercules type AU carbon fibers coated by electropolymerization of a series of different monomer systems were measured. Systems were selected to include representatives of various types of monomers in both aqueous and nonaqueous solvent-electrolyte systems encountered during the screening process. Composites prepared from Hercules type AU untreated, and type AS, commercially treated graphite fibers without further treatment were tested for comparison. It was seen that incorporation of a polymer interlayer on graphite fibers prior to embedding them in an epoxy matrix caused the strength of the composite, to vary over the range 45 to 82 MPa at $V_f = 50\%$, which is significant in relation to the standard deviations (less than ±5 MPa) involved in each set of measurements. Although more detailed studies are needed to standardize the electropolymerization technique in order to obtain optimum results, composite mechanical properties were found to be dependent upon monomer, solvent, electropolymerization time, and post treatment of the coated fibers.

Since significantly different shear strengths are obtained using different types of monomers in electropolymerization, it would appear that the shear strength is quite sensitive to the chemical and structural properties of the polymer interphase. The types of polymers formed varied from very flexible vinyl terminated butadiene-nitrile copolymer to rigid (acrylonitrile) polymers, and from those which had functional groups capable of reacting with the epoxy matrix to those which had none such. However, from the limited amount of data available, no correlation could be made of the trend in shear strength with the chemical or mechanical properties of the polymer coating. The properties of the polymer film formed on the graphite fiber need to be characterized before any such correlation can be made.

The results of impact tests on notched specimens provided further evidence that the carbon fiber-polymer matrix interface could be modified by electropolymerization. As was observed in shear tests, significant differences occur in the impact strength as well as in strength variation with volume fraction of fibers that had undergone different treatments.

When the trends of impact strength are compared to trends of shear strength, it was seen that usually an increased impact strength resulted whenever a decreasing shear strength was observed. This is in agreement with the general observation in composites that excessive interfacial bonding causes brittle failure and lowered impact strength. However, in exceptional instances, an increase in both shear and impact strength was observed.

It is useful to point out here that the composite specimens were prepared from fibers which were coated under one set of electropolymerization conditions, and that these conditions were not optimized with respect to composite properties. The results thus indicate the potential of electropolymerization for interphase modification in graphite fiber composites and the need to standardize electropolymerization conditions and monomer systems to control polymer film properties and, through them, composite properties.

The electrodeposition technique also has proved effective for interphase tailoring in carbon fiber composites (2,5). The copolymers employed for anodic electrodeposition on carbon fibers were a series of polymers with carboxyl groups, viz., copolymers of maleic anhydride with styrene, α-olefins and methyl vinyl ether. The introduction of the polymer interphase results in significant improvements in composite properties, and the extent of improvement is controlled by the nature of the polymer interphase. The molecular weight, chemical composition, and crosslinking of the interphase polymer are some of the molecular parameters modifying the effects observed. The most striking feature of the results was that, in most cases, increase in shear strengths was not accompanied by a corresponding loss in impact strength as is generally observed in various methods of surface treatment.

N-(2-hydroxyethyl)ethyleneimine.

The results obtained with HEEI are presented in Figure 1. The electropolymerization of HEEI was conducted at 12 VDC, from a solution of the monomer in DMF containing $NaNO_3$ as supporting electrolyte (0.1g HEEI per ml of 0.2N $NaNO_3$/DMF). The polymer coated fibers were washed with acetone, and vacuum dried at 60°C for 18 hours before preparation of composite specimens.

A number of interesting general features described above are seen to be illustrated in Figure 1. Polymer coating of Hercules AU fibers by electroinitiation of HEEI, at the anode, clearly raises the ILS of the epoxy composite to values comparable to those of specimens prepared from commercially available surface treated Hercules AS fibers.

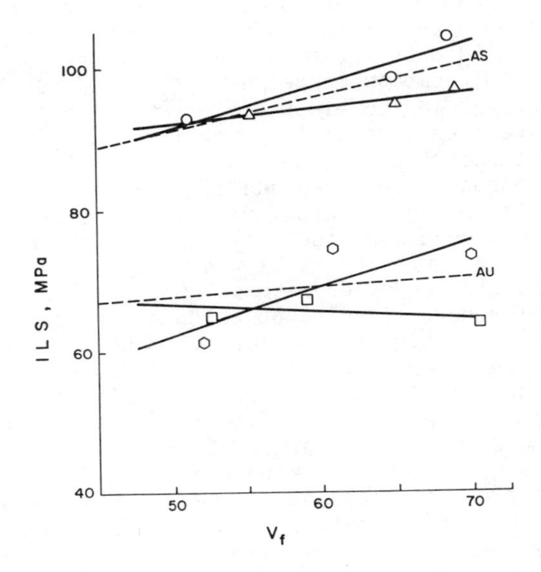

Figure 1. The interlaminar shear strengths (ILS) of composites prepared from Hercules AS and AU graphite fibers, and AU fibers coated by electropolymerization of N-(2-hydroxyethyl) ethyleneimine (HEEI) under conditions indicated, at various fiber volume fractions (V_f): (○) HEEI, 3 sec, anode; (△) HEEI, 10 sec, anode; (⬡) HEEI, 3 sec, cathode; (□) HEEI, 10-sec dip, no current passed

The duration of polymerization is seen as an important variant, changing from a 3 s to 10 s polymerization time has produced noticeable changes in ILS values.

Dipping the fibers in HEEI monomer for 10 s without electro-initiation does not raise the ILS of composites above that of untreated Hercules AU fibers. This observation is interesting since it emphasizes the importance of the interphase polymer, in this case poly-HEEI, in improving the ILS of the composite. Monomer HEEI carries the aziridine functional group which can readily react with carboxyl groups present on the fiber surface as follows:

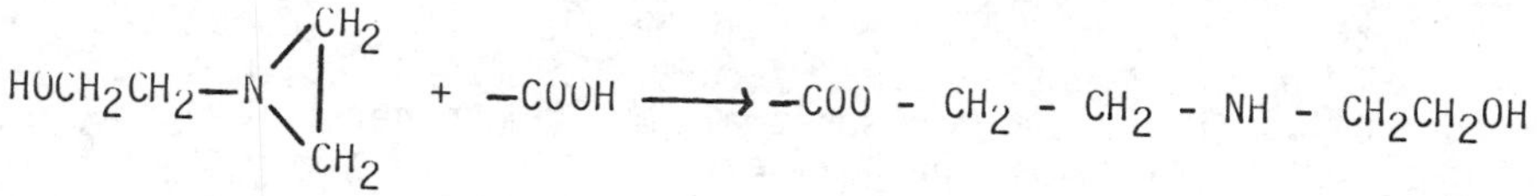

$$HOCH_2CH_2-N\begin{matrix}CH_2\\|\\CH_2\end{matrix} + -COOH \longrightarrow -COO - CH_2 - CH_2 - NH - CH_2CH_2OH$$

The resulting functional groups, amino and hydroxyl, are both capable of interacting with the epoxy matrix, thus forming a molecular bridge from the fiber surface to the matrix resin. Apparently, such a surface modification to produce reactive functional groups, without the formation of a polymer interphase, is not the cause of the observed improvement in ILS.

The absence of any significant change in ILS when HEEI monomer is present at the fiber cathode (Fig. 1) also supports the role of the polymer interphase in the observed property improvement. It has been observed in other experiments not reported here (6,16), that HEEI is polymerized only at the anode, and not at the cathode. The anodic polymerization proceeds by a cationic mechanism, leading to the polymer of the structure:

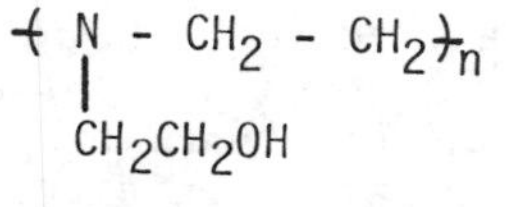

$$\left(\underset{\underset{CH_2CH_2OH}{|}}{N} - CH_2 - CH_2 \right)_n$$

The observed effectiveness of the anodic treatment and the lack of it in cathodic treatment (Fig. 1), must be attributed to the formation of the polymer of HEEI in the former case, and the absence of it in the latter. The structure of the polymer formed carrying tertiary amino, and hydroxyl functional groups, is favorable for subsequent participation in the epoxy curing reaction, and also for adhesion to the epoxy matrix formed.

Polyimides.

In view of the above results, and considering the importance of high temperature resistant graphite composites, the electropolymerization of HR-600, the oligomeric polyimide intermediate was conducted on graphite fibers. A polymer was readily formed by cathodic polymerization, as indicated by the weight increase of fibers illustrated in Figure 2. Since a variety of functional groups were found to be polymerized by electroinitiation (1), it was sought to establish the participation of the acetylenic terminal groups in the polymerization reaction. Using phenylacetylene as model monomer, it was found that it readily polymerized in $NaNO_3$/DMF in the cathode compartment to yield a polymer of average molecular weight 3000 (17). Carbon-hydrogen analyses agreed well with calculated values for polyphenylacetylene (calculated for C_8H_6: C,94.08; H,5.92. Found: C,92.14-93.05; H, 5.75-5.84). The ir and nmr spectral data and other evidence confirmed the polymer to be a linear polymer with a polyene structure formed by anionic addition of C≡C bonds.

It is therefore to be expected that the polyimide coating formed on carbon fiber during electropolymerization of HR-600 is the result of addition of acetylenic terminal groups. The compression molding of electrocoated fibers, with or without additional impregnation with polyimide oligomer, was adopted to produce composite specimens. The preparation of specimens has proved more difficult than with an epoxy matrix probably because of the flow properties of polyimide. Therefore, composite specimens were prepared from the polyimide coated Hercules-AU graphite fibers using an epoxy matrix. The results shown in Figure 3, confirm the modification of shear strength by the electropolymerized coating (3 s polymerization of HR-600 in 0.2N $NaNO_3$/DMF at 24 VDC). It is surprising to observe the large change in shear strength with fiber volume fraction. Perhaps it can be attributed to the fact that with increasing "volume fraction" of the polyimide coated fibers the matrix changes from a predominantly epoxy matrix to one containing increasing amounts of polyimide.

As indicated already, the polymerization of nitrile groups is also relevant to composite research in view of the nitrile crosslinked polyquinoxalines being developed (14). Polymerization of benzonitrile has shown that a linear polymer is formed with the repeating unit being $\leftarrow C_6H_5C = N \rightarrow$. The conjugated structure formed by addition of C N groups was supported by infrared and mass spectral data which showed the absence of the cyclic trimer 2,4,6-triphenyl-*s*-triazine (17,18). From these results it can be expected that the polymerization of nitrile terminated polyimide and polyquinoxaline intermediates can also be electroinitiated on graphite fibers. Electropolymerizations through C≡C and C≡N bonds have great significance therefore to high temperature resistant polyimide and polyquinoxaline composites and improvement of their shear

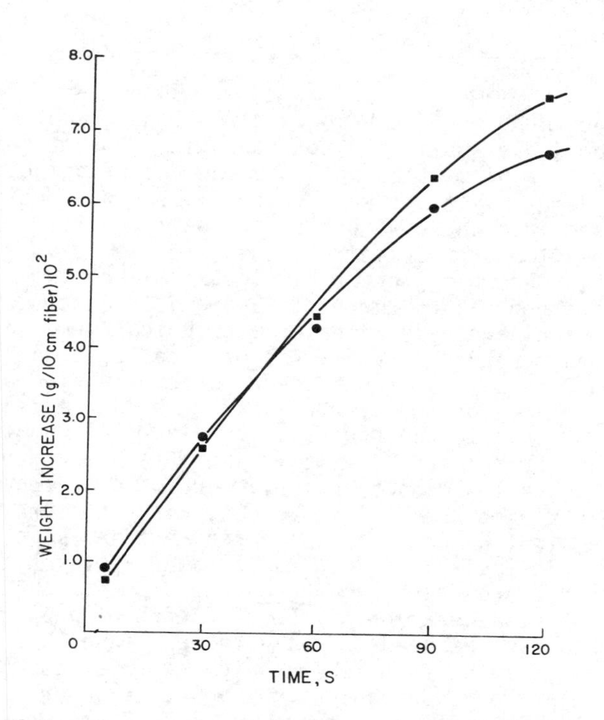

Figure 2. Electropolymerization of acetylene terminated polyimide (HR-600) on Fortafil carbon fibers: HR 600/DMF–$NaNO_3$ (24 VDC); (●) Fortafil 3U; (■) Fortafil 5U

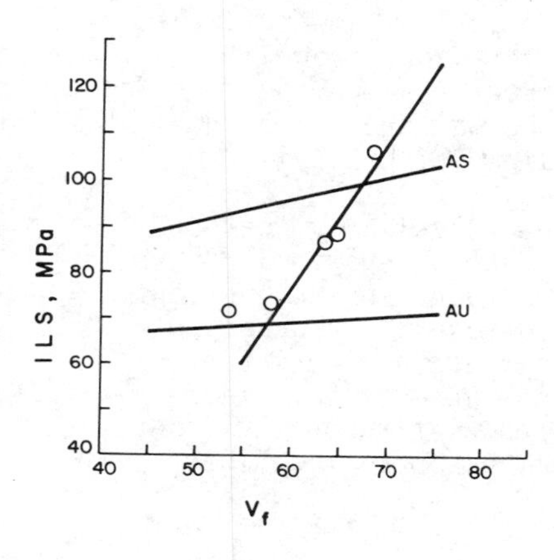

Figure 3. The interlaminar shear strengths (ILS) of composites prepared from Hercules AS and AU graphite fibers, and AU fibers coated by electropolymerization of HR-600, (○) at various fiber volume fractions (V_f).

strength and toughness.

Electrical Hazards of Conductive Fiber Fragment Release.

The electrochemical techniques of coating graphite fibers described above are applicable to addressing another important potential problem attendant upon the use of composites. The problems that may arise when carbon fibers are accidentally released into the environment are well publicized by now. The Department of Commerce issued a press release in January 1978, concerning potential electrical problems associated with carbon fibers currently being used in composites (19). NASA simultaneously released a technical memorandum concerning the observed effects on electrical systems of airborne carbon fibers (20). In these reports, the high electrical conductivity of the carbon fibers was identified as the prime factor in their effects on electrical equipment, with other properties such as small fiber diameter, generally short length, and low density being important contributory factors. Because of their light weight, carbon fibers can float in the air like dust particles and, if they come to rest on electrical circuits, can cause power failures, blackouts, shorts, or arcing that can damage equipment.

A serious consideration of these apsects of graphite composites emphasizes the need to evaluate new techniques of devising organic coatings for carbon fibers which can provide, in case of fire, a relatively nonconducting layer of char or other material that might result in fiber "clumping" preventing fiber release or, in the event of fiber release, prevent electrical contact with electrical components. In this context, research was conducted to apply the techniques of electropolymerization and electrodeposition developed for interphase modification of carbon fiber composites toward a solution to the problems of airborne carbon fiber fragments (6). The behavior of electrolytically formed organophosphorous and polyimide coatings at high temperature was therefore investigated (6). The first results obtained with polyimide coatings are presented here.

The use of graphite fibers, electrolytically coated with polyimide resins, as reinforcement for epoxy matrix resin is seen to offer potential advantage. On combustion, such a composite would be expected to result in increased fiber clumping because of the high temperature resistant polymer coating and thus show reduced tendency to release fiber fragments. The residual charred coating on the fibers is also likely to be less conductive than graphite fiber fragments.

In order to evaluate these possibilities, polyimide coatings on graphite fibers were formed not only from acetylene terminated polyimides as described earlier, but also from polyamic acids by electrodeposition. The thermal oxidative behavior of the fibers, matrix resins and composites was investigated by thermogravimetric analysis based on a study by Wentworth and coworkers (21) on the TGA of graphite fiber composites. The ability of different types of precursor coatings to reduce the potential for accidental release of carobn fibers was sought to be compared.

The weight increase of carbon fibers as a function of time at constant applied voltage during electrodepositions and electropolymerizations are shown in Figs. 2 and 4. It can be seen from these results that the amount of polyimide incorporated into a coating is a function, of the applied voltage and the exposure time of the treatment. Little difference is observed between fibers having different elastic moduli (Fortafil 3U and 5U, E=210 and 330 GPa, respectively).

Dynamic TGA of Resins and Mixtures.

The thermal oxidative behavior of the cured polyimide resin and polyimide-epoxy resin mixtures is shown in Fig. 5. As seen Fig. 5, the neat epoxy has three major breaks in the TGA curve. One starting at 275°C and another at 350°C, corresponding to resin decomposition to char, and a third starting about 450°C for the oxidation of the char residue. As expected, the polyimides, are clearly shown to be more thermally stable than the epoxy resin. In fact, the polyimides did not show any major decomposition below 500°C. At this temperature, the char from the epoxy resin had already begun to decompose. In the context of the purpose of this study, this would imply that a polyimide coating on the carbon fibers could survive to a higher temperature. That is, in the composite, the epoxy matrix resin and the resulting char could be completely consumed before the polyimide coating would begin to decompose. This would not only result in holding the fibers together, but would also provide an insulating layer on any released fibers, thereby preventing electrical contact.

The results of the thermal decomposition of the polyimide coated fibers lead to an interesting observation. Fibers coated with polyimides appear to decompose more rapidly than the untreated fiber. The results are shown in Fig. 6. HR 600 coated fibers showed a decomposition for the polyimide, beginning at about 500°C, followed by a very rapid fiber decomposition at about 760°C. Since HR 600 was electropolymerized in a $NaNO_3$-DMF solution, it is possible that a salt, codeposited with the HR 600, catalyzed the the oxidation of the fibers. Pyre-ML, Fig. 6, likewise appears to influence the fiber decomposition. The decomposition of

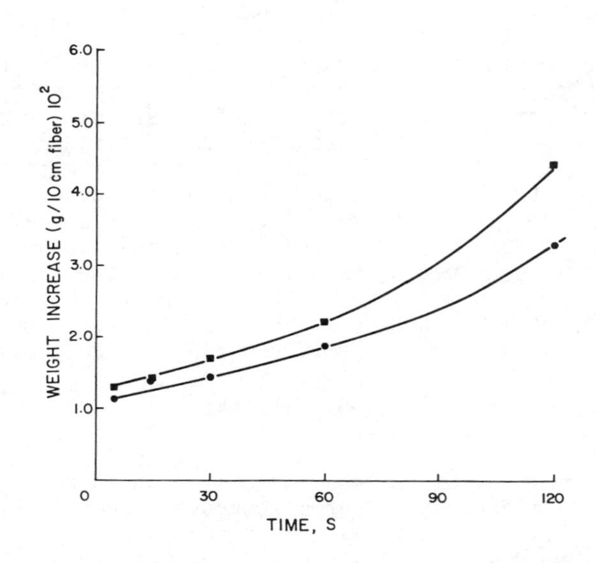

Figure 4. Electrodeposition of polyamic acid Pyre-ML on Fortafil carbon fibers: PYRE-ML (24 VDC); (■) Fortifil 5U, (●) Fortafil 3U

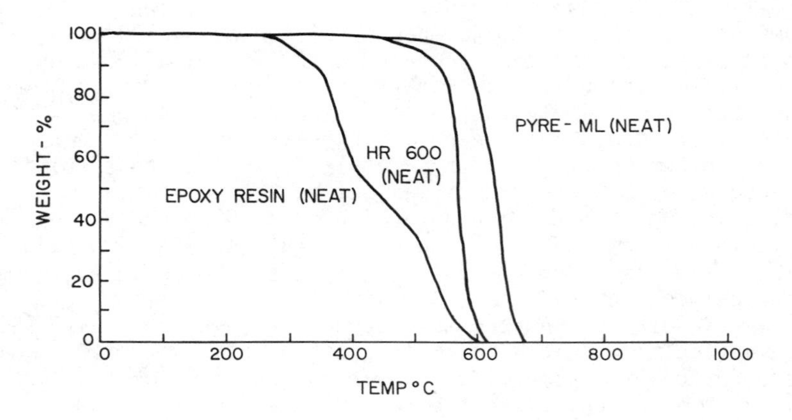

Figure 5. Dynamic TGA of epoxy and polyimide resins

the Pyre-ML, beginning about 550°C does not show any clear break before fiber oxidation begins. Instead, a smooth line is observed. It appears the polyimide has some catalytic effect on fiber oxidation, decomposing the fiber at a lower temperature than is observed for the untreated fiber.

Finally, the thermal behavior of composites prepared from polyimide coated fibers is shown in Fig.7. Several observations are worth noting. The fiber decomposition in both polyimide-coated fiber composites occur at temperatures somewhat lower than is observed in the untreated fiber composite. This again suggests that the decomposition of the polyimides in some manner catalyzes the oxidation of the fiber, as was observed with the coated fibers themselves. On the other hand, HR 600 and Pyre-ML resins correspond, almost exactly, to the observed resin decompositions in the composites. This observation adds support to the occurrence of an interaction between the polyimide resin and carbon fiber decomposition behavior.

In summary, the thermal oxidative behavior of the neat resins, coated fibers, and composites have shown that electrochemical treatments resulting in the decomposition of polyimides have a significant effect on the behavior of carbon fiber-epoxy matrix composites. Polyimides are not only more thermally stable than the epoxy resins, but also, appear to reduce the thermal stability of the carbon fiber' substrate. Other similar and significant effects of the coatings on the decomposition of the carbon fibers were observed with organophosphorus coatings also (6). All of these effects could be seen as either preventing the release of carbon fibers into the environment or, perhaps, as resulting in a fiber having a reduced conductivity, thereby preventing electrical contact once released.

It is obvious that a great deal of further study will be necessary to answer the questions raised by this work. Future study should be directed toward determining the effect these coatings have on the electrical conductivity of the fiber, both before and after thermal decomposition of the matrix resin. Also, since the coated fibers are to be used in composites, a study of the composite mechanical properties would be essential. Further study of the mechanical and thermal oxidative properties of carbon fibers treated by electrochemical means would be very useful in gaining an understanding of the behavior of the fiber in the composite.

Summary.

Interphase modification through electropolymerization and electrodeposition, such as that of N-(2-hydroxyethyl)

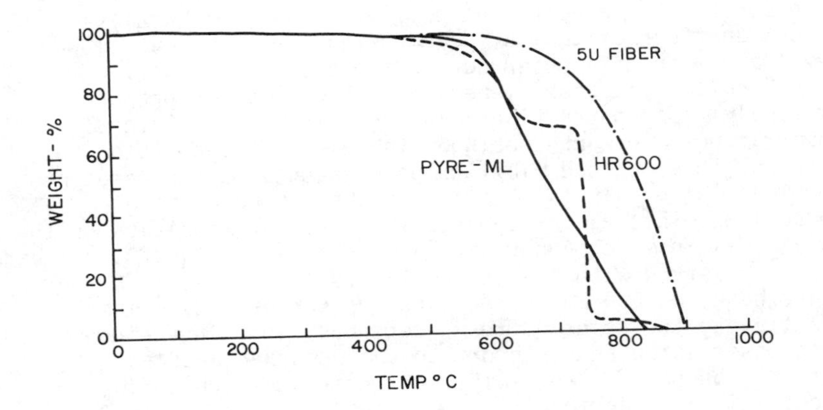

Figure 6. Dynamic TGA of untreated, Pyre-ML coated, and HR-600 coated Fortafil 5U carbon fibers

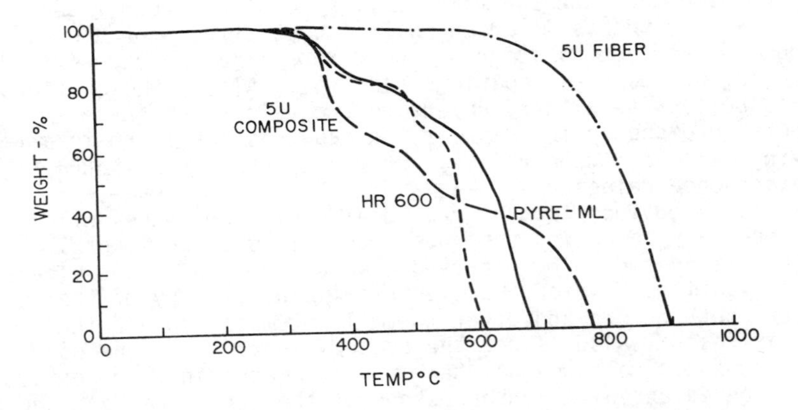

Figure 7. Dynamic TGA of Fortafil 5U fiber and epoxy composites prepared from untreated, HR-600 coated, and Pyre-ML coated Fortafil 5U fibers

ethyleneimine, has been shown to be effective in improving composite shear strength and toughness. Studies with model compounds such as phenylacetylene and benzonitrile confirm the occurrence of electroinitiated polymerization on graphite fibers through the C≡C or C≡N bonds in nitrile or acetylene terminated polyimide intermediates. The concept of interphase modification has been expanded by the application of electrochemical coating of carbon fiber to reduce the potential for release of conductive fiber fragments from graphite composites. Polyimide precursors were formed electrochemically on carbon fibers. Thermogravimetric analysis was used to measure the significant effects of the coatings on the thermal oxidative behavior of the system components. Electrochemical polymerization thus offers a new route for coating carbon fibers prior to embedding them in a polymer matrix. The potential value of these techniques to composite property modification was amply demonstrated.

Acknowledgement

This research was supported by grants from the Office of Naval Research and from the Washington State University Research Foundation.

Literature Cited.

1. R. V. Subramanian, James J. Jakubowski, Polym. Eng. Sci., 18. 590 (1978).

2. R. V. Subramanian, V. Sundaram, A. K. Patel, 33rd Ann. Tech. Conf. SPI Reinf. Plastics/Comp. Inst., 20F (1978).

3. R. V. Subramanian, James J. Jakubowski, F. D. Williams, J. Adhesion, 9, 185 (1978).

4. James J. Jakubowski, "Interface Modification of Carbon Fiber Composites by Electropolymerization" M.S. Thesis, Materials Science and Engineering, Washington State University, Pullman, WA (1976).

5. V. Sundaram, "Electrodeposition of Polymers on Carbon Fibers: Effects on Composite Properities" Thesis, Materials Science and Engineering, Washington State University (1977).

6. James J. Jakubowski, "Electrochemical Polymerization and Deposition on Carbon Fibers", Ph.D. Thesis, Materials Science and Engineering, Washington State University, Pullman, WA (1979).

7. A. S. Kenyon, H. J. Duffey, Polym. Eng. Sci., 7, 189 (1967).

8. A. S. Kenyon, J. Colloid Interface Sci., 27, 761 (1968).

9. G. J. Fallick, H. J. Bixler, R. A. Marsella, F. R. Garner, E. M. Fettes, Mod. Plast. 45, 143 (1969).

10. J. L. Kardos, F. S. Cheng, T. L. Tolber, Polym. Eng. Sci., 13, 455 (1973).

11. M. Xanthos, R. T. Woodham, J. Appl. Polym. Sci., 16, 381 (1972).

12. N. Bilow, A. L. Landis, Proc. 8th National SAMPE Conf. Seattle, WA (1976), p. 94.

13. R. F. Kovar, G. F. L. Ehlers, F. E. Arnold, AFML-TR-76-71 (June 1976); ibid, p. 106.

14. A. Wereta, AFML-TR-75-214 (1976).

15. W. B. Alston, Proc. 8th National SAMPE Conf. Seattle, WA (1976), p. 114.

16. J. Jakubowski and R. V. Subramanian, Polymer, (communicated).

17. R. V. Subramanian, J. J. Jakubowski, B. K. Garg, U. S. NTIS AD-Rep, AD-A047492 (1977), Chem. Abstr. 88, 170537q, (1978).

18. R. V. Subramanian and B. K. Garg, Polymer Bull. 1, 421 (1979).

19. U. S. Department of Commerce News, ITA 78-13, Jan. 20, (1978).

20. NASA TM 78652, January (1978).

21. S. E. Wentworth, R. J. Shuford, and A. O. King, Eighth North East Regional Meeting, ACS, Boston, June (1978).

22. D. C. Phillips, J. Electrochem. Soc., 119, 1645 (1972).

RECEIVED January 8, 1980.

21

Dynamic Mechanical Characterization of Advanced Composite Epoxy Matrix Resins of Altered Composition

M. VON KUZENKO—GAF Materials Laboratory, Landshuterstrasse 70, 8058 Erding, West Germany

C. E. BROWNING—Air Force Materials Laboratory, Wright–Patterson Air Force Base, OH 45433

C. F. FOWLER—University of Dayton Research Institute, Dayton, OH 45469

The compositions of several advanced composite epoxy matrix resins were intentionally altered to assess acceptable limits of compositional variability. The objective of the work reported herein was to determine if these compositional variations could be related to the material's dynamic mechanical (DM) response. DM characterization of epoxy resins has been shown (1,2) to be useful in elucidating structure-property relationships.

In this work the DM behavior of these altered samples was studied in shear as a function of frequency, temperature and absorbed moisture. Absorbed moisture has been shown (3) to dramatically affect an epoxy resin's mechanical response.

Experimental

The epoxy resins investigated are representative of current high performance composite epoxy matrix resin systems. They contain multifunctional epoxy resins, an aromatic diamine curing agent (diaminodiphenylsulfone, DDS), and, in certain formulations, an organometallic catalyst (boron trifluoride complex). Compositional variations investigated are shown in Tables I, II, and III. Samples were prepared by casting the resin between glass plates. Cure was one hour at 177°C. This was followed by removal from the glass mold, machining into rectangular test specimens (5.08 cm x 1.27 cm x 0.3175 cm), and postcure at 177°C for eight hours.

DM testing was performed with a Rheometrics Mechanical Spectrometer, Model RMS-7200, in a torsion mode with shear moduli (storage and loss) and loss tangent being measured for each variation. The dynamic shear modulus is defined by:

$$G = G' + i G'' \qquad (1)$$

where G' is the storage modulus and G" is the loss modulus. The loss tangent, Tan δ, is defined by:

TABLE I. Epoxy Matrix Resin A Test Data Summary

Sample Code	Variation	Concentration (w/o)/ Viscosity			Test Condition	Storage Modulus Data					Loss Tangent Data					
						T(°C)			G'(10^{10}dynes/cm^2)		T(°C)					
		Epoxy	Curing Agent	Catalyst		Onset	G'(3x10^9)	G'(R)	Onset	G'(R)	α-peak Onset	α-peak Max	Tan δ= 10^{-1}	$W_{10^{-1}}$	β-peak Max	γ-peak Max
12kV11	Standard	N/N	N	N	Dry Wet Redried	170 70	224 164 215	272	1.20 0.98	0.015	130 50 102	234	208 114 151	41	51 -6	-78 -100
16kV1	High Viscosity Epoxide	N/High	N	N	Dry Wet Redried	160 80	218 145 214	273	1.10 1.10	0.012	126 66 100	228	200 104 150	54	62 2 62	-71 -101 -68
8kV2	Low Viscosity Epoxide	N/Low	N	N	Dry Wet Redried	160 80	226 146 217	272	1.20 0.92	0.016	130 60 95	234	208 106 155	43	44 -4 42	-80 -94 -70
8kV3	Brittle System	N/Low	0.9N	1.2N	Dry Wet Redried	160 80	222 142 206	268	1.10 0.90	0.016	110 66 90	231	205 105 146	43	48 3 54	-79 -104 -73
16kV4	Ductile System	N/High	1.1N	0.8N	Dry Wet Redried	160 80	220 146 216	271	1.10 1.00	0.012	120 78 110	231	198 106 148	56	73 -2 74	-62 -110 -68

N - Nominal

$W_{10^{-1}}$ - Peak Width at Tan δ = 10^{-1}

G'(3x10^9) - Storage Modulus Value is 3 x 10^9 dynes/cm^2

G'(R) - Storage Modulus at Onset of Rubbery Region

TABLE II. Epoxy Matrix Resin B Test Data Summary

Variation	Concentration (w/o)/Viscosity			Test Condition	Storage Modulus Data					Loss Tangent Data					
					$T(^{o}C)$			$G'(10^{10}dynes/cm^2)$		$T(^{o}C)$					
	Major Epoxy	Minor Epoxy	Curing Agent		Onset	$G'(3x10^9)$	G'(R)	Onset	G'(R)	α-peak Onset	α-peak Max	Tan δ= 10-1	W_{10}-1	β-peak Max	γ-peak Max
Standard	N/N	N/N	N	Dry Wet	170 80	245 211	>300	1.05 1.05	0.01	175 78	254	232 138	35	124	-70 -100
Under-Catalyzed	1.05N/Low	1.05N/N	0.95N	Dry Wet	156 80	242 208	>300	1.10 1.10	0.01	170 80	248	226 135	40	150	-74 -100
Over-Catalyzed	0.95N/High	0.95N/N	1.05N	Dry Wet	160 80	246 216	>300	1.05 1.10	0.01	154 77	254	234 138	37	---	-72 -95

N - Nominal

W_{10}-1 - Peak Width at Tan $\delta = 10^{-1}$

$G'(3x10^9)$ - Storage Modulus Value is 3×10^9 $dynes/cm^2$

G'(R) - Storage Modulus at Onset of Rubbery Region

TABLE III. Epoxy Matrix Resin C Test Data Summary

Variation	Concentration (w/o)/Viscosity					Test Condition	Storage Modulus Data					Loss Tangent Data					
							T(°C)			G'(10^{10} dynes/cm^2)		T(°C)					
	Epoxy 1	Epoxy 2	Epoxy 3	Curing Agent	Catalyst		Onset	G'($3x10^9$)	G'(R)	Onset	G'(R)	α-peak Onset	α-peak Max	Tan δ= 10^{-1}	$W_{10}-1$	β-peak Max	γ-peak Max
Standard	N/N	N/N	N	N	N	Dry Wet	110 70	205 142	250	1.2 1.1	0.015	70 43	217	194 108	50	56	-78 -87
Brittle	N/Low	0.9/N	1.1N	0.9N	1.2N	Dry Wet	120 70	205 139	240	1.2 1.1	0.017	82 55	218	195 110	50	65	-68 -92
Ductile	N/High	1.1N/N	0.9N	1.1N	0.8N	Dry Wet	110 70	205 143	246	1.2 1.1	0.014	60 40	215	194 110	52	30	-72 -78

N - Nominal

$W_{10}-1$ - Peak Width at Tan δ = 10^{-1}

G'($3x10^9$) - Storage Modulus Value is 3×10^9 dynes/cm^2

G'(R) - Storage Modulus at Onset of Rubbery Region

$$\text{Tan } \delta = \frac{G''}{G'} \quad (2)$$

where δ is the phase angle between the sinusoidal stress and strain waves. A strain is applied to the sample and the resultant stress is measured. Test results will be discussed in terms of the storage modulus, G', and the loss tangent, Tan δ.

The DM testing was performed in the temperature range of -150°C to +300°C using a programmed incremental heating rate. At each temperature increment the instrument automatically subjects the specimen to the four test frequencies - 15.8 m Hz, 158.5 m Hz, 1.585 Hz, and 15.85 Hz.

Wet test specimens were prepared by subjecting dry specimens to a near equilibrium absorption of moisture in a 71°C/100% R.H. environment. Redried test specimens are defined as wet specimens DM tested from 5°C to 95°C followed by drying at 80°C *in vacuo* to a constant weight.

Results

Epoxy Resin A. Epoxy resin A contains a single multifunctional epoxide resin (tetraglycidylmethylene dianiline, TGMDA), DDS, and a boron trifluoride complex catalyst.

Typical G' and Tan δ curves are shown in Figure 1 using the reference standard 12KV11 as an example. There are three transitions observable - one is the α-transition at 234°C associated with Tg and the other two are glassy state dispersions, a β-transition at 51°C and a γ-transition at -78°C. A total of 11 compositional variations were run in a dry condition at all four frequencies over the temperature range of -150°C to +300°C with the result that the most significant detectable differences were attributable to those variations involving the base epoxide's viscosity. For this reason the discussion will be limited to those variations shown in Table I, which comprise extremes in viscosity, brittleness/ductility, and the reference standard/nominal material. Also, the data and curves will be limited to the lowest frequency (15.9 m Hz). The most important parameters taken from G' and Tan δ curves are shown in Table I.

In the dry samples, two parameters consistently differentiated between high and low viscosity epoxide resins - G' at the onset of the rubbery region, G'(R), and the α-peak width at a reference Tan δ value of 10^{-1}, $W_{10}{-1}$. In all cases the low viscosity epoxide samples gave higher G'(R) and smaller $W_{10}{-1}$ values than the high viscosity samples (Table I). This is consistent in that low viscosity epoxide corresponds to low epoxide equivalent weight which, in turn, would correspond to a lower molecular weight between crosslinks, M_c, resulting in a higher G'(R). The converse would be true for a high viscosity epoxide composition. This G' behavior is shown in Figure 2 for 8KV2 vs. 16KV4. Assuming that the equilibrium moduli represent crosslinked network response, values of M_c were calculated from:

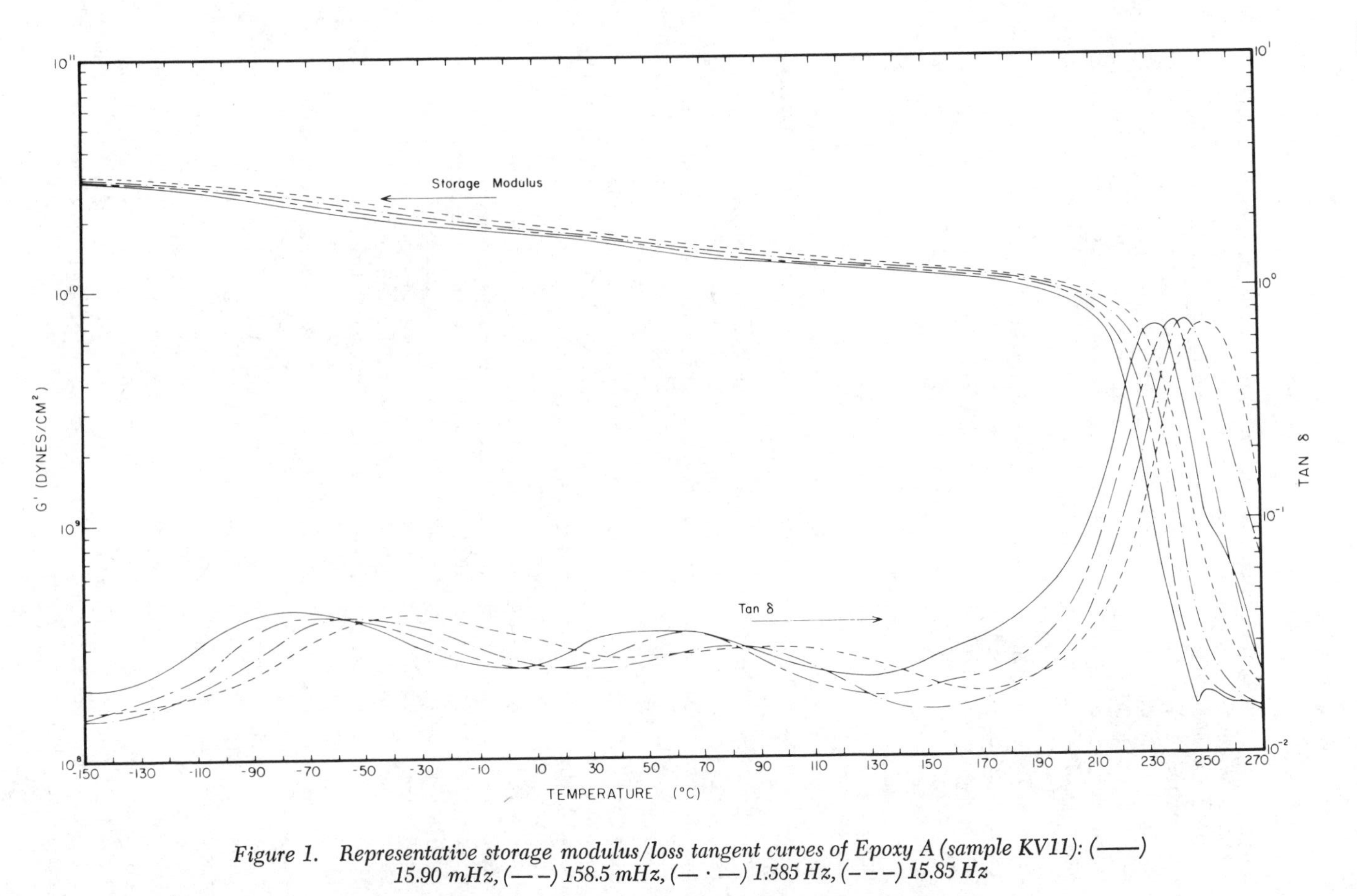

Figure 1. Representative storage modulus/loss tangent curves of Epoxy A (sample KV11): (——) 15.90 mHz, (— —) 158.5 mHz, (— · —) 1.585 Hz, (- - -) 15.85 Hz

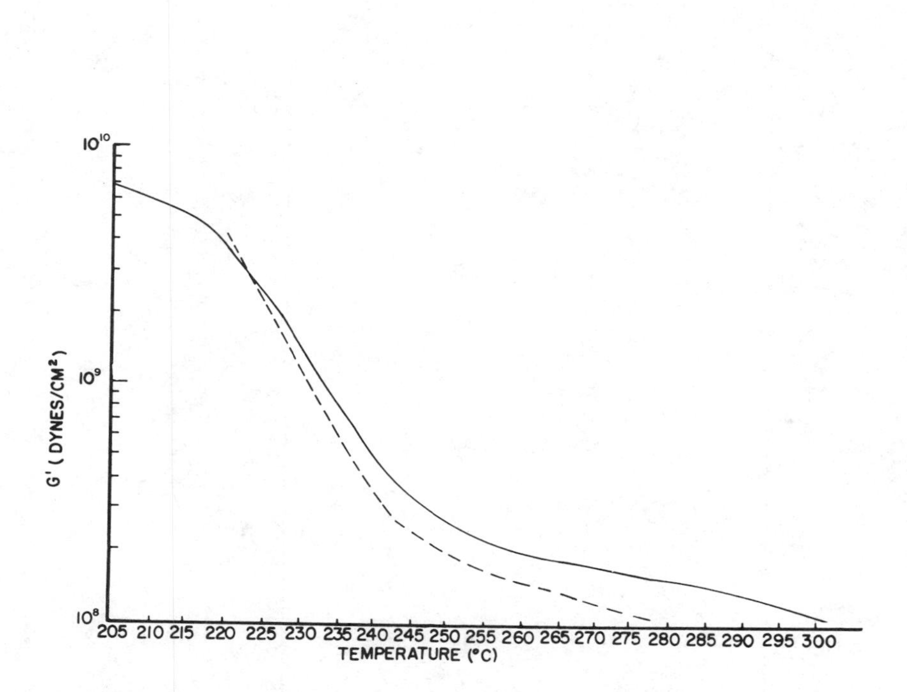

Figure 2. Storage modulus curves at rubbery onset for (——) low viscosity (8KV2) and (– – –) high viscosity (16KV4) Epoxy A samples

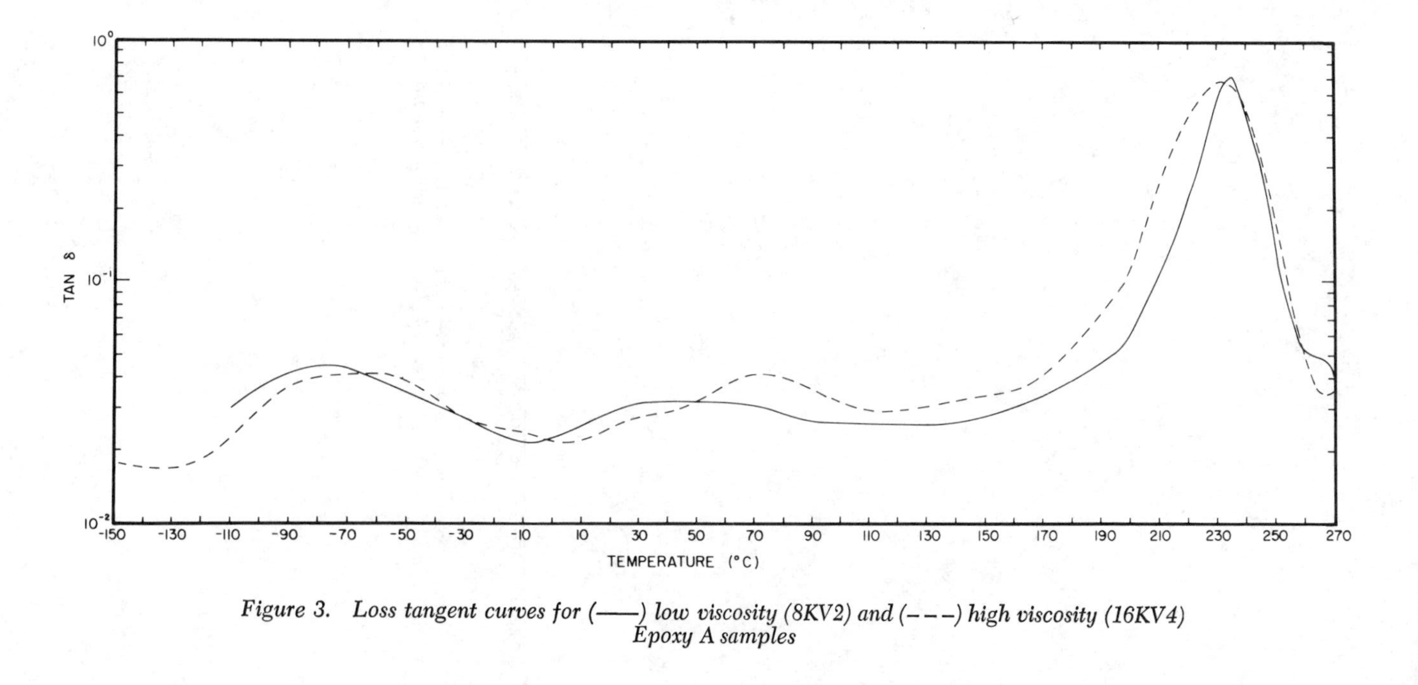

Figure 3. Loss tangent curves for (——) low viscosity (8KV2) and (- - -) high viscosity (16KV4) Epoxy A samples

$$M_c = \frac{3 \rho RT}{E} \quad (3)$$

where values of E were calculated from the G' values in Table I. M_c values of 90.3 and 120.4 were found for 8KV2 (low viscosity) and 16KV4 (high viscosity), respectively, which agrees with the observed response.

The broad Tan δ dispersion in the glass transition region corresponds to a continuous unfreezing of a spectrum of molecular mobilities. It could be inferred that under constant conditions, a narrow dispersion would reflect a smaller variety of available molecular motions and vice versa. Therefore, it could be expected that the low viscosity material would give a smaller $W_{10^{-1}}$ than the high viscosity material as was found (Table I). This is illustrated in Figure 3 for 8KV2 vs. 16KV4. Other Tan δ differences were found for the β and γ transitions. The high viscosity samples consistently gave higher β and γ-peak maximum temperatures.

Failure modes at very high temperatures (>250°C) were in all cases characterized by the growth of large cracks in the specimens.

With respect to proposed engineering applications of the material (use temperatures ≤150°C), no significant differences were detected in the DM response of all variations under dry conditions.

Data taken from curves of the DM response of wet specimens (6.5% weight gain) are tabulated in Table I. Typical effects of moisture on G' behavior can be seen in Figure 4 (wet vs. dry 8KV3). Moisture exhibits a significant plasticization effect as previously reported (3), shifting G' at Tg onset 80°C to lower temperatures. G' in the glass transition region at a reference value of 3 x 10^9 dynes/cm^2 was also shifted 80°C to lower temperatures. G' at the onset of Tg was lowered from a nominal 1.2 x 10^{10} to 0.9 x 10^{10} dynes/cm^2.

Similar moisture induced plasticization was observed in the Tan δ results as shown in Table I and Figure 5 (wet vs. dry 8KV3). The onset temperature of the α-peak was shifted 60°C to lower values, and the temperature at a reference Tan δ value of 10^{-1} was shifted 100°C to lower values. Discussion of results at temperatures above 100°C is not useful because of the onset of significant moisture loss (drying); however, it can be observed that moisture causes the α-dispersion to become very broad and irregular.

Moisture also shifts the β and γ-peaks to lower temperatures (Table I) and broadens them such that the valley between them in the dry condition becomes less and less distinct. Moisture has also been observed in all variations to induce a cracking-type failure mode at very high temperatures (>250°C) characterized by the splitting-out of discs from the specimen surface.

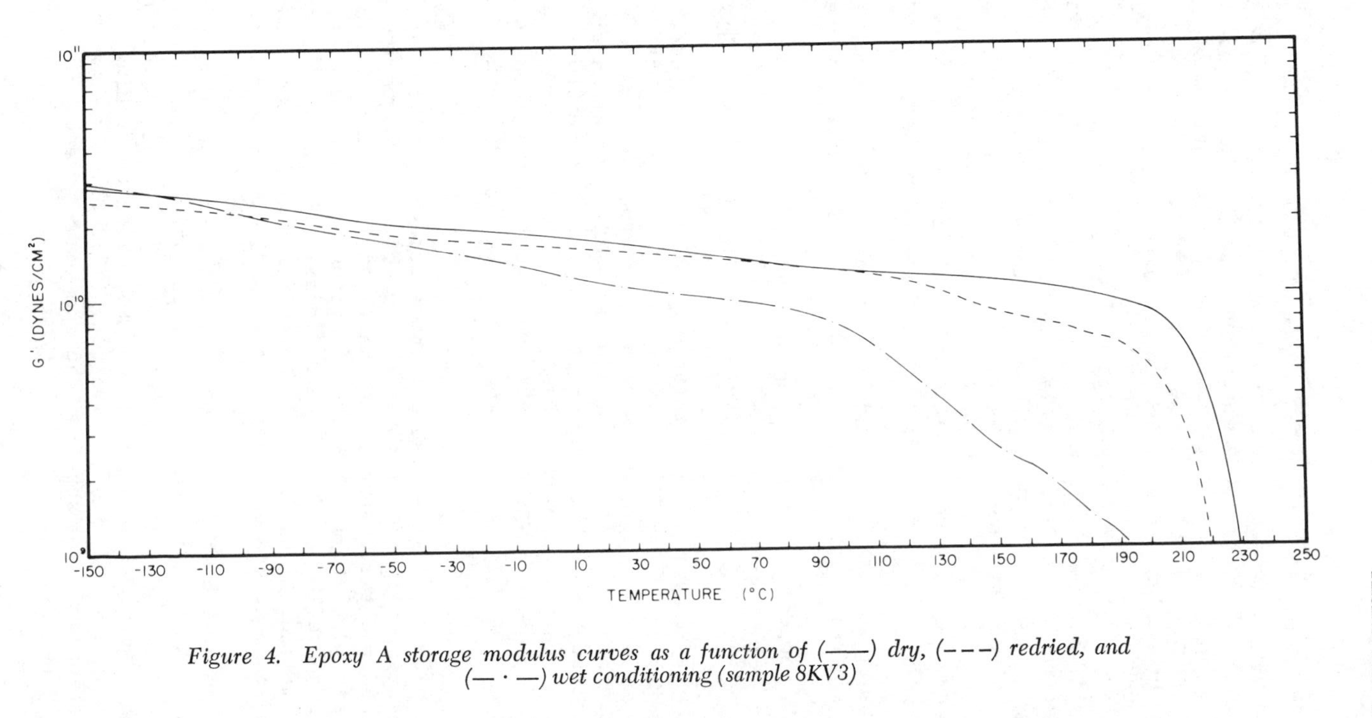

Figure 4. Epoxy A storage modulus curves as a function of (——) dry, (– – –) redried, and (— · —) wet conditioning (sample 8KV3)

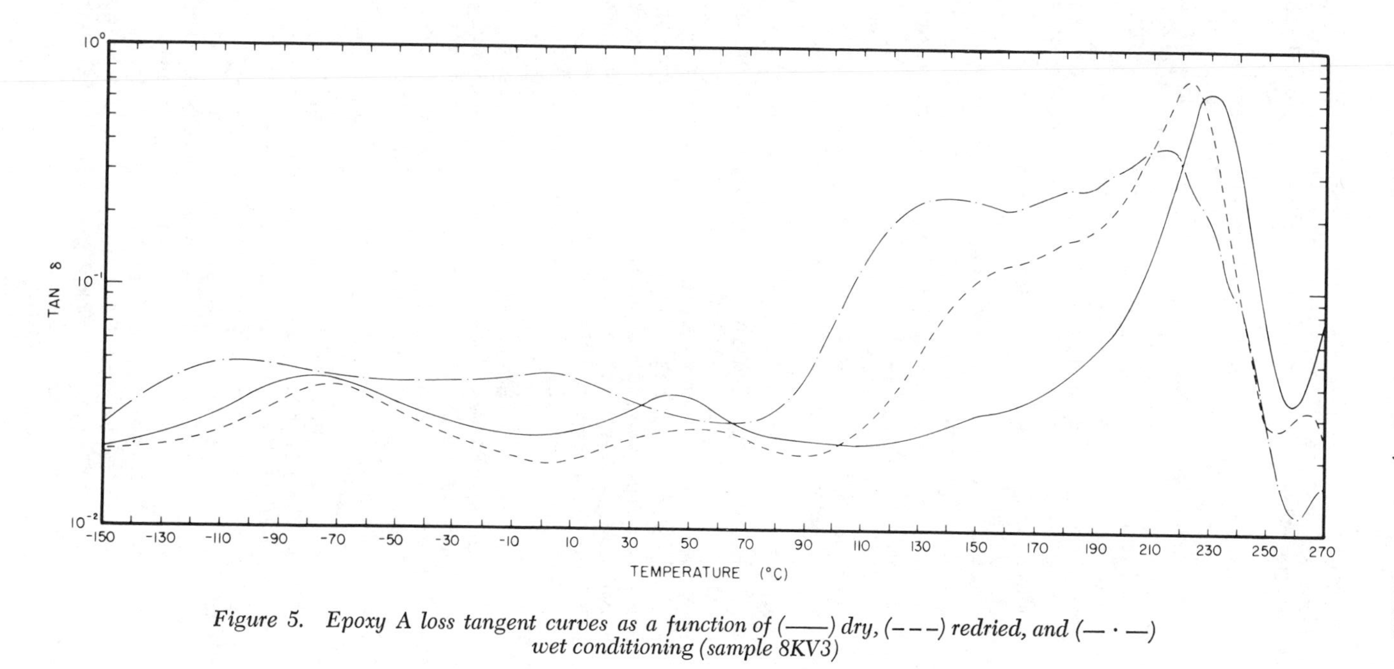

Figure 5. Epoxy A loss tangent curves as a function of (——) dry, (– – –) redried, and (— · —) wet conditioning (sample 8KV3)

Overall, moisture does not exhibit an effect such that compositional variations can be discriminated. Rather, it exhibits such an overpowering plasticization effect that potential composition-induced effects are masked.

Redried specimens were DM characterized to determine if moisture could induce permanent, irreversible damage in the material. Curves for 5-95°C tests were identical to the wet curves. However, after drying, an anomalous behavior was observed (Figures 4 and 5). Each variation investigated in a redried condition exhibited a characteristic inflection in their G' curves at about 130°C (Figure 4), irrespective of composition. Similar irregular Tan δ behavior was also observed (Figure 5) in the same temperature region. The mechanism producing this behavior has not been defined and will be the subject of future work.

Additional irreversible property losses were observed. Redried G' values at the reference 3×10^9 dynes/cm^2 were always lower than those of dry samples and Tan δ values were always shifted to lower temperatures (Table I).

In summary, the redried samples did not exhibit a response indicative of composition. Rather, the DM response discussed above was observed for all samples independent of composition.

Epoxy Resin B. Epoxy resin B contains two multifunctional epoxide resins. The principal resin is TGMDA, the minor epoxide is a glycidyl ether of a bisphenol A novolac, and the curing agent is DDS. Formulations representing the standard composition, an overcatalyzed composition, and an undercatalyzed composition (Table II) were characterized in the same manner as Epoxy A, with the most important parameters extracted from the storage modulus and loss tangent curves tabulated in Table II.

The moduli and loss tangent curves of this resin were similar to those of resin A as shown in Figures 6 and 7 for the standard composition. The curves for the standard material are representative of all three compositions. None of the compositions exhibited significant variations or trends compared to themselves as did Epoxy A compositions. Use of G'(R) values for differentiation was not possible because the temperatures at G'(R) exceeded the instrument's 300°C capability. Values shown in Table II are extrapolated estimates.

As with Epoxy A, moisture (5.3% weight gain) exhibited a significant plasticization effect, shifting G' at Tg onset 70-80°C to lower temperatures. G' in the glass transition region at a reference value of 3×10^9 dynes/cm^2 was shifted 30-34°C. This behavior is shown in Figure 6 for the standard resin. In the case of the loss tangent data, the onset temperature of the α-peak was shifted 80-95°C to lower values, and the temperature at the reference Tan δ value of 10^{-1} was shifted the same amount. This behavior is illustrated in Figure 7 for the standard resin.

As was the case with Epoxy A, compositional variations do not exhibit a discriminating moisture-induced response. Rather, the

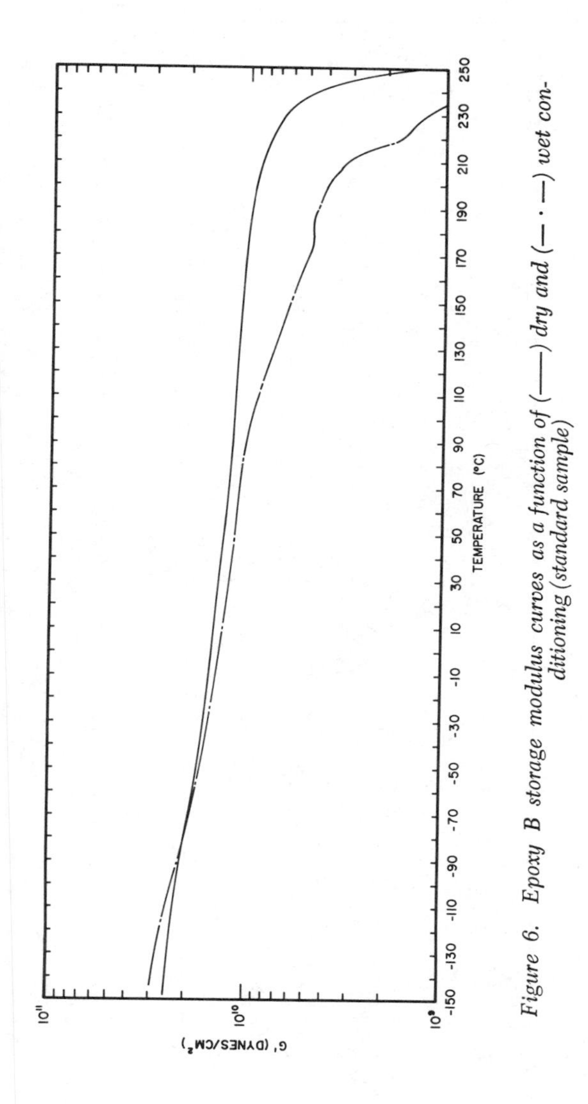

Figure 6. Epoxy B storage modulus curves as a function of (——) dry and (— · —) wet conditioning (standard sample)

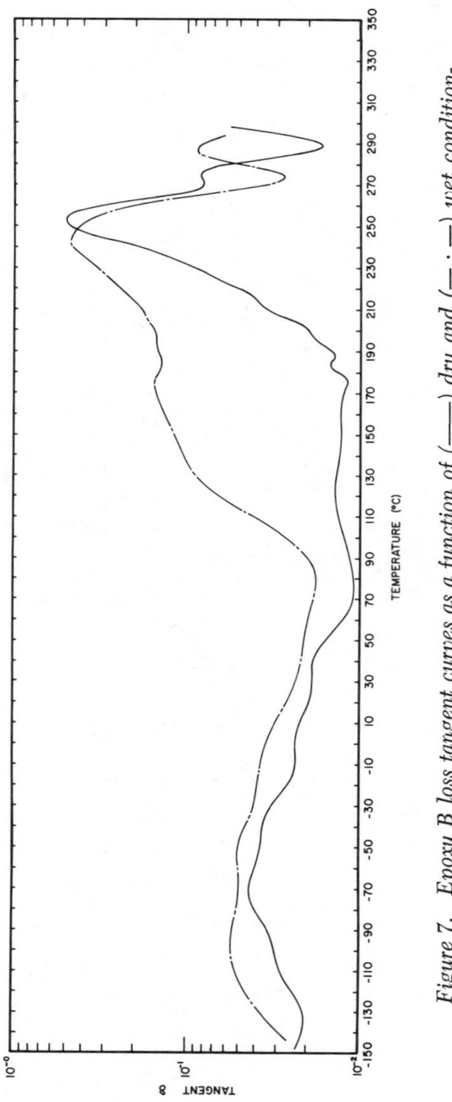

Figure 7. Epoxy B loss tangent curves as a function of (——) dry and (— · —) wet conditioning (standard sample)

plasticization effect is so overpowering that any potential composition-induced effects are masked.

A comparison of G' data for Epoxy B vs. Epoxy A shows that while the values of the onset temperature and corresponding moduli of both are similar (dry and wet), the temperatures at a reference G' value of 3 x 10^9 dynes/cm^2 show Epoxy B to have a much higher temperature capability than Epoxy A.

A comparison of loss tangent data shows that the width of the Tan δ peak at the reference value of 10^{-1} is smaller for Epoxy B vs. Epoxy A. In terms of the discussions presented under Epoxy A, this narrower peak width response reflects a smaller variety of available molecular motions, indicating that Epoxy B may have a more uniform, homogeneous network.

Epoxy Resin C. Epoxy resin C contains three epoxy resins - TGMDA (principal epoxide, Epoxy 1 in Table III), a cycloaliphatic diepoxide (a minor epoxide, Epoxy 2 in Table III), and an epoxy cresol novolac (a minor epoxide, Epoxy 3 in Table III). The curing agent is DDS and the catalyst is a boron trifluoride complex. Formulations representing the standard composition, a brittle resin composition, and a ductile resin composition were characterized in the same manner as Epoxy resins A and B. Resultant data are tabulated in Table III.

The shear moduli and loss tangent curves are similar to those of Epoxies A and B as illustrated in Figures 8 and 9 for the representative standard composition. In a dry condition, G' at the onset of the rubbery region, G'(R), showed the same discriminating behavior as with Epoxy A. The brittle formulation containing the low viscosity major epoxide gave a higher G'(R) value while the ductile formulation containing the higher viscosity major epoxide gave a lower G'(R) value (Table III). This behavior is in accordance with the discussions presented under Epoxy A.

Moisture (6.12% weight gain) exhibited a significant plasticization effect as with the other two resin systems. G' at the onset of Tg was shifted 40-50°C to lower values and G' at the reference value of 3 x 10^9 dynes/cm^2 was shifted 60°C to lower values (shown in Figure 8 for the standard composition). In the case of the loss tangent, the onset temperature of the α-peak was shifted 20-30°C to lower temperatures and the temperature at the reference Tan δ value of 10^{-1} was shifted 85°C to lower values (Figure 9). As was the case with Resins A and B, the plasticization effect of moisture is so strong that any composition-induced effects are masked.

The G' and Tan δ data for Epoxy C vs. Epoxies A and B show that this resin has the lowest temperature capability of the three resin systems. This would be expected because of the presence of a lower functionality epoxide in the formulation. In further comparing C to A and B, the α-peak width at the reference Tan δ value of 10^{-1} for resin C is larger than for A or B in the case of the standard compositions. Furthermore, the total α-peak

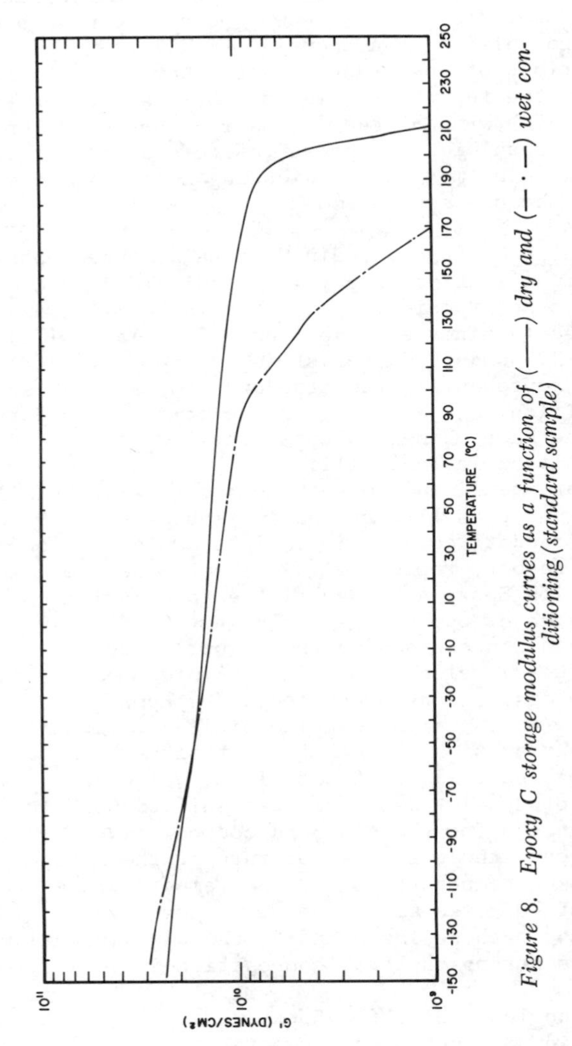

Figure 8. Epoxy C storage modulus curves as a function of (——) dry and (— · —) wet conditioning (standard sample)

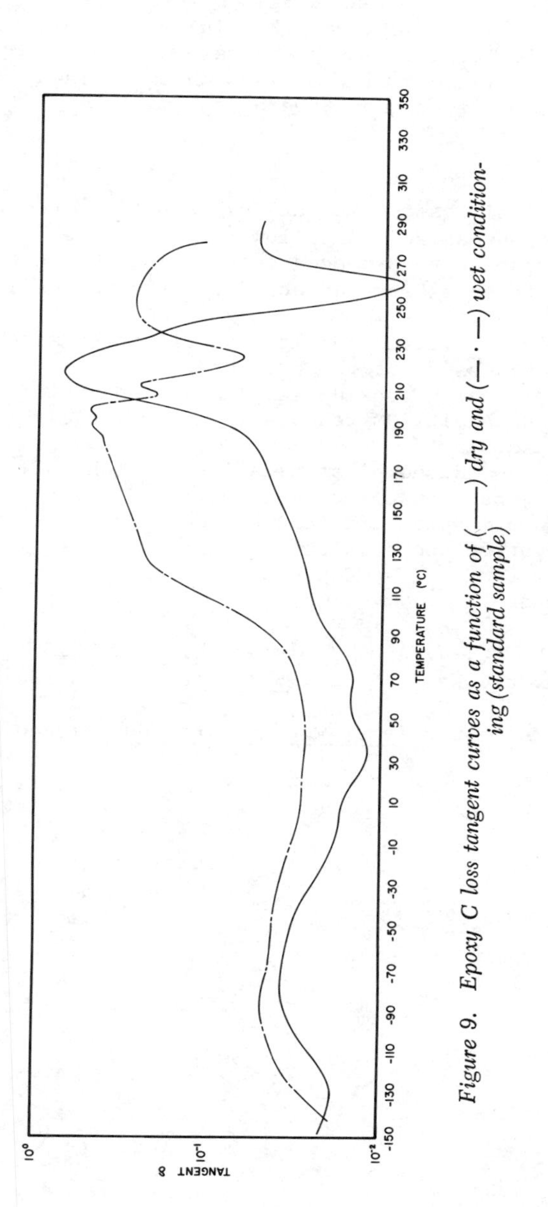

Figure 9. Epoxy C loss tangent curves as a function of (——) dry and (— · —) wet conditioning (standard sample)

(Figure 9) of resin C from onset temperature and above is much broader than the α-peaks of A and B. In keeping with previous discussions, this broader peak width response reflects a larger variety of available molecular motions which is in accordance with a more complex, multicomponent resin system.

Conclusions

The most significant observations resulting from this investigation can be summarized as follows:

(1) With respect to proposed engineering applications, no significant differences were detected for the compositional variations investigated.

(2) Dry, as-fabricated samples exhibit detectable differences related to the viscosity of the base epoxide.

(3) High viscosity epoxide samples are characterized by a lower cross-link density DM response, and conversely for low viscosity epoxide samples.

(4) Moisture-induced plasticization substantially alters the DM response, independent of composition.

(5) Moisture induces irreversible degradation of DM properties, independent of composition.

Literature Cited

1. Kaelble, D. H., J. Appl. Polymer Sci., 9, 1213 (1965).
2. May, C. A. and Weir, F. E., SPE Transactions, July, 211 (1962).
3. Browning, C. E., Polymer Engineering and Science, 18, 16 (1978).

RECEIVED January 17, 1980.

22

Anisotropic Measurements on Single-Ply Lamina Composites

WAYNE J. MIKOLS and JAMES C. SEFERIS

Department of Chemical Engineering, University of Washington, Seattle, WA 98195

Environmental and chemical effects significantly influence the physical properties of composite materials (1,2). Environmental effects cause property alterations due to various hygrothermal histories and states of the composite (3). Chemical effects manifest themselves through structural considerations. Such things as resin/catalyst stoichiometry, matrix impurities, fiber surface impurities, and other process related phenomena cause chemical alterations in the composite which affect the macroscopic and/or microscopic structural nature of the system (4,5).

Many investigations have focused upon the influence of either chemical or environmental effects upon various composite system properties. It should be apparent that these two effects are strongly coupled. Physical properties of composites are of course dependent upon the ultimate physio-chemical structure of the system. Consequently, this basic structure is an overriding consideration which is responsible for the nature and extent to which environmental effects can influence system properties. By examining basic anisotropic viscoelastic properties of a composite system, this study relates moisture absorption to property and structural considerations affecting the absorption process.

The particular graphite-epoxy composite system employed in this study was made with T-300 carbon fibers (Union Carbide) and 5208 epoxy (Narmco). Unidirectional continuous single-ply T-300/5208 laminates were used. Sample materials were exposed to either desiccated or soaked environments prior to testing.

In a composite system, properties can be viewed on a macroscopic or microscopic scale. Macroscopically, composites exhibit anisotropic behavior due to the nature in which carbon fibers are oriented in the matrix material. Dynamic mechanical experiments were performed on both soaked and desiccated samples. These measurements on the two sets of samples were used to study the macroscopic behavior of the composites at 0°, 15°, 30°, and 45° directions with respect to the fiber direction. Similar experi-

0-8412-0567-1/80/47-132-293$05.00/0

ments were carried out using sonic modulus measurements. Microscopically, the composite matrix material exhibits viscoelastic behavior that can strictly be assigned to the crosslink structure of the resin. Graphite reinforcement only serves to damp out viscoelastic transitions of the resultant composite. By combining dynamic mechanical data with moisture diffusion studies, a fundamental understanding of the nature and mechanisms involved in the moisture absorption (and property degradation) processes can be inferred.

Experimental

The composite studied in this work was made with a commercial epoxy (Narmco 5208) whose primary constituent is tetraglycidyl 4,4' diaminodiphenyl methane epoxy cured with 4,4' diaminodiphenyl sulfone (TGDDM-DDS). Continuous T-300 carbon fibers impregnated with epoxy provided a unidirectional prepreg tape having a 34.7% resin content and an aerial density of 95 g/m^2. Sheet lamina were prepared from a 30.48 cm wide role of prepreg tape by vacuum bag casting. Sheets of prepared prepreg were placed in a 120°C oven. The oven temperature underwent a linear ramp to 177°C over 30 minute interval. The sheets were held at this elevated temperature for 2 hours. The oven was permitted to return to room temperature. Cured samples were removed and allowed to set at ambient laboratory conditions for at least one month (101.3 kPa, 22°C, 60% Relative Humidity).

Several sets of samples were cut from the cured lamina sheets which had a thickness of 0.02 cm. Portions of the material were cut having nominal dimensions of 7cm x 14cm. These samples were subjected to subsequent desiccation over anhydrous calcium sulfate and/or soaking in deionized water. An ambient laboratory temperature of 22°C was maintained throughout this portion of the experiment. Typical data from these tests are presented in Figure 1. Swelling measurements were collected at desiccated, 60% relative humidity, and soaked sample environments. Average swell dimensions for the sample of Figure 1 are illustrated in Figure 2. Length and width swell dimensions correspond to respective fiber and transverse fiber composite directions. These measurements were obtained using calipers accurate to $\pm$0.001 cm (i.e. less than 0.01% of sample dimensions). Lamina thickness measurements were obtained using a dished plate micrometer accurate to $\pm$0.0002 cm (i.e. approximately 1% of sample dimensions).

Several strip samples having approximate dimensions of 0.4cm x (10-17cm) x 0.02 cm were also cut from the cured sheet material at various directions to the fiber direction. These angles were 0, 15, 30, and 45 degrees to the fiber direction. It was found that strips cut at angles greater than 45 degrees could not sustain any load for testing. Extreme care was taken to insure that no visible microcracks remained along sample edges

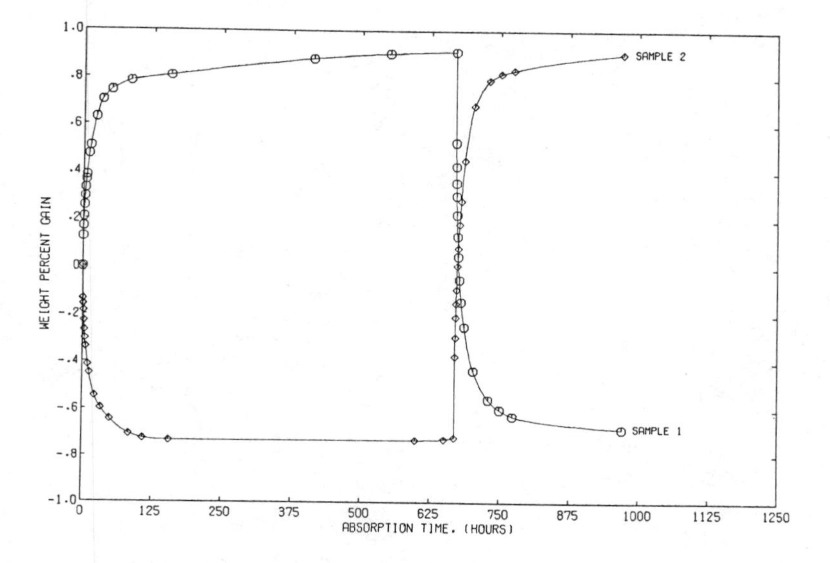

Figure 1. Soaking and desiccation of T-300/5208 composite sheets; percentage gain or loss as a function of time with respect to initial equilibrium at 60% relative humidity

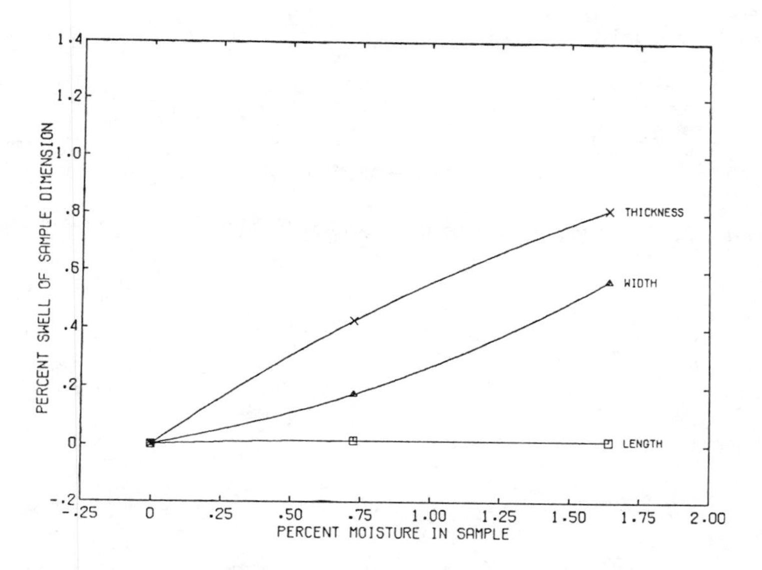

Figure 2. Percentage swell of linear sample dimensions as a function of moisture content

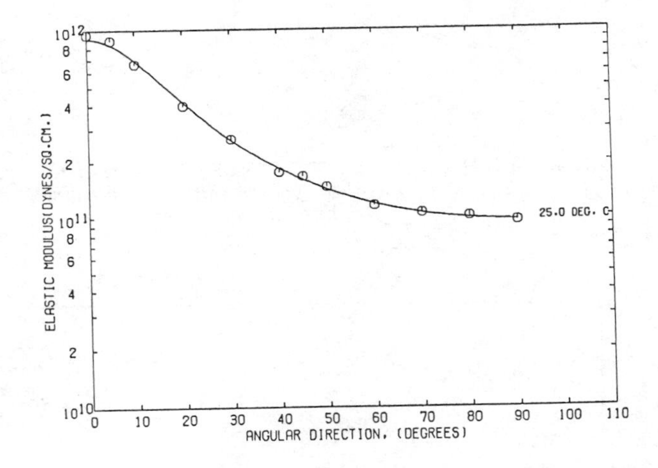

Figure 3. Sonic moduli at various angles, ϕ, to fiber direction for a composite sample

prior to desiccation or soaking. All cut sample edges were prepared by first sanding with 240, then 400 mesh sand paper. Final polishing of the edge was completed using crocus polish cloth.

Samples to be soaked were sealed in deionized water at 22°C and allowed to absorb at least 95% of their estimated theoretical maximum water uptake. This was determined from preliminary soaking experiments on sheet samples (ref. Fig. 1) in conjunction with information supplied by Carter (6). Using a similar procedure, desiccated samples were estimated to have less than 0.05% by weight water at the time they were used in dynamic mechanical and sonic modulus experiments. Sonic modulus experiments were conducted on the 7cm x 14cm sheet samples as well as the 0.4cm wide strip samples using a commercially available apparatus (PPM-5, H. M. Morgan Inc., Norwood, Mass.). The sonic modulus tester, which employs a piezoelectric transducer to measure a 10 kHz sonic pulse, was used to evaluate desiccated, 60% relative humidity, and soaked samples. Typical anisotropic sonic modulus data for a 60% relative humidity composite samples are illustrated in Figure 3. Samples which were removed from their respective environments for sonic test (usually lasting less than 0.5 hours) were reinserted in their environmental chambers for at least two additional weeks prior to dynamic mechanical testing.

The dynamic mechanical properties of all strip samples described above were obtained for sample lengths of at least 6.5cm. The temperature dependence of the dynamic mechanical properties were obtained in tension with the Rheovibron DDVII which was modified to prevent sample slippage and ambient moisture condensation on the sample at low temperatures (7, 8). All experiments were performed at a frequency of 11 Hz and over the temperature range -160°C to 320°C with a heating rate of one degree (°C) per minute. The samples in the Rheovibron were maintained in an atmosphere of dry nitrogen. The data obtained were recorded and plotted in the form of the traditional tan δ, and dynamic moduli (E' and E'').

Initially some concern was expressed over the moisture removal from the samples during the sample mounting and data collection periods. Generally, sonic modulus data was collected within minutes after removal from the sample's environment. Bulk water lost or gained by the sample during this period of time represents a negligible change in the bulk modulus. However, the sample mounting and calibration operations for dynamic mechanical experiments generally caused about a two-hour delay between sample environment removal and initial data point collection at -160°C. Even longer periods existed before dynamic mechanical property data were collected at elevated temperatures. Furthermore, experimental conditions exposed the samples to a dry nitrogen atmosphere for a near 8-10 hour data collection period. This experimental procedure introduced negligible environmental changes for desiccated composite samples. On the other hand,

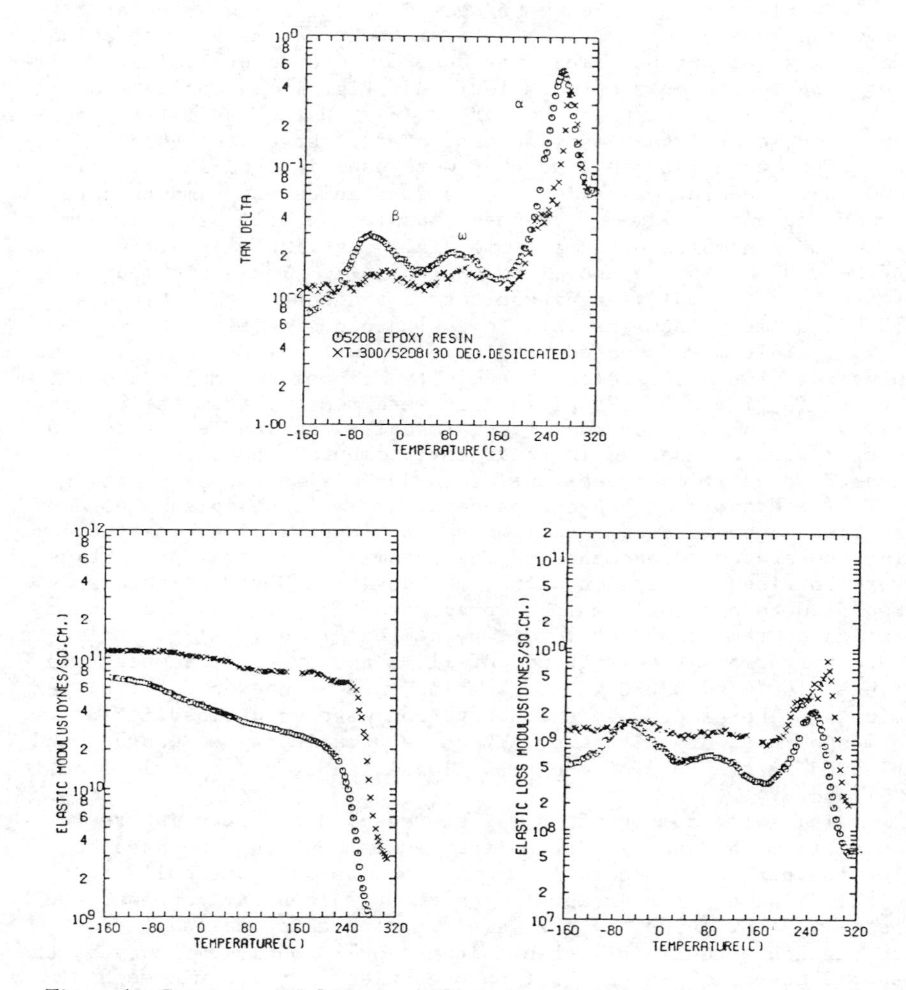

Figure 4. Comparison of dynamic mechanical properties of a 30° composite and neat epoxy resin showing β, ω, and α transitions

water did escape from soaked samples during this period. Our experimental results suggest that not all of the water affecting composite properties can ever be rapidly removed until high temperatures are reached. This consequence provides new insight into the nature of matrix plasticization (and/or degradation) as well as moisture absorption kinetics.

Discussion

Viscoelastic Behavior. Typical dynamic mechanical spectra for a carbon fiber composite are shown next to a similar curve for neat resin in Figure 4. The fibers in the composite samples are oriented at 30° with respect to the direction of 11 Hz dynamic mechanical oscillations. Three major transitions are evident. These correspond to the familiar β, ω, and α transitions of the neat 5208 epoxy resin (4).

The low temperature β transition is centered near -50°C. It is primarily attributed to minor localized motion in the polymeric matrix network. The ω transition is linked to unreacted molecular segments and/or inhomogeneities arising from dissimilar crosslink densities. This transition occurs in the 100°C temperature vicinity. Measurements on 15°, 30°, and 45° soaked and desiccated samples did exhibit variations in the vicinity of the β and ω temperature transitions. However, such measurements were made near the sensitivity limits of the instrumentation. High sample rigidity contributed to data scatter. General trends with respect to these two transitions were noted but need further investigation before they can be fully discussed.

The high temperature α peak in Figure 4 is clearly attributed to the glass transition of the matrix material. The peak maximum was observed near 275°C. All of the other noted transitions have been previously observed in our studies of neat resin viscoelastic properties (4). However, the magnitudes of these transitions are substantially reduced as a result of carbon fiber structural reinforcement. Figure 4 demonstrates how the low temperature transitions are masked by the presence of fiber reinforcement. Previous work demonstrates that the low temperature ω transition is affected by absorbed moisture in the neat resin (4). The fact that no discernible difference in the ω (and β) transitions, other than a significant reduction in magnitude, is observed for composites over neat resins suggests that the presence of fibers do not alter the network structure of the resin. As might be expected, a composite takes on an increasingly more matrix-like response as load is applied in transverse fiber directions. Figure 5 illustrates this by plotting the dynamic elastic modulus, E', at four separate angular directions of single ply T-300/5208 composite, as well as for the neat epoxy resin.

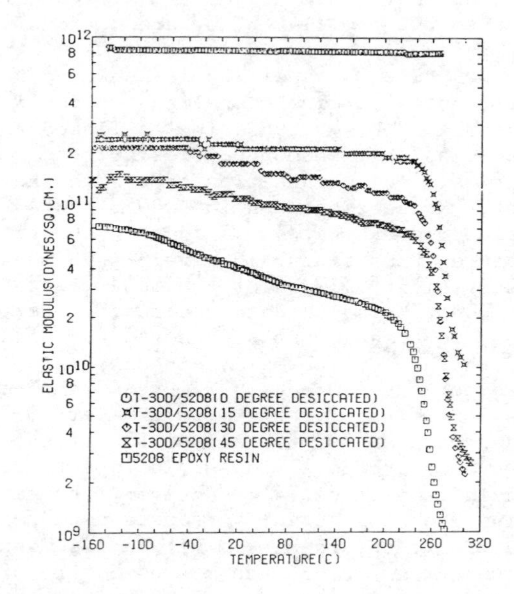

Figure 5. Dynamic moduli, E′, of the composite at various angles, φ, to fiber direction. Neat epoxy data is also plotted for comparison.

Water Adsorption. One interesting comparison of dynamic mechanical properties is between soaked and desiccated samples. Figure 6 illustrates one such comparison for a 30° sample. The 30° sample has been selected here for illustration. Similar plots for other angles demonstrate how viscoelastic properties for soaked samples return to those of desiccated samples in the vicinity of 240-245°C. This behavior is not peculiar to composite materials. Similar conclusions were obtained by Keenan and Seferis (4) on soaked and desiccated neat resin samples. These temperatures are well above 100°C temperature at which one might expect all water to leave the composite material. This is particularly unusual behavior for single-ply laminates with an extremely thin cross section and a high percentage of exposed cross-sectional edge. Experimental design required the thin soaked composite samples to remain in a dry nitrogen atmosphere for at least 9 hours prior to measurement of the dynamic mechanical properties at 240°C. Figure 1 suggests that at best, these circumstances would leave only 40% of the water originally contained in the sample. (60% would be lost to the dry nitrogen atmosphere during the preceding 9 hours required to reach 240°C). Probably the actual percentage is significantly less than this since actual diffusion at elevated temperatures would be accelerated well above the 22°C case employed in Figure 1.

Available literature data suggest that water is retained in the 5208 matrix due to hydrogen bond formation. This formation is proposed by Carter and Kibler (6) to explain some of the anomalous moisture behavior of the epoxy systems (9, 10). They propose that diffusion of water in the 5208 resin matrix is governed by two diffusing states of water; one mobile state and one strongly bound state. Preliminary analysis (11) of the relative magnitudes of the β and ω transitions, as they are influenced by moisture uptake, lends support to Carter's model (6).

Sonic Data. Sonic modulus data does not show any marked difference between sheet and strip or between soaked and desiccated composite samples. Keenan (12) demonstrated that both sheet and strip samples give comparable moduli when care is taken to isolate discontinuities which occur with data taken near the edge of sheet composite samples or with close probe spacings. The fact that negligible differences are observed between soaked and desiccated material sonic moduli is not surprising in view of the fact that low temperature dynamic mechanical data (which are equivalent to high temperature sonic modulus data) demonstrated near similar behavior. This result concurs with the work of Kaelble and Dynes (13).

Conclusions

The presence of carbon fiber reinforcement in a 5208 epoxy matrix tends to override the ω and β transitions normally ob-

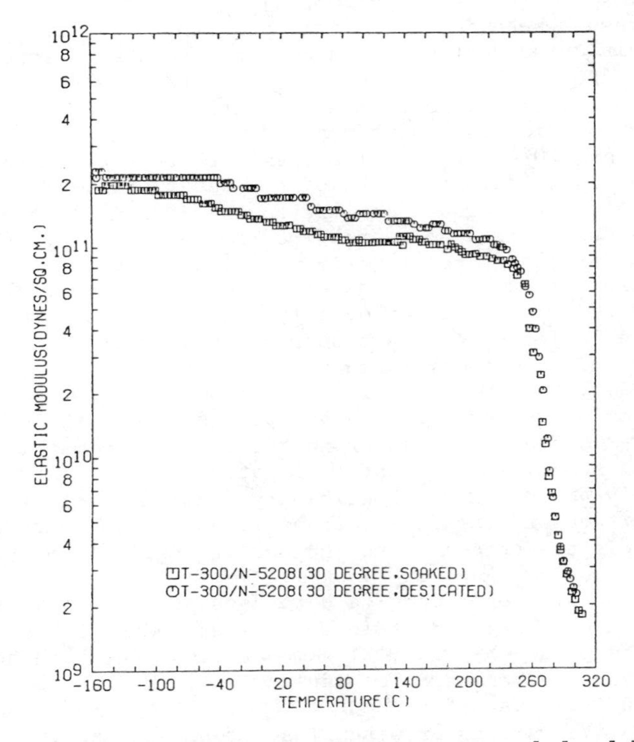

Figure 6. Comparison of dynamic moduli, E′, for a 30° soaked and desiccated composite sample

served in the neat resin material. The α transition however, which is associated with the glass transition of the matrix material, prevails strongly inspite of fiber reinforcement.

In general, moisture serves to plasticize the composite material. Data suggest that moisture diffusion in the composite matrix can be divided into at least two distinct groups: mobile and bound species. Bound species remain relatively fixed in the matrix at low temperatures for long periods of time due to steric considerations. In general, this bound water constitutes a small fraction of the total absorbed water. This bound water is rapidly released from the matrix material by heating the sample to the initial stages of its glass transition.

Acknowledgments

The authors express their appreciation to the Boeing Commercial Airplane Company for providing financial assistance for this work. We also thank the late Mr. Lonnie Hilliard for preparing the composite material.

Abstract

It is well known that the physical properties of composite materials are highly anisotropic. Consequently, factors that affect the properties of composite materials will manifest themselves differently depending on the direction of experimentation. In this work the influence and nature of moisture absorption on the anisotropic dynamic mechanical and sonic modulus properties for single ply lamina composites has been examined. These experiments provide further elucidation of the interaction between fiber and matrix components of the composite.

Literature Cited

1. Hung, C. and Springer, G. S., J. Composite Materials, 11, 2, Jan., (1977).
2. Morgan, R. J. and O'Neal, J. E., Polym.-Plast. Technol. Eng., 10 (1), 49, (1978).
3. Adams, D. F. and Miller, A. K., J. Composite Materials, 11, 285, July, (1977).
4. Keenan, J. D., Seferis, J. C., and Quinlivan, J. T., in press, J. Appl. Polym. Sci.
5. Fleming, G. J. and Rose, T., AD-770-407, Nov., (1973).
6. Carter, H. G. and Kibler, K. G., J. Composite Materials, 12, 118, April, (1978).
7. Seferis, J. C., McCullough, R. L., and Samuels, R. J., Appl. Polym. Symp., 27, 205, (1975).
8. Wedgewood, A. R. and Seferis, J. C., in press, J. Polym. Sci.
9. Shirrell, C. D., ASTM Symposium, Dayton, Ohio, Pg. 21, Sept. 29-30, (1977).

10. Whitney, J. M. and Browning, C. E., Advanced Composite Materials Environmental Effects, Vinson, J. R., Ed., ASTM STP 658, p. 43, (1977).
11. Mikols, W. J. and Seferis, J. C., in preparation.
12. Keenan, J. D., Masters Thesis, Department of Chemical Engineering, University of Washington, Seattle, WA, 98195, (1979).
13. Kaelble, D. H. and Dynes, P. J., Materials Evaluation, p. 103, April, (1977).

RECEIVED October 19, 1980.

23

Acoustic Fatigue Strength of Fiber-Reinforced Plastic Panels

T. FUJII and T. FUKUDA

Department of Mechanical Engineering, Osaka City University, Sumiyoshi, Osaka 558, Japan

S. IIDA and M. SANO

National Aerospace Laboratory, Chofu, Tokyo 182, Japan

Composite materials are used for many aircraft parts, including structural members which allow for the possible development of an all composite aircraft (1). The increased power of aircraft engines over the past three decades has resulted in a major social and scientific problem - aeronautical noise. Acoustic fatigue of the aircraft structure occurs due to the merciless hammering of the fluctuating sound pressures (2). In Japan fatigue studies of composite materials is limited, although a program is in progress at the University of Southhampton, ISVR, which is concerned with the acoustic fatigue of carbon fiber reinforced panel-type structures (3). Consequently, an immediate need exists to develop fundamental data based upon acoustic fatigue tests for composite materials.

Four kinds of FRP panels with three layers consisting of roving glass cloth and/or glass-fiber mat reinforced unsaturated polyester resin are acoustically excited and fatigue test data obtained. An observation was made from the tests that the glass-fiber mat reinforced test panel may be more isotropic in the failure pattern than the roving glass cloth reinforced test panels. The experimental results indicate that the acoustic fatigue strength of FRP test panels is considerably lower than the fatigue strength of FRP materials tested under constant stress. Making certain assumptions the acoustic fatigue life of test panels can be predicted from the experimental data of fatigue strength of FRP obtained under constant stress testing. Measuring any increase of acoustic fatigue strength required bonding CFRP on the middle part of a panel or laminating carbon fiber woven tapes simultaneously during panel fabrication. The latter stiffening method proved more effective.

Acoustic Fatigue Tests

Test Panel. As shown in Figure 1, the shape of each test panel is rectangular and identified with one bay of the aircraft body. Table I shows properties of test panels which are laminated

0-8412-0567-1/80/47-132-305$05.00/0

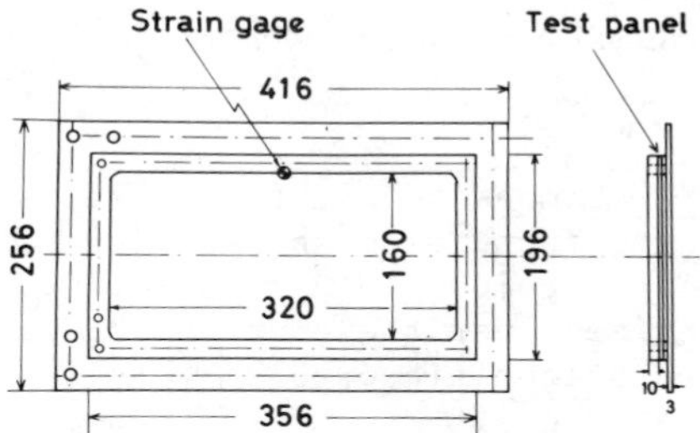

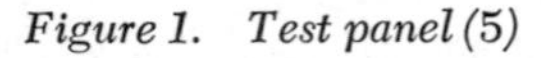
Figure 1. Test panel (5)

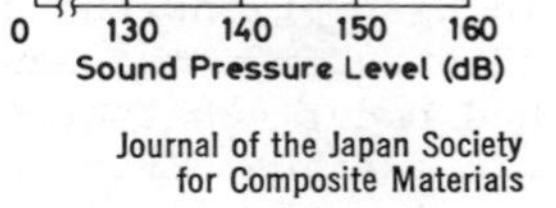
Journal of the Japan Society for Composite Materials

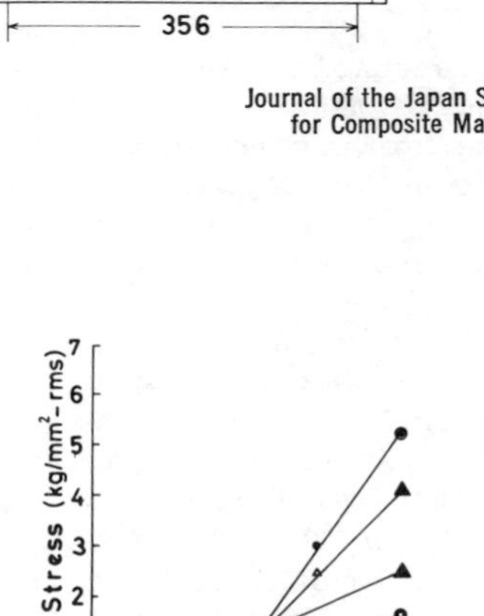

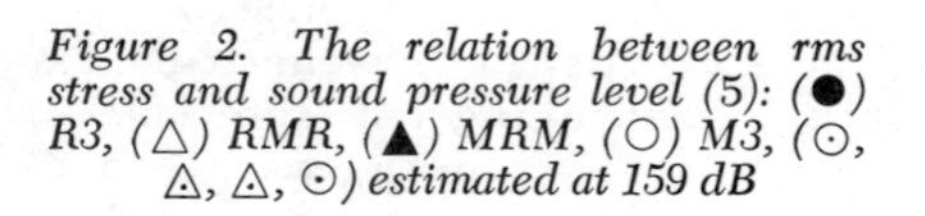
Figure 2. The relation between rms stress and sound pressure level (5): (●) R3, (△) RMR, (▲) MRM, (○) M3, (⊙, △, △, ⊙) estimated at 159 dB

Journal of the Japan Society for Composite Materials

by hand lay-up method and cured for 24 hours at 60°C. The four kinds of test panels are R3, RMR, MRM, and M3 where R is the glass roving cloth, M the glass-fiber mat, 3 the number of layers, and specific identification represents the sequence of layers. In order to monitor and measure strain response, a strain gage is mounted as in Figure 1 where maximum stress will occur in the test panel.

Table I

Test panel	t	E	W_f	f_n
R3	1.0	3240	46.5	206
RMR	1.0	3050	55.3	206
MRM	1.2	2480	52.5	217
M3	1.1	1600	57.6	165

R; roving glass cloth E; tensile modulus (kg/mm^2)
M; glass-fiber mat
t; thickness (mm) W_f; glass content by weight
f_n; fundamental natural freq. (Hz)

Journal of the Japan Society for Composite Materials

Relation Between rms Stress and Sound Pressure Level. The relationship between load and stress (or strain) in the acoustic fatigue test is different from that observed in the usual fatigue test under constant stress. The acoustic load is a random noise whose intensity is expressed by a sound pressure level (SPL, unit: dB) and measured by the rms volt-meter, while the spectrum of strain response of the test panel shows multiple resonances characterized by vibration mode of the panel. The resonance whose intensity peaks in the spectrum significantly contributes to the fatigue damage of the panel.

Figure 2 shows the relationship between rms value of stress and the sound pressure level for each test panel. It should be noted that rms stress at SPL of 159dB is estimated by extrapolating linearity between load and stress since the strain gage is damaged at SPL values of more than 148dB.

Results of Test and Discussions

Failure Pattern. Figures 3-6 depict the failure patterns observed on different panels when exposed to strong acoustic noise. Figure 3 shows the failure of an unsaturated polyester resin 3mm in thickness. Failure of the isotropic panel occurs after a few minutes exposure; the crack starting at point A or B and propagat-

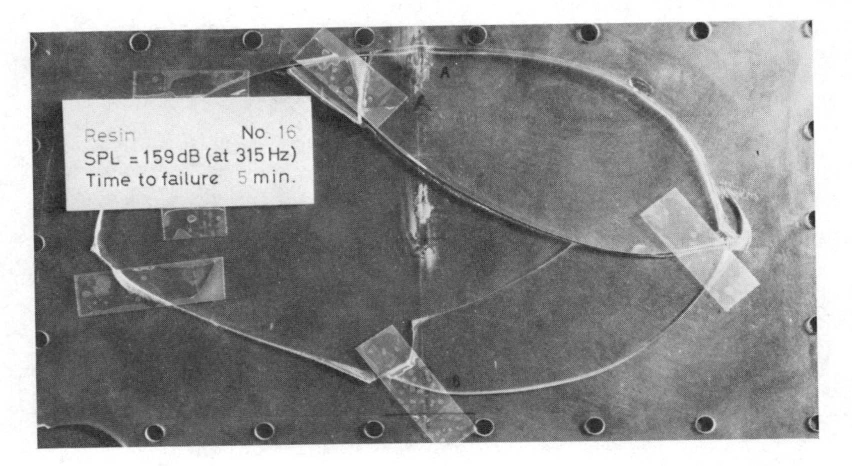

Journal of the Japan Society for Composite Materials

Figure 3. Failure pattern for unsaturated polyester resin (5)

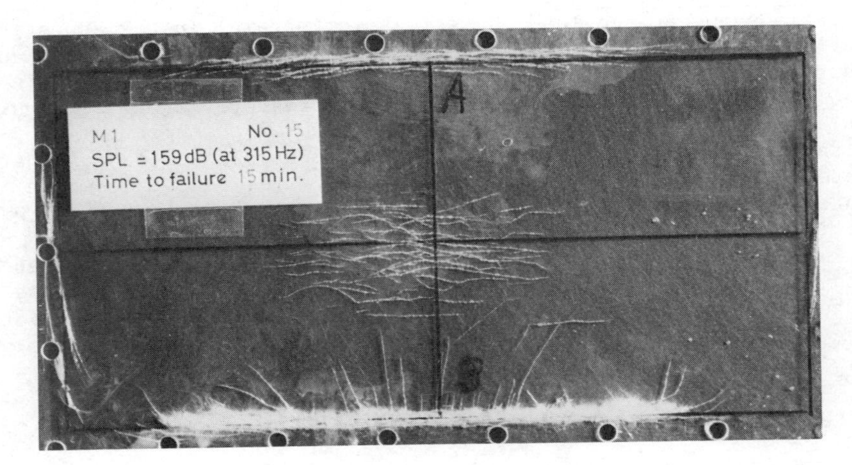

Figure 4. Failure pattern for test panel with a glass-fiber mat layer

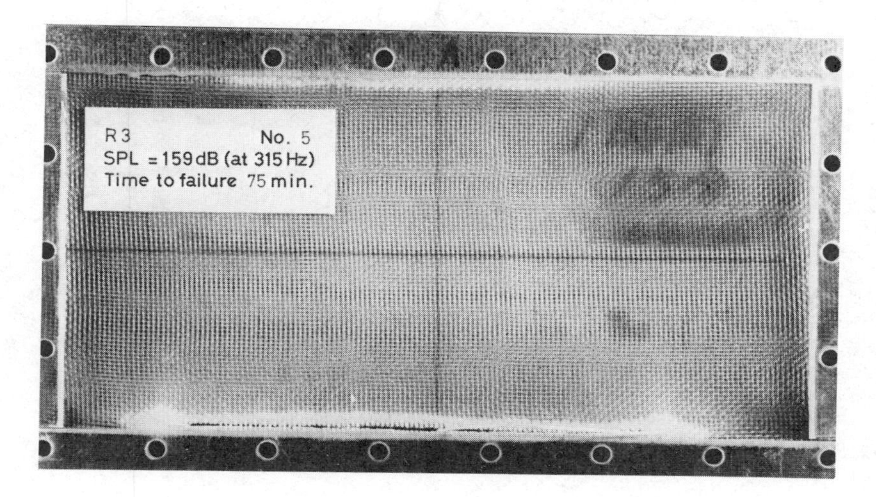

Journal of the Japan Society for Composite Materials

Figure 5. Failure pattern for R3 test panel (5)

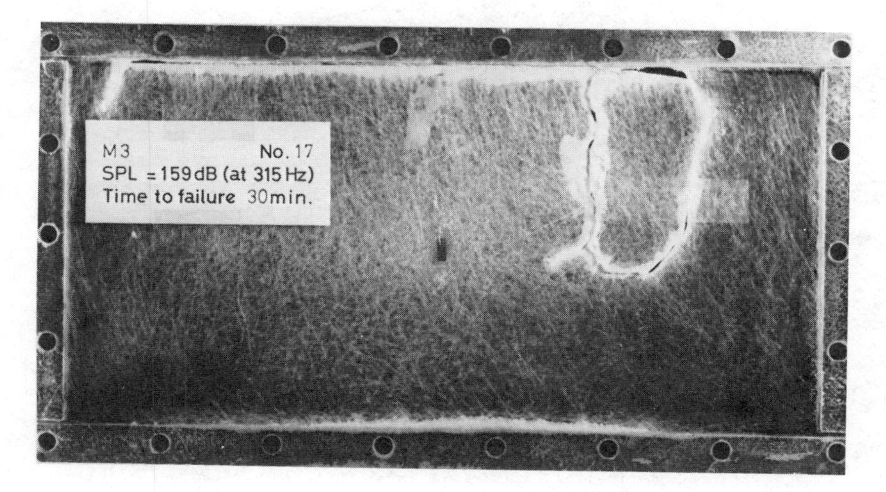

Journal of the Japan Society for Composite Materials

Figure 6. Failure pattern for M3 test panel (5)

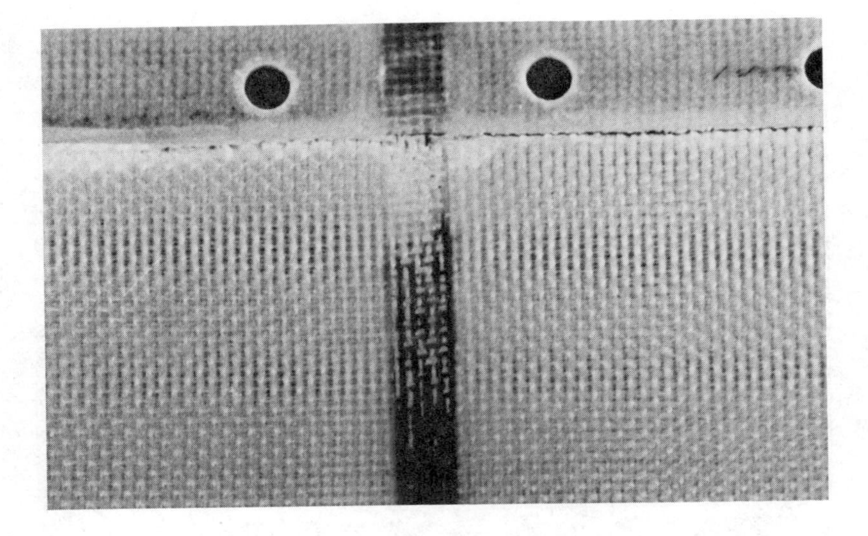

Figure 7. Failure pattern for RC-B test panel

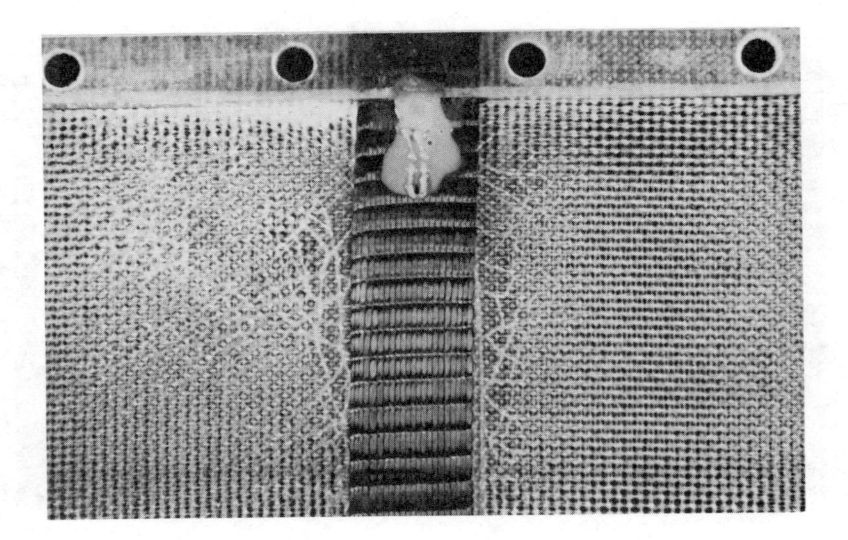

Figure 8. Failure pattern for RC-L test panel

ing along an elliptical locus. Figure 4 shows the effect of adding one layer of glass-fiber mat as reinforcement. The effect of the fibers reinforcing the matrix is evident from how and where the cracks develop.

Figures 5 and 6 show the failure patterns of R3 and M3 test panels. The results indicate that glass-fiber mat reinforced panels are more isotropic than glass cloth roving reinforced ones. The differences in failure patterns is probably due to differences in reinforcing mechanisms between glass cloth roving and glass-fiber mat. Additionally, the M3 test panel is prone to catastrophic failure compared to the R3 test panel. Although photographs for RMR and MRM test panels are omitted, their failure patterns are almost intermediate between R3 and M3.

The investigation on acoustic failure included methods to increase the acoustic fatigue strength of FRP. A simpler stiffener was made by bonding CFRP on the middle of a panel (RC-B) or laminating fiber woven tapes during the molding process (RC-L). Their respective failure patterns are shown in Figures 7 and 8. The best stiffening method was in RC-L where delamination could not take place while the stiffener for RC-B debonded when exposed to acoustic noise.

Acoustic Fatigue Strength. Test results for acoustic fatigue strength of each panel are plotted in Figure 9 along with fatigue strength under constant stress of specimens having similar static strength and glass fiber contents. In Figure 9 the left side ordinate is in rms stress while the right side is in rms peak stress. By multiplying test data by the $\sqrt{2}$ to achieve rms peak stress, the acoustic fatigue strength of R3 and M3 can be compared on the same chart as their fatigue strength under constant stress. Cycles to failure plotted on the abscissa are calculated so that they will be approximately equal to the values of fundamental natural frequency multiplied by failure time of each test panel.

It is found that the acoustic fatigue strength for R3 and M3 is considerably lower than the fatigue strength obtained from constant stress testing. It is considered that this fact is caused by multiple resonant phenomenon of the panel at very high cyclic frequencies where the fatigue strength of FRP, in general, reduces due to the generation of heat (4).

In order to estimate the acoustic fatigue life of test panels, it is assumed that the acoustic load used is a random Gaussian noise. Peak values of stresses taking place in panels are supposed to be Rayleigh distributed having the probability density function

$$f(\sigma) = (2\sigma/\sigma_{rms}^2)e^{-\sigma^2/\sigma_{rms}^2} \qquad (1)$$

The second assumption is concerned with the fundamental S-N curve obtained from the fatigue test with constant stress. In this case, we can get

$$N\sigma^{m} = C \quad , \tag{2}$$

where C and m are material constants.

If the modified Miner's rule is assumed to be applicable to FRP materials, the cumulative damage during fatigue of FRP is given by

$$\begin{aligned}\Sigma(n/N) &= (N_f/C)\int_0^{\infty} \sigma^{m} f(\sigma)d\sigma \\ &= (N_f/C)\sigma_{rms}^{m} D(m) \quad ,\end{aligned} \tag{3}$$

where N_f means the number of cycles to failure of FRP panels under acoustic loadings and σ_{rms} is the rms peak stress in Eq.(1). And D(m) is called the damage function in the form of the following integral

$$D(m) = \int_0^{\infty} 2\sigma^{m+1} e^{-\sigma^2} d\sigma \equiv \Gamma(1+m/2) \quad , \tag{4}$$

where Γ is the Gamma function.

Thus, using Eq.(3), we can predict the acoustic fatigue strength of FRP from the experimental data of fatigue strength of FRP obtained from constant stress testing. Computational results are given in Figure 9 which have the good agreement with their experimental data.

Improvement of Acoustic Fatigue Strength

Acoustic fatigue strength for glass cloth roving reinforced FRP, whose inherent strength is the highest among tested panels, is improved by making the following type laminate. RC-B panel is made by bonding CFRP as a sandwich on both sides of an R3 panel. RC-L panel is laminated so that a carbon fiber woven tape is centered between two layers of glass cloth roving. Table II lists data on RC-B and RC-L test panels including width and thickness dimensions for the stiffness. Figure 10 shows the increase of acoustic fatigue due to stiffening effects. It should be noted that the slopes of plotted data for RC-B and RC-L are greater than the slope of the S-N curve for R3 under constant stress.

Table II

	Thickness (mm)	Width (mm)	Tensile modulus (kg/mm)
RC-B	0.23 x 2	10.0	6000
RC-L	0.5	20.0	8700

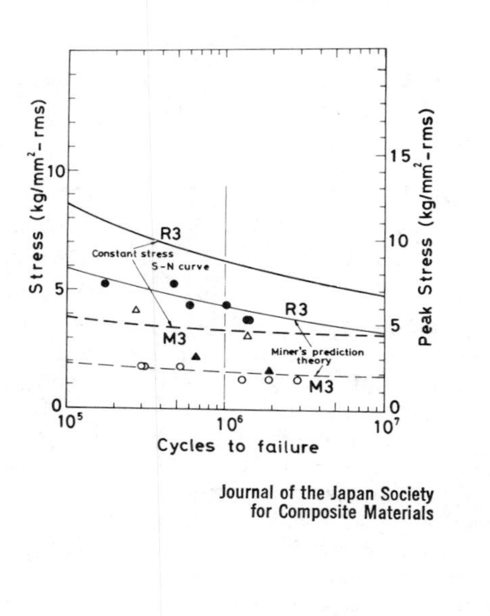

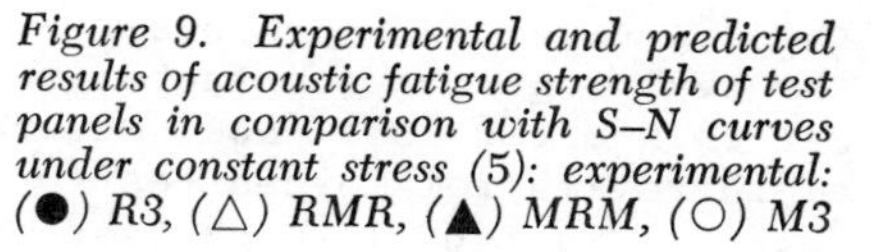

Figure 9. Experimental and predicted results of acoustic fatigue strength of test panels in comparison with S–N curves under constant stress (5): experimental: (●) R3, (△) RMR, (▲) MRM, (○) M3

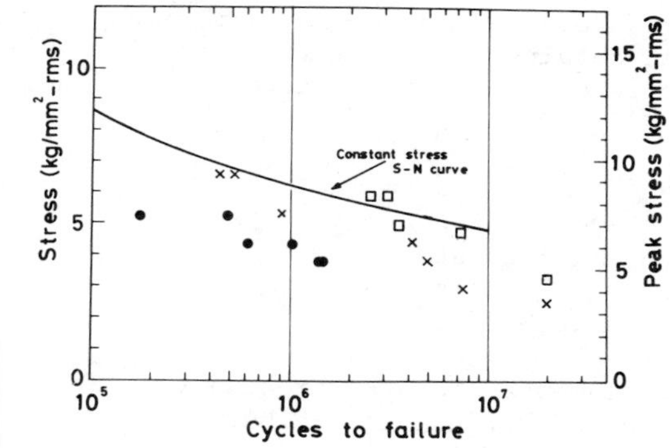

Figure 10. Increase of acoustic fatigue strength by CFRP stiffeners: (○) R3, (×) RC-B, (□) RC-L

Conclusions

- Acoustic fatigue strength of FRP panels is considerably lower than their fatigue strength obtained from constant stress testing.
- Making a few assumptions acoustic fatigue strength can be predicted from experimental fatigue strength data with reasonably good agreement.
- The better method for panel stiffening is lamination of carbon fiber woven tapes.

Abstract

Four kinds of FRP panels with three layers consisting of glass cloth roving and/or glass mat reinforced unsaturated polyester resin were acoustically excited and fatigue tested. From the experiments, it is observed that the laminated structure of the test panel may characterize the failure pattern and the glass mat reinforced test panel may be more isotropic than the glass cloth roving reinforced test panel. In order to predict the acoustic fatigue life of test panels, it is assumed that the acoustic load is a random noise with narrow frequency band-width and the modified Miner's rule is applicable to FRP materials. Using these assumptions, the acoustic fatigue life of test panels is estimated with good agreement from experimental data of FRP fatigue strength obtained under constant stress. Simple stiffening methods are investigated to achieve improved acoustic fatigue strength.

List of Symbols

C, m	Material Constants
D( )	Damage Function
E	Tensile Modulus
f( )	Probability Density Function
f_n	Fundamental Natural Frequency
M	Chopped Strand Glass-Fiber Mat
N, N_f	Cycles to Failure
R	Roving Glass Cloth
t	Thickness of Panel
W_f	Glass Content by Weight
$\Gamma( )$	Gamma Function
σ	Stress Amplitude
σ_{rms}	rms Peak Stress
$\Sigma(n/N)$	Cumulative Damage

Acknowledgments

The authors thank Mr. K. Itami for his help in preparations of test panels and experiments. They are also indebted to Nippon Glass Fiber Co. for glass-fibers, Takeda Chemical Industries Ltd. for resins and Toray Industries Inc. for CFRP.

Literature Cited

1. Fujii, T. J., Society of Materials Science, Japan, 1976, 25, (277).

2. Richards, E. J., Mead, D. J., Ed. "Noise and Acoustic Fatigue in Aeronautica"; John Wiley & Sons Ltd., London, 1968, p. vi.

3. White, R. G., Memo of ISVR, Univ. of Southampton, 1974, 552.

4. Fujii, T., Fukuda, T., Yoshida, H., Preprint of the 7th FRP Symp., Japan, 1978, 59-62.

5. Fujii, T., Fukuda, T., Iida, S., Sano, M. J., Japan Society for Composite Materials, 1978- 4, (4).

RECEIVED February 15, 1980.

24

A New Polyester Matrix Resin System for Carbon Fibers

ROBERT EDELMAN and PAUL E. McMAHON

Celanese Research Company, 86 Morris Avenue, Summit, NJ 07901

Epoxy resins are the primary matrix materials used in carbon fiber composites today. Prepreg systems made with these resins are used in recreation and aerospace applications. The epoxies were chosen for their good handling characteristics as well as their good ambient and elevated temperature composite properties. However, epoxy resin systems have certain deficiencies in that the prepreg materials must be kept refrigerated resulting in limited out time. Cost is also higher than other available thermoset materials and cure time is often long.

In the aerospace industry, these deficiencies have been tolerated when high levels of performance that could be achieved only with epoxy materials, were desired for certain applications. However, as additional applications for composites are considered, where lower property levels are acceptable, it is likely that other types of prepreg could be used in place of the epoxy systems currently available, particularly the "250°F" curing epoxy systems. Polyester resin prepregs should be considered as a potential alternative to these epoxy materials, since they do not have their negative features.

Most polyester systems that are used today contain glass reinforcement and use styrene as the reactive monomer. Styrene is the most frequently used monomer since it provides good ambient temperature properties at low cost. Elevated temperature properties (100°C and above) are frequently poor. Styrene does have a significant drawback in that it is quite volatile at room temperature leading to environmental control problems. In addition, styrene monomer undergoes very high shrinkage and exotherming on cure necessitating the use of extensive filler addition to minimize warpage and cracking. (1,2) This use of fillers frequently results in a distinct lowering of the properties of the system.

It is obvious then that a significantly different polyester resin system must be considered as a possible replacement for an epoxy material. There are commercially available materials that appear to meet the necessary requirements. A formulation that contains diallyl phthalate monomer and a high performance unsat-

0-8412-0567-1/80/47-132-317$05.00/0

TABLE I

PROPERTIES OF DIALLYL ORTHOPHTHALATE (DAP)

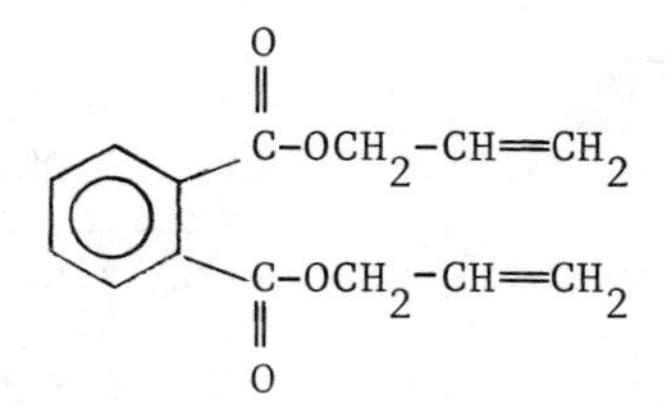

VAPOR PRESSURE AT ROOM TEMPERATURE	LOW (0.1 Torr at 90°C)
VOLUME SHRINKAGE	11.8% (Styrene = 17%)
GELATION TIME	Very long at temperatures up to 82°C
	Rapid at temperatures greater than 149°C (1-2 minutes at 177°C when catalyzed)
EXOTHERM DURING CURE	LOW (93°C)
	Comparable styrene system exotherms to 204°C
THERMAL RESISTANCE	Very good at temperatures up to 121°C
MOISTURE RESISTANCE	Very good at temperatures up to 95°C

urated polyester produces an unexpectedly high level of composite mechanical properties. Diallyl phthalate is a commonly available product that is used in critical electrical/electronic applications requiring high reliability under long-term adverse environmental conditions. (3) The reasons why diallyl phthalate was chosen can be seen in Table I which shows the excellent balance of physical and chemical properties of the material. The vapor pressure of the monomer is low contributing to reduced environmental hazards. Shrinkage of the material is the lowest of the commonly available monomers. This tends to result in lower warpage in the final part. Gel time behavior is ideal for a prepreg resin. At temperatures up to 82°C virtually no cure will occur in the catalyzed material. This important property removes the need for refrigeration which is a necessary feature with the use of epoxy prepregs. This particular mode of behavior is related to the activity level of the stable allyl radical. At lower temperatures, despite the availability of free radicals from an initiator, the reaction rate of the allylic species is simply too sluggish for significant advancement to occur. Thus, it is necessary to use higher temperature initiators to catalyze allylic systems. In addition, the exothermic heat released during the cure cycle is low compared to other monomers, resulting in significantly less cracking in the cured part. (4) Finally, physical properties such as thermal and moisture resistance are good because of the tight crosslinked network that can be developed by the difunctional monomer. Ester linkages are not readily available for attack by water because of the latter feature.

Results and Discussion

A carbon fiber prepreg was prepared from a polyester matrix resin system formulated with diallyl phthalate and a high performance unsaturated polyester.

An initial set of room temperature composite properties was obtained by molding the prepreg in a compression mold using a 177°C press. Time in the mold was fifteen minutes at a pressure of 500 psi (3.5 MPa). The data obtained are shown in Table II. The fiber used in all of the work discussed is Celanese's Celion 6000 carbon fiber. For purposes of comparison, key properties were obtained on some prepregs prepared from commercially available styrene-polyester, vinyl ester as well as an epoxy product. The values obtained are shown in Table III. It is apparent that in the key area of interlaminar shear strength, the diallyl phthalate system and the epoxy material are virtually identical. The other two materials are distinctly inferior. However, flex and tensile properties are quite similar for most of the materials examined. Fiber volumes obtained with the vinyl ester system were quite low and this resulted in a reduction of tensile and flex properties. Unfortunately, the

TABLE II

MECHANICAL PROPERTIES OF CELION 6000/DAP-POLYESTER COMPOSITES (22°C)[1]

Tensile[2]	Strength, Ksi	(MPa)	222	(1531)
	Modulus, Msi	(GPa)	20.3	(140)
	% Elongation		1.1	
Flex[2]	Strength, Ksi	(MPa)	273	(1882)
	Modulus, Msi	(GPa)	17.7	(122)
Interlaminar Shear Strength,	Psi	(MPa)	13,600	(93.8)
Compressive	Strength, Ksi	(MPa)	151	(1041)
	Modulus, Msi	(GPa)	17.7	(122)
	% Elongation		1.6	
Transverse Tensile	Strength, Psi	(MPa)	5,100	(35.2)
	Modulus, Msi	(GPa)	1.3	(8.96)
	% Elongation		0.39	

1 Compression molded cure at 177°C, 500 Psi (3.5 MPa) for fifteen minutes.

2 Tensile and flex values (excluding transverse tensile) are normalized to 62% fiber volume.

TABLE III

COMPARATIVE MECHANICAL PROPERTIES OF CELION 6000 COMPOSITES (22°C)[1]

		STYRENE POLYESTER	VINYL ESTER	DAP - POLYESTER	"250°F" EPOXY
Tensile[2]	Strength, Ksi (MPa)	229 (1579)	208 (1434)[3]	220 (1517)	221 (1524)
	Modulus, Msi (GPa)	20.1 (139)	18.2 (125)	20.2 (139)	20.0 (138)
	% Elongation	1.1	1.1	1.1	1.1
Flex[2]	Strength, Ksi (MPa)	303 (2089)	224 (1544)[4]	272 (1875)	287 (1979)
	Modulus, Msi (GPa)	20.8 (143)	16.9 (117)	17.7 (122)	19.3 (133)
Interlaminar Shear Strength, Psi (MPa)		10,350 (71.4)	8,170 (56.3)	13,400 (92.4)	14,000 (96.5)

1. Compression molded at 177°C 500 Psi (3.5 MPa) for fifteen minutes.

2. Tensile and flex properties normalized to 62% fiber volume for all systems except vinyl ester.

3. 49.6% fiber volume.

4. 53.8% fiber volume.

TABLE IV

COMPARATIVE MECHANICAL PROPERTIES OF CELION 6000 COMPOSITES (82 °C) [1]

	STYRENE POLYESTER		VINYL ESTER		DAP - POLYESTER		"250°F" EPOXY	
Flex Strength, Ksi (MPa) [2]	161	(1110)	147 [3]	(1013)	239	(1648)	289	(1993)
Interlaminar Shear Strength, Psi (MPa)	6,600	(45.9)	5,720	(39.4)	8,200	(56.5)	11,100	(76.5)

1 Compression mold cured at 177°C, 500 Psi (3.5 MPa) for fifteen minutes.

2 Flex strength normalized to 62% fiber volume for all systems except vinyl ester.

3 53.8% fiber volume.

polyester used in the styrene and diallyl phthalate systems is not the same. The polyester in the former system contains significant levels of adipic acid which would tend to have an adverse affect on flex and shear strength, particularly at elevated temperatures.

In Table IV, we see a comparison of key properties at elevated temperatures. Here, the flex and shear strengths show a greater spread of values between the DAP/polyester system and the epoxy.

Out Time Of The DAP-Polyester System. Carbon fiber prepreg prepared with the DAP-Polyester system was stored in a polyethylene bag at room temperature for a seven and a half month period. At the end of this period the prepreg had adequate tack and drape. Panels prepared from the material using a short cure cycle of two minutes at 163°C in a compression mold followed by a free standing fifteen minute post cure at 177°C gave very good mechanical properties. These are shown in Table V. The epoxy prepreg system used in this work for comparative purposes has a maximum out time of one month.

DAP-Polyester Short Cure Cycles. The previous data relating to the new polyester prepreg was generated using a fifteen minute compression molding cure cycle at 177°C and 500 psi (3.5 MPa). In Table VI, composite properties are shown as a function of cure cycle. The top row shows the properties obtained using a long cure cycle. The next two rows show that a two to five minute cure at temperatures of 138-163°C followed by a short "off the tool" post cure results in virtually optimum properties. The last two rows indicate that a short cure of one or two minutes at 177°C without any post cure gives quite acceptable properties.

DAP-Polyester Temperature Performance. In a separate set of experiments, the performance of Celion 6000/DAP-polyester composites was characterized as a function of temperature (the longer cure cycle was employed). It can be seen (Table VII) that the shear strength at 82°C is reduced to 60% of its ambient value. However, the high initial value still keeps this parameter reasonable at 82°C. At 121°C, a further decrease in shear strength has occurred and the flex strength has also decayed to 60% of its initial level.

In a new formulation that we have recently evaluated, elevated temperature performance has been improved compared to the original system. In Table VIII, properties are shown as a function of post cure time and temperature. A short cure cycle of two minutes followed by brief free standing post cures of five and twenty minutes at 177°C result in high levels of flex and shear properties.

In Table IX, properties obtained as a function of tempera-

TABLE V

COMPOSITE MECHANICAL PROPERTIES OF CELION 6000/DAP-POLYESTER AFTER 7½ MONTHS OF ROOM TEMPERATURE AGING[2,3]

TEST TEMPERATURE	FLEXURAL[1] STRENGTH	FLEXURAL[1] MODULUS	INTERLAMINAR SHEAR STRENGTH
(°C)	KSI (MPa)	MSI (GPa)	PSI (MPa)
22	294 (2027)	20.6 (142)	13,400 (92.4)
82	270 (1862)	--	9,900 (68.3)
121	198 (1365)	--	6,800 (46.9)

1 Flex values are normalized to 62% from 56.9%.

2 Samples were compression mold cured at 163°C and 500 PSI (3.5 MPa) for two minutes. Post cure was done free standing at 177°C for fifteen minutes.

3 Prepreg was kept in a polyethylene bag during this period.

TABLE VI

CELION 6000/DAP-POLYESTER COMPOSITE PROPERTIES AS A FUNCTION OF CURE CYCLE

COMPRESSION MOLD CURE CYCLE	POST CURE	TEST TEMPERATURE (°C)	FLEX[1] STRENGTH KSI (MPa)	FLEX[1] MODULUS MSI (GPa)	INTERLAMINAR SHEAR STRENGTH PSI (MPa)
FIFTEEN MINUTES AT 177°C AND 500 PSI (3.5 MPa)	NONE	22	273(1882)	17.7(122)	13,600(93.8)
TWO MINUTES AT 163°C AND 500 PSI (3.5 MPa)	NONE	22	262(1806)	17.4(120)	12,900(88.9)
		82	287(1979)	16.2(112)	8,300(57.2)
	FIVE MINUTES AT 325°F	22	247(1703)	17.7(122)	13,200(91.0)
FIVE MINUTES AT 138°C AND 500 PSI (3.5 MPa)	NONE	22	281(1937)	17.9(123)	12,500(86.2)
	FIFTEEN MINUTES AT 350°F	22	274(1889)	18.3(126)	13,300(91.7)
TWO MINUTES AT 177°C AND 500 PSI (3.5 MPa)	NONE	22	276(1903)	18.0(124)	12,900(88.9)
ONE MINUTE AT 177°C AND 500 PSI (3.5 MPa)	NONE	22	263(1813)	17.7(122)	12,800(88.3)

1 Flex values normalized to 62% fiber volume.

TABLE VII

PROPERTIES OF CELION 6000/DAP-POLYESTER COMPOSITES
AS A FUNCTION OF TEMPERATURE

TEST TEMPERATURE (°C)	FLEX STRENGTH Ksi (MPa)	FLEX MODULUS Msi (GPa)	INTERLAMINAR SHEAR STRENGTH Psi (MPa)
22	247 (1703)	17.6 (121)	13,400 (92.4)
82	240 (1655)	15.9 (110)	8,200 (56.5)
121	155 (1069)	16.3 (113)	5,000 (34.5)

1 Cured at 177°C, 500 psi (3.5 MPa) for fifteen minutes.

2 Flex values normalized to 62% fiber volume.

TABLE VIII

CELION 6000/DAP-POLYESTER COMPOSITE PROPERTIES AS A FUNCTION OF POST CURE TIME[1]

Post Cure[3] Time (min.)	Test Temperature (°C)	Flex[2] Strength Ksi (MPa)	Flex[2] Modulus Msi (GPa)	Interlaminar Shear Strength Psi (MPa)
0	22			11,000(75.8)
	82			4,800(33.1)
	121			3,100(21.4)
5	22	258(1779)	19.2(132)	12,700(87.6)
	82	192(1324)	18.8(130)	9,400(64.8)
	121	184(1269)	18.2(125)	7,200(49.6)
20	22	250(1724)	19.2(132)	12,800(88.3)
	82	206(1420)	19.0(131)	9,300(64.1)
	121	182(1255)	18.7(129)	8,000(55.2)

(1) Compression mold cured at 163°C, 500 psi(3.5MPa) for two minutes.

(2) Flex values normalized to 62% fiber volume.

(3) Post cure is done free standing at 177°C.

TABLE IX

COMPARATIVE COMPOSITE PROPERTIES OF
CELION 6000/DAP-POLYESTER AND CELION 6000/EPOXY

		TEST TEMPERATURE (°C)	DAP POLYESTER[1]	"250°F" EPOXY[2]
FLEXURAL[3]	STRENGTH KSI (MPa)	22	258(1779)	284(1958)
		121	184(1269)	142(979)
	MODULUS MSI (GPa)	22	19.2(132)	19.8(137)
		121	18.2(125)	--
INTERLAMINAR SHEAR STRENGTH PSI (MPa)		22	12,700(87.6)	14,100(97.2)
		121	7,200(49.6)	7,300(50.3)

(1) Compression mold cured at 163°C, 500 PSI (3.5 MPa) for two minutes followed by a five minute free standing post cure at 177°C.

(2) Compression mold cured at 154°C, 500 PSI (3.5 MPa) for fifteen minutes. No post cure was done.

(3) Flex values are normalized to 62%.

ture are compared with those obtained using a "250°F" curing epoxy system. The polyester and epoxy prepreg materials offer similar performance.

The composites prepared from the DAP-Polyester system can be used at temperatures as high as 150°C for short excursions. An extended post cure of at least two hours should be done if this level of temperature will be experienced. A longer post cure of about five hours is preferred. In Table X, ambient and 150°C properties are compared after a two hour post cure.

Moisture Resistance of the DAP-Polyester System. Another set of samples has been used to check the moisture sensitivity of the mechanical properties of Celion® 6000/DAP-Polyester composites. The results obtained are summarized in Table XI, wherein the excellent retention of dry "as is" properties is seen after a 24 hour soak in 71°C water. Flex and shear strengths are regarded as the key properties since they are most sensitive to changes at the interface.

Vacuum Bag - Autoclave Molding Of The DAP-Polyester Prepreg. For aerospace applications, the preferred technique used to prepare composite parts would be vacuum bag autoclave molding. A polyester prepreg containing the new formulation referred to above was used in this work. Panels were prepared using a total cure cycle of seventy minutes. The lay up was consolidated using a vacuum bag under 7-14 KPa (2-4 in. Hg) vacuum for fifteen minutes at room temperature. The bag was then placed in an autoclave, and a pressure of 85 psi (0.59 MPa) was applied. Vacuum was maintained throughout the entire cure. Heating was done at a rate of 2.2°C/minute until a temperature of 65.6°C was reached. Dwell time at this temperature was ten minutes. Heating was then continued at a rate of 3.9°C/minute until a temperature of 163°C was reached. Dwell time at this temperature was fifteen minutes. The part was then quickly cooled under full pressure and vacuum. The cure cycle used is shown graphically in Figure 1. Composite properties obtained on the panels are shown in Table XII. It should be mentioned that it is very likely that a straight up cure cycle could also be used with this system.

Conclusions

The properties shown for the composites made from the DAP-Polyester carbon fiber prepreg system are similar to those obtained with a "250°F" curing epoxy system. The material can be cured rapidly using compression molding procedures or by vacuum bag autoclave molding. Thermal and moisture resistance are very good. Cost of the resin system is low and refrigeration is not necessary. Out time of the product is in excess of six months. Prepreg handleability is excellent as evidenced by

TABLE X

COMPOSITE MECHANICAL PROPERTIES OF CELION 6000/DAP-POLYESTER AFTER EXTENDED POST CURE

		TEST TEMPERATURE (°C)	
FLEXURAL	STRENGTH KSI (MPa)	22	292(2013)
		150	107(738)
	MODULUS MSI (GPa)	22	20.2(139)
		150	17.4(120)
INTERLAMINAR SHEAR STRENGTH PSI (MPa)		22	13,200(91.0)
		150	4,500(31.0)

(1) Flex values are normalized to 62% from 53.5%.

(2) Samples were compression mold cured at 163°C, 500 PSI (3.5 MPa)for two minutes. Post cure was free standing in a 177°C circulating air oven for two hours.

(3) Three point modulus values. Flexural strength values are four point measurements.

TABLE XI

EFFECT OF MOISTURE ON MECHANICAL PROPERTIES OF CELION 6000/DAP-POLYESTER COMPOSITES[1]

SAMPLE CONDITIONING	TEST TEMPERATURE (°C)	FLEX[2] STRENGTH Ksi (MPa)	FLEX[2] MODULUS Msi (GPa)	INTERLAMINAR SHEAR STRENGTH Psi (MPa)
CONTROL	22	263(1813)	18.3(126)	13,700(94.5)
	82	240(1655)	15.7(109)	8,300(57.2)
AGED 24 HOURS IN 160°F WATER TESTED WET	22	272(1875)	17.7(122)	13,200(91.0)
	82	248(1710)	16.6(114)	7,300(50.3)

(1) Compression mold cured at 177°C, 500 psi (3.5MPa) for fifteen minutes.

(2) Flex values normalized to 62% fiber volume.

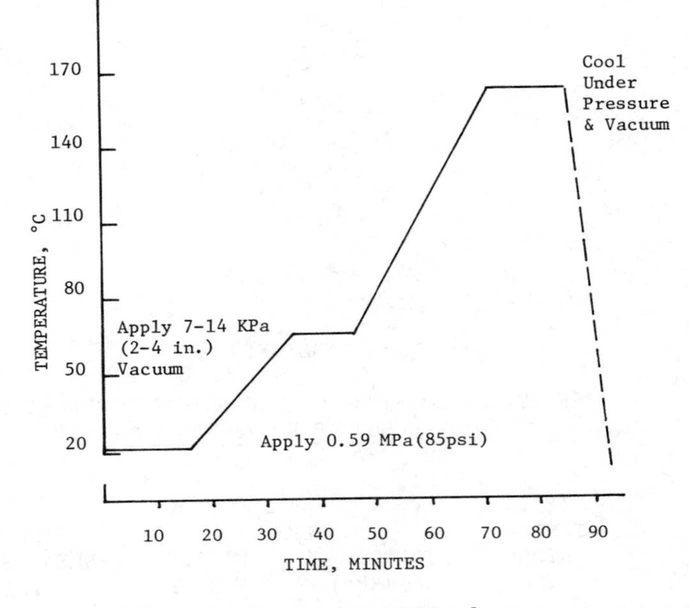

Figure 1. Cure cycle for DAP polyester

TABLE XII

MECHANICAL PROPERTIES OF CELION 6000/DAP-POLYESTER COMPOSITES - VACUUM BAG AUTOCLAVE MOLDED[1]

TEST TEMPERATURE	FLEX[1] STRENGTH KSI(MPa)	FLEX[1] MODULUS MSI(GPa)	INTERLAMINAR SHEAR STRENGTH PSI(MPa)
22	259(1786)	18.4(127)	12,500(86.2)
82	228(1572)	16.6(114)	9,300(64.1)

(1) Flex values normalized to 62% fiber volume.

good tack and drapeability. Environmental hazards are minimal because of the absence of volatile monomers. For these reasons the material should be considered for certain applications where epoxies are now currently being used.

Acknowledgments

We would like to acknowledge the extremely valuable assistance of Hector Zabaleta who fabricated the panels used in this work. Ray Steele ably assisted in this effort.

Howell Peterson and Val Stanton prepared all of the prepreg materials. Physical testing measurements were carried out by George Brenn, Stan Urbanski and Joe Cabanas. Dick Dzejak assisted in formulation preparation and in obtaining viscosity data.

Literature Cited

1. Boenig, H. V., "Unsaturated Polyesters," Elsevier Publishing Co., New York, N.Y., 1964, p. 140.
2. Raech Jr., H., "Allylic Resins and Monomers," Reinhold Publishing Corporation, New York, N.Y., 1965, p. 28.
3. Thomas, J. L. in "Modern Plastics Encyclopedia," Volume 55, McGraw Hill and Co., New York, N.Y., 1978-79, p. 9.
4. Raech Jr., H., "Allylic Resins and Monomers," Reinhold Publishing Corporation, New York, N.Y., 1965, p. 7.

RECEIVED January 8, 1980.

INSTRUMENTAL CHARACTERIZATION TECHNOLOGY

25

Nuclear Magnetic Resonance Characterization of Some Polyphthalocyanine Precursors

C. F. PORANSKI, JR. and W. B. MONIZ

Naval Research Laboratory, Washington, DC 20375

Many organic polymers have found aerospace applications as adhesives, sealants and coatings, and as matrix resins for fiber reinforced composites. During the past few decades polymer chemists have devoted a considerable amount of effort toward the development of improved organic polymers for such applications. These programs have had various goals, such as increased thermal or hydrolytic stability, higher strength, easier processing, lower costs, etc. A dozen years ago, Helminiak and Gibbs discussed the place of characterization of materials in such synthetic efforts.(1) They divided characterization into three categories: chemical identification, utility evaluation and structure-property relationships.

These three elements were incorporated in a recent Naval Research Laboratory program in the development of high performance adhesives and composites for vertical and short takeoff/landing aircraft (V/STOL).(2) As part of that program, our efforts were directed to the characterization of a variety of organic matrix materials used in graphite-fiber reinforced composites. We aimed primarily at determining the chemical composition of the polymer systems. We found carbon-13 and proton nuclear magnetic resonance spectroscopy (nmr) to be extremely useful techniques for this purpose.(3-6)

One of the organic materials studied in this program is N,N'-bis(3,4-dicyanophenyl) decane diamide, structure IA in Figure 1. This compound is one of a series, synthesized by Griffith and coworkers, which condense upon heating at about 200°C to form thermosetting polymers given the name, polyphthalocyanines.(7,8) The thermomechanical properties of these resin systems have been studied extensively.(9,10,11) In this Symposium, Griffith and Keller have reported the synthesis of a new class of phthalonitrile monomers, in which the dicyanophenyl groups are connected by dialkoxy or diphenoxy linkages.(12) The structures of three of these new materials are given in Figure 1.

This paper discusses the chemical characterization, by carbon-13 and proton nmr, of the compounds shown in Figure 1. In

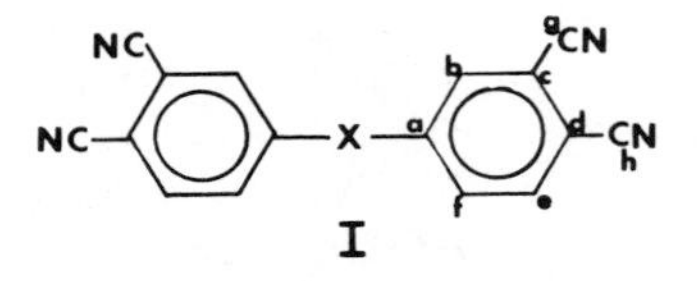

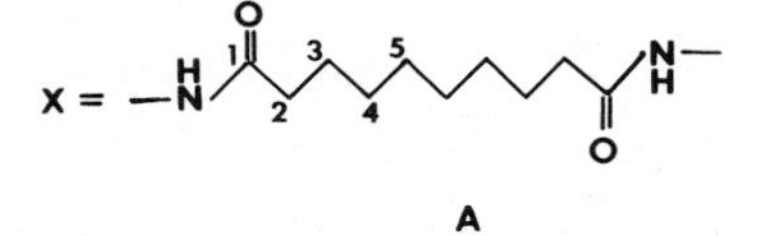

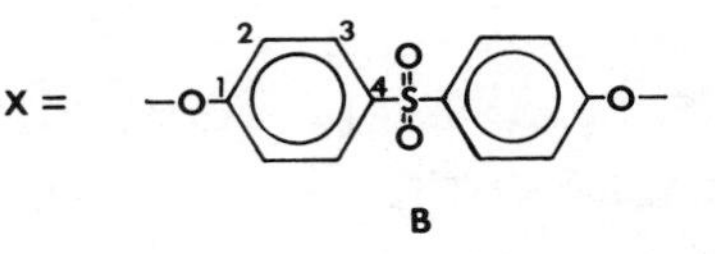

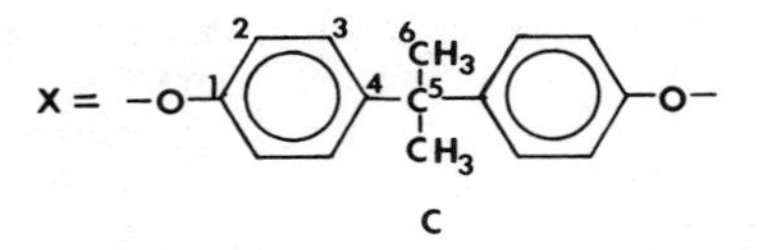

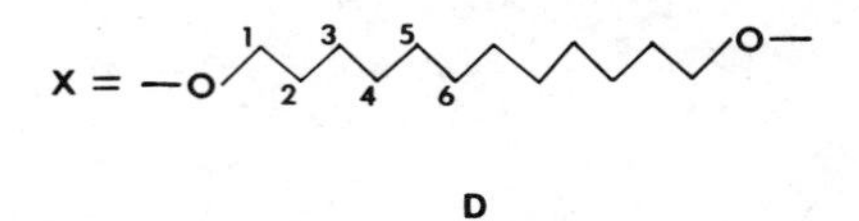

Figure 1. Structures of compounds IA–ID (9). The letters and numerals annotating the various carbon atoms are for use with Table I. Note that symmetry obviates annotating every carbon atom in each structure.

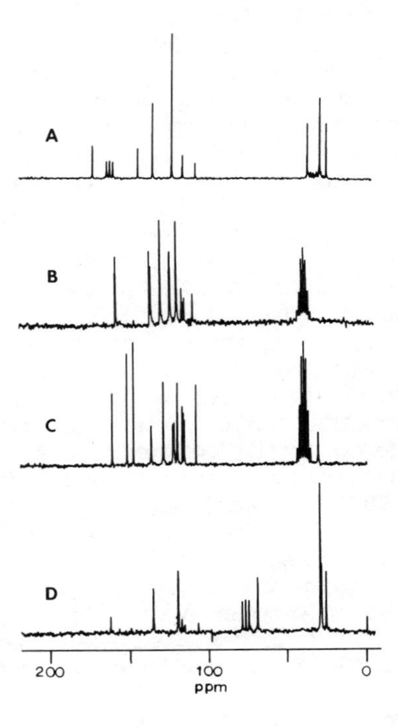

Figure 2. 15-MHz Carbon-13 NMR spectra of compounds IA–ID (9). Extraneous lines attributable to the solvents are located as follows: for IA, the triplet of the carbonyl carbon of dimethylformamide-d_7 is centered at 163 ppm, while the two multiplets from the carbons of the nonequivalent methyl groups contribute to the baseline noise around 35 ppm; for IB and IC, the septet from the methyl carbons of dimethylsulfoxide-d_6 is centered at 39.5 ppm; for ID, the triplet from chloroform-d is centered at 77 ppm.

the cases of compounds IB, IC and ID, we wanted only to verify the proposed structures. In the case of compound IA, however, we had a large number of samples from various sources and we were able to assess the quality control potential of carbon-13 and proton nmr for this material.

Carbon-13 NMR

Figure 2 gives the carbon-13 nmr spectra of compounds IA to ID. At first glance the spectra seem quite different. We will show, however, that there are similarities based on the common structural feature of the phthalonitrile ring. Furthermore, the differences which do occur are the keys to identification of the various X linkages in the four compounds. The solvent peaks in each spectrum are identified. Interference occurs only in the case of IC, where the line due to the bridging neo carbon of the bisphenol-A moiety is buried in the septet arising from the solvent, perdeutero-dimethyl sulfoxide, DMSO-d_6. The line can easily be located when the multiplet is expanded, however. Note that the spectra were not obtained under quantitative conditions. Therefore, in any given spectrum the intensities of the lines are not a measure of the relative numbers of each type of carbon nucleus in the molecule.

Table I gives the chemical shifts of the lines observed in the carbon-13 nmr spectra of the four compounds. Assignments of these lines to the various carbon atoms were made using carbon substituent chemical shift calculations (13) and comparisons with model compounds. In some cases there are ambiguities; these are noted in the table.

Two spectral features are characteristic of the phthalonitrile portion of these compounds. We have assigned the lines near 116 ppm to the cyano carbons and to the cyano-substituted aromatic carbon meta to the X substituent. Also there is a line near 108 ppm in each of the four spectra which we have assigned to the other cyano-substituted aromatic carbon, the one para to the X substituent.

Analysis of the 140 to 175 ppm regions of the four spectra shows significant differences related to the nature of the X group. In the spectrum of IA, the line at 173.4 ppm arises from the amide carbonyl carbon. In the phthalonitrile rings the aromatic carbons to which the substituents are attached have different shifts in the four compounds. In IA, the line from the nitrogen-substituted carbon appears at 144.8 ppm. The lines assigned to the oxygen-substituted carbons in the phthalonitrile rings of IB, IC and ID are at 158.7, 160.9 and 162.3 ppm, respectively. Compounds IB and IC each have a second oxygen-substituted aromatic carbon (#1 in the structures). In IB, the assignments of carbons C-1 and C-a may be interchanged. For IC the assignment of the 151.5 ppm peak to C-1 is consistent with the result of our substituent effect calculations. The assign-

Table 1

Carbon-13 Chemical Shifts Assignments for Compounds IA-ID

Carbon Chemical Shift[1] (ppm from TMS)

Compound	Solvent	a	b	c	d	e	f	g	h	1	2	3	4	5	6
IA	DMF	144.8	123.4	116.7†	108.5	135.6	123.4	116.5†	116.5†	173.4	37.5	25.6	29.6*	29.7*	-
IB	DMSO	158.7*	124.2‡	116.9†	110.0	136.4	124.4‡	115.7†	115.2†	158.9*	120.2	130.3	137.1	-	-
IC	DMSO	160.9	122.6*	116.6†	108.0	136.2	121.8*	115.8†	115.3†	151.5	119.8	128.6	147.4	42.0	30.5
ID	$CDCl_3$	162.3	119.6	117.4	106.9	135.3	119.8	115.8†	115.4†	69.4	29.8*	25.8	29.2*	29.5	29.5

1. Refer to Figure 1 for labelling of carbon atoms.

*, †, ‡: Assignments among these lines are uncertain.

ments of the remaining lines in the 120-140 ppm region are as given in Table I.

The several lines between 25 and 30 ppm in the spectra of IA and ID arise from the aliphatic chains in the X group. The main difference between IA and ID is in the position of the lines of the terminal methylene carbons of the chains; the α-methylene carbon resonance of the amide, IA, occurs at 37.5 ppm, while that of the ether, ID, is at 69.4 ppm. The two lines from the methyl and neo carbon of the isopropylidine group in IC appear at 30.5 and 42.0 ppm respectively.

Proton NMR

The proton nmr spectra obtained for these four compounds are given in Figure 3. Compounds IA and ID have only three aromatic protons and, in the aromatic region, their spectra show patterns typical of spin-spin coupling in ABC spin systems. The aromatic regions of the spectra of IB and IC are more complex because of additional lines from the 4 protons of the *para* disubstituted benzene rings in the linking groups. These patterns could be completely analyzed in terms of proton chemical shifts and coupling constants should one so wish.

Compounds IA, IC and ID also have aliphatic protons. The methyl protons of the bis-phenol A structure of IC give a broad, featureless line at 1.71 ppm. Both IA and ID have a band between 1.3 and 1.7 ppm typical of $(-CH_2-)_n$. In ID the protons of the methylene group α to the ether oxygen give rise to a triplet at 4 ppm. The triplet due to the methylene protons α to the carbonyl group in IA appears at 2.3 ppm.

Finally, IA has amide protons which, in dimethyl formamide solution, give a broad peak at 10.6 ppm.

Several points are worth noting about the proton spectra. First, although several different solvents were used, there are sufficient differences in these spectra to allow qualtitative distinction among the compounds. Of course, to develop a "fingerprint" type of quality control standard, all of the compounds would have to be run in the same solvent.

Second, proton FT nmr spectra are more easily quantified than carbon-13 spectra. Hence integration of the proton spectra can give a proton count to verify a proposed structure. Third, proton nmr is more sensitive than carbon-13 nmr (by $\sim$100 times) and has a higher potential for revealing the presence of impurities. This last point will be illustrated later.

Compound IA

Of the various alkane diamides originally synthesized by Griffith and co-workers, the so-called C_{10} diamide, IA, was selected for detailed study in the NRL V/STOL program. A relatively large number of batches of various sizes were available;

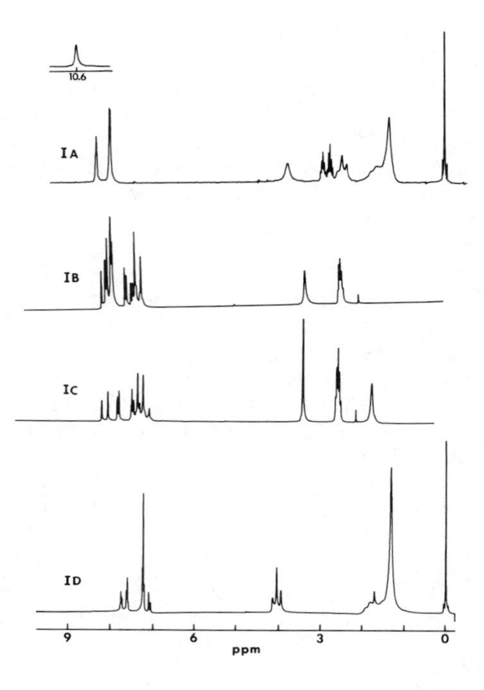

Figure 3. 60-MHz proton NMR spectra of compounds IA–ID (9). Extraneous peaks in each spectrum are noted as follows: in IA—4 ppm (H_2O), 2.6 and 2.9 ppm (DMF-d_7); in IB and IC—3.5 ppm (H_2O), 2.5 ppm (DMSO-d_6), 2.1 ppm (acetone). The insert in IA shows the peak due to the amide protons of IA.

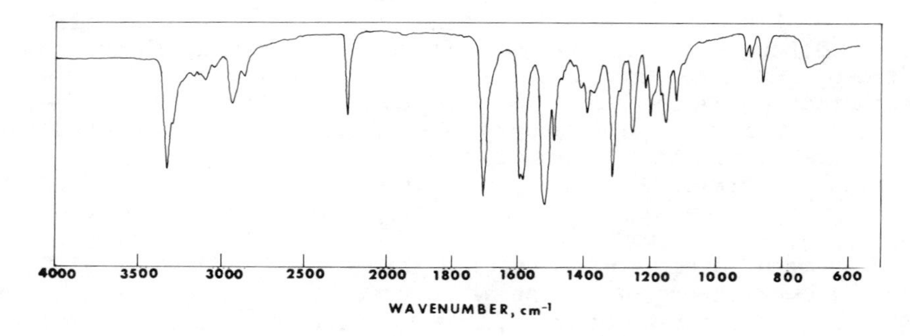

Figure 4. IR spectrum of compound IA (KBr pellet)

eleven batches prepared at NRL and five batches produced commercially. We were interested in how carbon-13 and proton nmr would fare in comparison with another potential method of quality control, infrared spectroscopy.

Figure 4 shows the infrared spectrum obtained from a sample of IA synthesized at NRL. We have assigned some of the more prominent bands as follows: 3330 cm^{-1} (N-H), 3100 cm^{-1} (aromatic C-H), 2940 and 2860 cm^{-1} (aliphatic C-H), 2240 cm^{-1} (C≡N), 1710 cm^{-1} (C=O), 1590, 1520, 1490 cm^{-1} (C=C), 1340 cm^{-1} (PhN-H), 1255 cm^{-1} (amide III). Inspection of the infrared spectra of the various batches of IA showed that all of them were similar to the spectrum in Figure 4. However, variations occurred in some bands which could not be correlated with melting point or method of purification of the batch. The two major variations were changes in the width and shape of the carbonyl band, and the occasional appearance of a moderately strong band at 1410 cm^{-1} which overwhelmed the three small bands in that region. Clearly significantly more work is required before infrared spectroscopy could function as a quality control vehicle in any way other than gross identification of IA.

The quality control role of carbon-13 nmr is much the same as for infrared spectroscopy, but for a different reason. It is sensitivity, rather than precision or reproducibility, which limits the potential of carbon-13 nmr. For example, the carbon-13 chemical shifts listed in Table I for compound IA are the average of those measured for over 16 different samples of IA. The largest standard deviation found for any line is 0.3 ppm. There are no obvious problems due to the previous histories of the samples. However, carbon-13 nmr is hard pressed to detect impurities at low levels, say below 5%, without resorting to time-consuming accumulations impractical for routine use. The primary reason for this low sensitivity is the 1% natural abundance of the carbon-13 isotope. In the case of IA, the problem is compounded by its low solubility: this results in lowering the effective sensitivity towards impurities and leads to dynamic range problems in the detection system due to the large solvent signals.

Proton nmr, on the other hand, does not suffer from low sensitivity. It provides, therefore, a quick method for identification as well as for determining impurities at the 1-5% level. In spite of the low solubility of IA in dimethyl sulfoxide (DMSO-d_6) (∿ 9 mg/ml) we were able to detect in one sample ∿2% dimethyl formamide (DMF), a solvent used in the commercial preparation of IA (Figure 5). In more concentrated solutions in DMF-d_7 (∿ 160 mg/ml) we have observed in the proton nmr spectrum of several samples of IA, traces (∿ 1%) of acetone or ethyl alcohol, solvents which may be used to wash IA after synthesis.

Proton nmr has also given us some insight into the chemistry of IA systems. For example, we used proton nmr to study the reaction product formed when IA was B-staged with a stoichiomet-

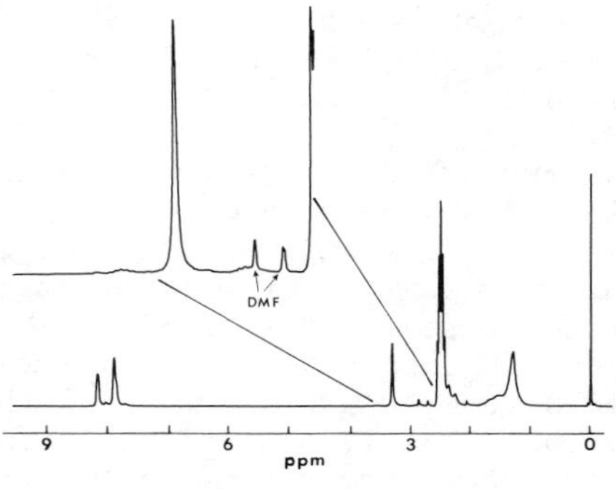

Figure 5. 60-MHz proton NMR spectrum of IA in DMSO-d_6 (9). The expanded insert shows the two peaks attributable to residual dimethyl formamide. The peak attributable to the amide protons of IA at 10.6 ppm is not shown.

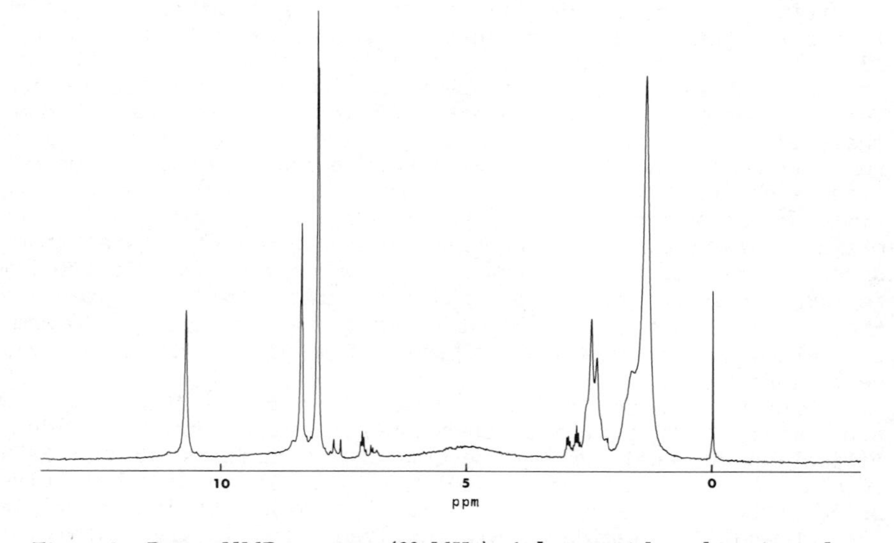

Figure 6. Proton NMR spectrum (60 MHz) of the material resulting from the B-staging of a stoichiometric mixture of IA and $SnCl_2 \cdot 2H_2O$

ric amount of $SnCl_2 \cdot 2H_2O$. It had been found that although this system polymerizes, the high temperature properties of the resulting polymer are not as good as those of the polymer formed by curing IA alone. The proton nmr spectrum of the prepolymer from IA and $SnCl_2 \cdot 2H_2O$ is given in Figure 6. The spectrum contains two features not found in the spectrum of IA alone. First, the triplet at 2.0 ppm, arising from the protons of the methylene group adjacent to the amide carbonyl group, appears to be an unsymmetrical quartet. Second, there are a number of additional lines between 6.5 and 7.0 ppm due to aromatic protons. Subsequent analysis showed that the new aromatic lines were due to 4-aminophthalonitrile (Figure 7). This led us to conclude that during early stages of curing IA/$SnCl_2 \cdot 2H_2O$ mixtures, the $SNCl_2 \cdot 2H_2O$ may catalyze the hydrolysis of IA as shown in Scheme I.

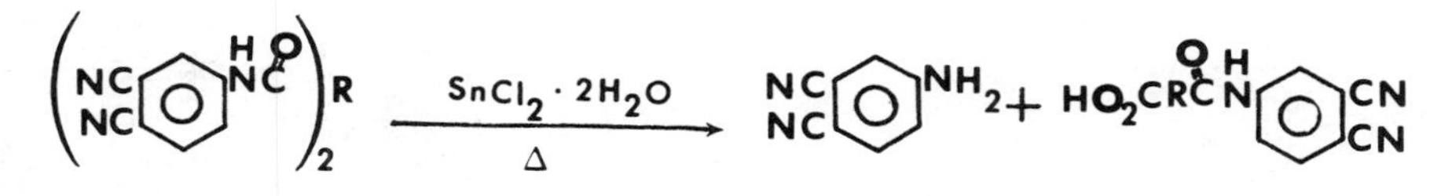

Scheme I

Experimental

Proton decoupled carbon-13 spectra were run on either of two spectrometers; a JEOL FX-60Q or a Varian HA-100 modified for pulse FT operation at 25.15 MHz and equipped with a Nicolet data system for accumulation and Fourier transformation. In both cases the sweep width was 250 ppm, the data block was 8192 points and the pulse angle was 90°. The number of accumulations varied according to the needs of the sample. Proton spectra were obtained on a Varian HA-100, JEOL PS-100 or FX-60Q spectrometer. All chemical shifts are expressed in parts per million, ppm, from the reference compound, tetramethylsilane (TMS). Infrared spectra were run of KBr pellets containing 1% by weight of compound on a Perkin Elmer Model 267 spectrophotometer.

Materials

Single samples of IB, IC, and ID were run in DMSO-d_6, DMSO-d_6, and $CDCl_3$ respectively, at concentrations of 100 mg/ml. Eleven samples of NRL prepared, and 5 samples of commercially prepared IA were used as received and run in DMSO-d_6 (9 mg/ml) and DMF-d_7 (160 mg/ml). Most carbon-13 spectra of IA samples were obtained with normal DMF as the solvent.

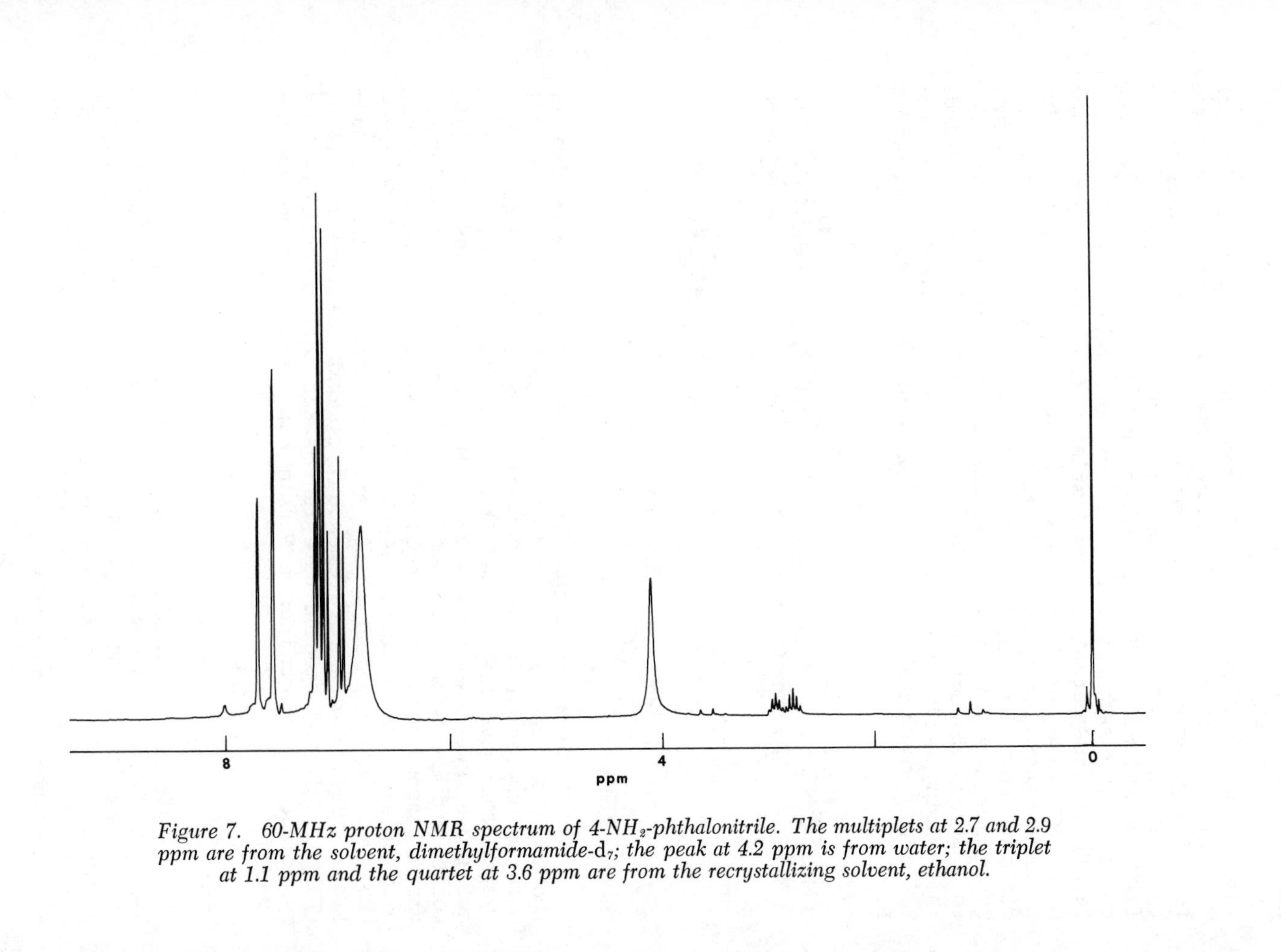

Figure 7. 60-MHz proton NMR spectrum of 4-NH_2-phthalonitrile. The multiplets at 2.7 and 2.9 ppm are from the solvent, dimethylformamide-d_7; the peak at 4.2 ppm is from water; the triplet at 1.1 ppm and the quartet at 3.6 ppm are from the recrystallizing solvent, ethanol.

Literature Cited

1. Helminiak, T.E., and Gibbs, W.E., Polymer Preprints, 1967, 8 (2), 934.
2. "High Performance Composites and Adhesives for V/STOL Aircraft, Third Annual Report," L.B. Lockhart, Ed., Naval Research Laboratory Memorandum Report 4005, May 1979.
3. Poranski, Jr., C.F., and Moniz, W.B., J. Coatings Tech., 1977, 49, No. 632, 57.
4. Poranski, Jr., C.F., Moniz, W.B., and Giants, T.W., Preprints, ACS Div. Org. Coat. Plas. Chem., 1978, 38, 605.
5. Moniz, W.B., and Poranski, Jr., C.F., Preprints, ACS Div. Org. Coat. Plas. Chem., 1978, 39, 99.
6. Poranski, Jr., C.F., Moniz, W.B., Birkle, D.L., Kopfle, J.T., and Sojka, S.A., "Carbon-13 and Proton NMR Spectra for Characterizing Thermosetting Polymer Systems," Vol. 1, Naval Research Laboratory Report 8092, June, 1977.
7. Griffith, J.R., O'Rear, J.G., and Walton, T.R., in "Copolymers, Polyblends and Composites," N.A.J. Platzer, Ed., Adv. Chem. Series No. 142, ACS, Washington, D.C., 1975, p. 458.
8. Griffith, J.R., Proc. Nat. SAMPE Symp., 1975, 20, 582.
9. Gillham, J.K., Preprints, ACS Div. Org. Coat. Plas. Chem., 1979, 40, 866.
10. Gillham, J.K., Poly. Sci. Eng., 1979, 19, 319.
11. Bascom, W.D., Bitner, J.L., Cottington, R.L., and Henderson, C.M., in "High Performance Composites and Adhesives for V/STOL Aircraft; Third Annual Report", L.B. Lockhart, Jr., Ed., Naval Research Laboratory Memorandum Report 4005, May, 1979, p. 13.
12. Keller, T.M., and Griffith, J.R., Preprints, ACS Div. Org. Coat. Plas. Chem., 1979, 40, 781.
13. Wehrli, F.W., and Wirthlin, T., "Interpretation of Carbon-13 NMR Spectra," Heyden & Son., Inc., New York, 1976. p. 47.

RECEIVED January 14, 1980.

26

Torsional Pendulum Analysis of the Influence of Molecular Structure on the Cure and Transitions of Polyphthalocyanines

JOHN K. GILLHAM

Polymer Materials Program, Department of Chemical Engineering, Princeton University, Princeton, NJ 08544

This report follows up an earlier study (1) in which an automated torsion pendulum was used to investigate the transformation (cure) of a "C_{10} diamide phthalocyanine" resin monomer to thermoset polymer using supported specimens [torsional braid analysis, TBA, (2,3).

The overall chemical reaction for polymerization of the C_{10} diamide monomer is shown in Figure 1. The bis aromatic ortho-dinitrile monomer reacts to form a network of C_{10} diamide segments linked by tetra functional phthalocyanine branching sites. In general thermosetting polymerization proceeds from liquid monomer (with a glass transition temperature below the initial melting temperature of the pure monomer), through gelation (as branched molecules of infinite molecular weight form), through the rubbery state (as network molecules form), and finally (when the temperature of reaction is below the maximum attainable glass transition temperature) to vitrification. Fabrication can involve use of tractable prepolymer (i.e. non-gelled oligomer) for molding *in situ* to intractable product.

Cure was examined directly from isothermal plots at different temperatures, and also from transition temperatures, which were obtained from thermomechanical plots of intermittently cooled specimens, versus time of isothermal cure. In the present work the influence of molecular structure on the transitions, particularly the glass transition (Tg) and a cryogenic transition (T_{sec}) associated with the flexible molecular segment of the monomer and of the resulting network, has been investigated using analytically pure monomers designated C_6-methyl, C_{10}, C_{14} and C_{22} diamide phthalocyanine resin monomers. A preliminary report has been published (4).

0-8412-0567-1/80/47-132-349$05.00/0

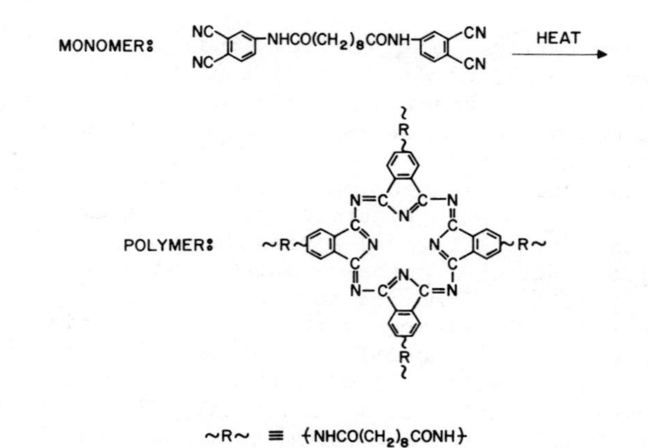

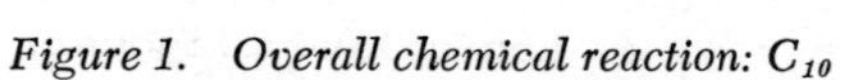

Figure 1. Overall chemical reaction: C_{10}

Experimental

Monomers.

The monomers, which differ in the linkage connecting the two aromatic ends, were obtained from the Naval Research Laboratory, Washington, D.C. The chemistry of their synthesis and polymerization has been reported (5). The monomers and their designations were:

"C_6-methyl diamide"
  i.e., N,N'-bis(3,4-dicyanophenyl) 3-methylhexanediamide, mp 203-206°C.

"C_{10} diamide"
  i.e., N,N'-bis(3,4-dicyanophenyl) decanediamide, mp 185-189°C.

"C_{14} diamide"
  i.e., N,N'-bis(3,4-dicyanophenyl) tetradecanediamide, mp 163-165°C.

"C_{22} diamide"
  i.e., N,N'-bis(3,4-dicyanophenyl) docosanediamide, mp 144-147°C.

Torsion Pendulum.

An automated, intermittently activated, freely vibrating torsion pendulum operating at about 1 Hz (2,3) was used for all experiments. (A version of the instrumental system is available from Plastics Analysis Instruments, Inc., P.O. Box 408, Princeton, NJ.) Each specimen was made by dipping a glass braid into a slurry of monomer in ethyl alcohol. After mounting a specimen in the apparatus at room temperature (RT) the temperature was raised to the isothermal cure temperature at 5°C/min. Cure was monitored both continuously at the isothermal temperature and, in a separate experiment, by intermittently cooling and then heating back (generally at 1.5°C/minute) to the isothermal cure temperature to record the thermomechanical spectra (which were used to assign transition temperatures). All experiments were performed in dry helium. Each specimen consumed about 20 mg of monomer.

The torsion pendulum plots display the relative rigidity ($1/P^2$, where P is the period in seconds of a freely damped wave) and logarithmic decrement Δ (= $\ln A_i/A_{i+1}$, where A_i is the amplitude of the *ith* oscillation of a wave) versus temperature (mV from an iron-constantan thermocouple) or time. These mechanical parameters relate directly to dynamic mechanical testing in that the relative rigidity is directly proportional to the in-phase shear modulus (G') and $\Delta \simeq \pi \tan\delta$, where δ is the phase angle between cyclic stress and strain. Transition temperatures were assigned using the peaks of the logarithmic decrement; intensities of transitions were assigned using values of Δ.

Results and Discussion

An isothermal temperature of 250°C was selected for studying structure-property relationships since higher temperatures lead to weak glass transitions and to degradation. The immediate discussion emphasizes Tg which was identified as the most dominant loss peak at high temperature in a thermomechanical spectrum. Other transitions are considered more completely later. The glass transition temperatures versus time of cure at 250°C are included in Tables I-IV and in Figures 2-5. The initial Tg for all of the monomers was less than 100°C. On heating, the glass transition temperature increased more rapidly and, after about 5 hours, reached a higher limiting temperature (designated ${}_{250}T_{g\infty}$) the shorter the length of the inter-aromatic molecular linkages. Values for ${}_{250}T_{g\infty}$ (see Table V) were: C_6-methyl > 244°C, C_{10} ∿ 202°C, C_{14} ∿ 185°C, and C_{22} ∿ 157°C. A plot of ${}_{250}T_{g\infty}$ versus the number of inter-aromatic C atoms, which is an index of segmental molecular flexibility, is shown in Figure 6. As the temperature of T_g increased with extent of cure, the intensity of Tg decreased for all four materials (Tables I to IV). This could lead to an uncertainty in designating Tg (see below).

Plots of isothermal cure for 5 hours at 250°C for the four monomers are shown in Figures 7A, 7B, 7C, and 7D. The corresponding subsequent thermomechanical plots (250°C → -190°C) are shown in Figures 8A, 8B, 8C and 8D. Comparison of each pair of isothermal and thermomechanical plots shows that two loss processes (peaks or shoulders) occur isothermally neither of which is assigned as the glass transition (vitrification), which is revealed in the corresponding thermomechanical plots at lower temperatures than the isothermal cure temperature. The first isothermal loss peak probably occurs at an isoviscous level and has been used to measure gelation times (1,2,3); the second isothermal "peak" presumably corresponds to the $T_{\ell\ell}$ (or T > T_g) relaxation (1,6) which occurs immediately above T_g in temperature scans of thermoplastic material (see later).

The procedure for assigning T_g is more convincingly demonstrated than in Figures 7A-7D and Figures 8A-8D by comparing the plot of isothermal cure at 220°C (Figure 9) and the subsequent thermomechanical plot (Figure 10) of the C_{14} material: there can be little doubt in this case that the intense relaxation below the temperature of isothermal cure is T_g. Vitrification was not observed as a third process on isothermal cure for the four materials since the temperatures (250°C, Figures 7A to 7D and 220°C, Figure 9) were above the maximum T_g attained in the time scale of the experiments. In the absence of thermal degradation full cure would be expected to lead to a network of short linkages (i.e. high crosslink density) which would give rise only to a weak glass transition perhaps better characterized as a secondary transition. If isothermal cure occurs at temperatures just below the glass transition temperature of the fully cured material, the loss peak associated with the rising glass transition temperature (i.e.

TABLE I. C_6-CH_3: Transitions (>RT) Versus Cure Time at 250°C.

Time at 250°C Min.	Plot No.	ΔT/Δt - or +	TRANSITIONS Tg(Δ) °C		T > Tg °C
	1	+			
5	2	-	92	(0.964)	111
5	3	+	92	(0.966)	110
23	4	-	138	(0.274)	246
23	5	+	139	(0.267)	223
41	6	-	186	(0.147)	>250
41	7	+	185	(0.141)	
83	8	-	212	(0.105)	
83	9	+	213	(0.0986)	
133	10	-	226	(0.0829)	
133	11	+	222	(0.0786)	
230	12	-	233	(0.0646)	
230	13	+	231	(0.0605)	
545	14	-	244	(0.0441)	

TABLE II. C_{10}: Transitions (>RT) Versus Cure Time at 250°C.

Time at 250°C Min.	Plot No.	ΔT/Δt - or +	TRANSITIONS Tg(Δ) °C		T > Tg
10	1	-	60.5	(2.587)	81.5
10	2	+	60	(2.648)	79.5
28	3	-	71	(1.801)	106
28	4	+	71	(1.782)	106.5
46	5	-	99.5	(0.818)	183.5
46	6	+	99	(0.804)	182
94	8	-/+	165.5	(0.362)	
131	9	-/+	187.5	(0.283)	
269	10	-/+	199	(0.208)	
635	11	-/+	202.5	(0.139)	

TABLE III. C_{14}: Transitions (> RT) Versus Cure Time at 250°C.

Time at 250°C Min.	Plot No.	ΔT/Δt - or +	TRANSITIONS Tg (Δ) °C		T > Tg (T' > Tg) °C	
0	1	+				
5	2	-	49	(2.218)	70	(∿ 86)
5	3	+	49	(2.222)	*	
5	4	-	50	(1.832)	75	(∿102)
5	5	+	50.5	(1.829)	*	
5	6	+	51	(2.499)	*	
23	7	-	64	(1.414)	132	(∿204)
23	8	+	65	(1.404)	131	(∿222)
33	9	-	100	(0.551)	202	
33	10	+	99	(0.534)	198	
66	11	-	150	(0.334)		
99	12	-	168	(0.257)		
99	13	+	169	(0.250)		
465	14	-	183	(0.137)		
465	15	+	185	(0.132)		
1005	16	-	176	(0.100)		
1005	17	+	175	(0.0986)		
1695	18	-	163	(0.0840)		
1695	19	+	161	(0.0830)		
2355	20	-	152	(0.0784)		

*Plots display multiple melting/crystallization transitions (e.g. see Figure 11 for Plot No. 5). Adjacent plots obtained on cooling (i.e. 2,4,7) display amorphous behavior with T > Tg and T' > Tg relaxations.

TABLE IV. C_{22}: Transitions (> RT) Versus Cure Time at 250°C.

Time at 250°C Min.	Plot No.	ΔT/Δt - or +	Tg (Δ) °C		TRANSITIONS T > Tg (T' > Tg) [Other] °C
5	2	-	87(?)	(0.243)	[T_{cry} 93]
5	3	+	90	(0.217)	[T_m 134]
23	4	-	58	(0.538)	124 (219)
23	5	+	60	(0.535)	129 [141/∿231]
41	6	-	71	(0.388)	184 (∿254)
41	7	+	71	(0.382)	183
89	8	-	108	(0.256)	233
89	9	+	111	(0.250)	∿226
131	10	-	128	(0.202)	
131	11	+	130	(0.200)	
296	12	-	145	(0.143)	
296	13	+	147	(0.145)	[∿50]
824	14	-	156	(0.101)	
2432	15	-	157	(0.100)	
2432	16	+	151.5	(0.0789)	[52/∿88]
5048	17	-	153	(0.0660)	
5048	18	+	148	(0.0685)	[∿48]

TABLE V. ${}_{250}T_{g\infty}$ Versus Effective Number of Inter-Aromatic Carbon Atoms.

Polymer	Effective Number of Carbon Atoms*	${}_{250}T_{g\infty}$ (°C)
C_6-CH_3	5.5	> 244
C_{10}	10	∿ 202
C_{14}	14	∿ 185
C_{22}	22	∿ 157

*Effective number of carbon atoms/inter-aromatic segment is taken as the number of unbranched C atoms between the amide N atoms in the monomer, except for C_6-CH_3 which is assigned 0.5 less than 6 because of the stiffening effect of a branched C.

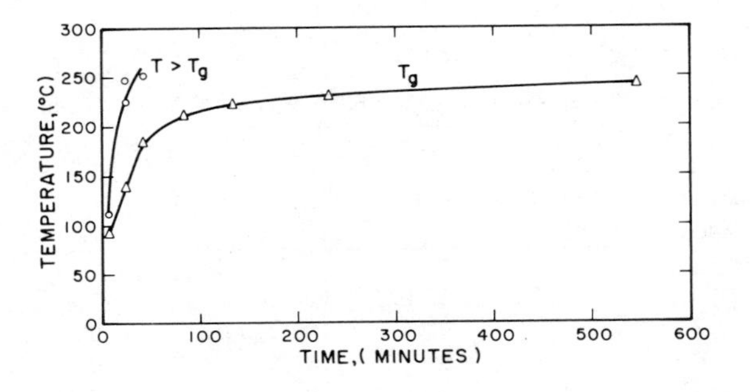

Figure 2. $C_{6\text{-}CH_3}$*: transitions (> RT) vs. cure time at 250°C*

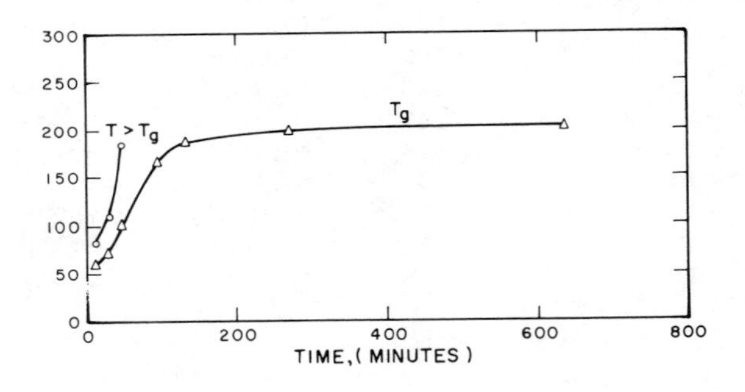

Figure 3. C_{10}*: transitions (> RT) vs. cure time at 250°C*

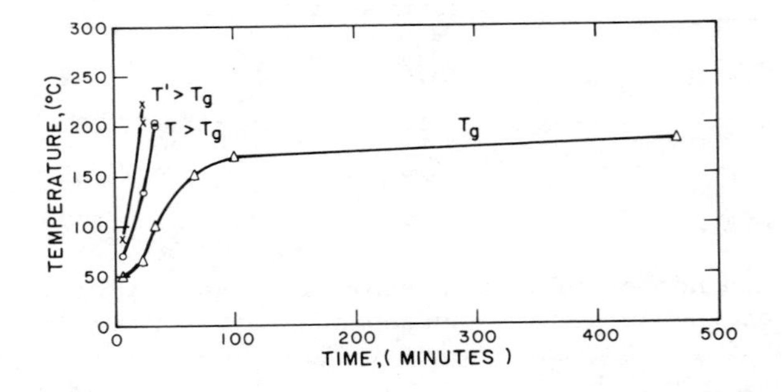

Figure 4. C_{14}*: transitions (> RT) vs. cure time at 250°C*

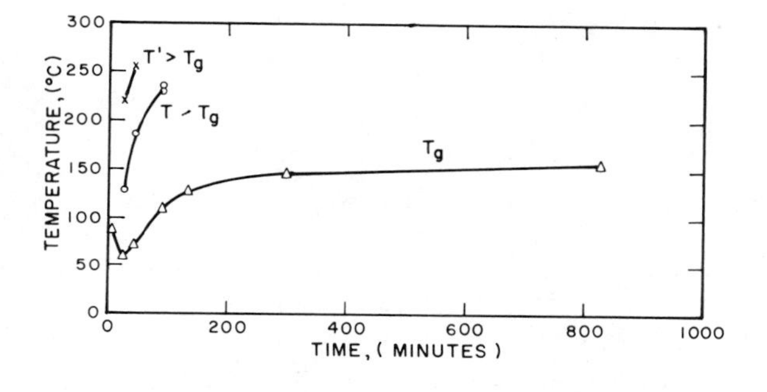

Figure 5. C_{22}: transitions (> RT) vs. cure time at 250°C

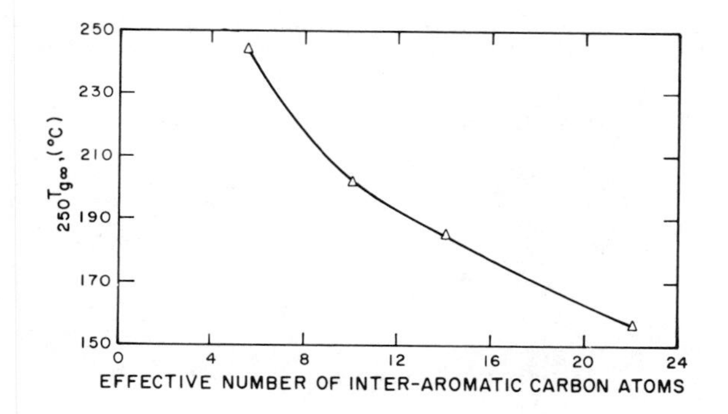

Figure 6. ${}_{250}T_{g\infty}$ vs. effective number of interaromatic carbon atoms

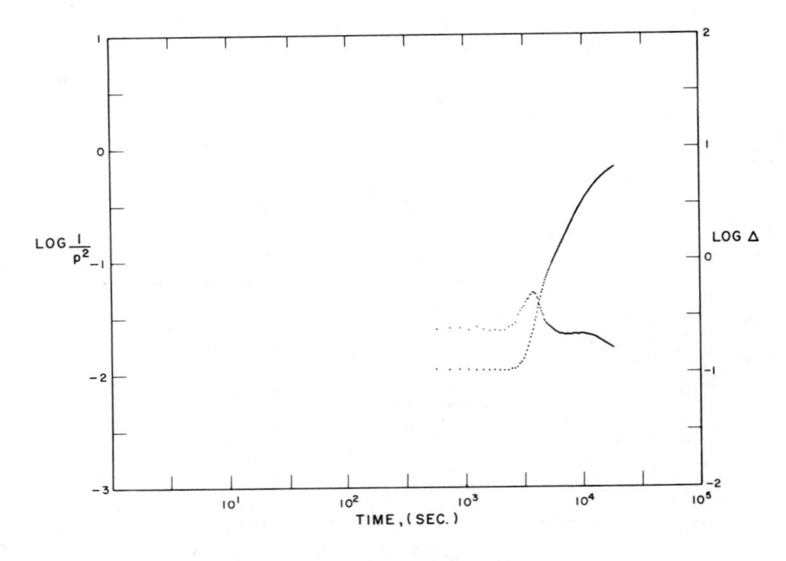

Figure 7A. Isothermal cure (250°C/5 hr): $C_{6\text{-}CH_3}$

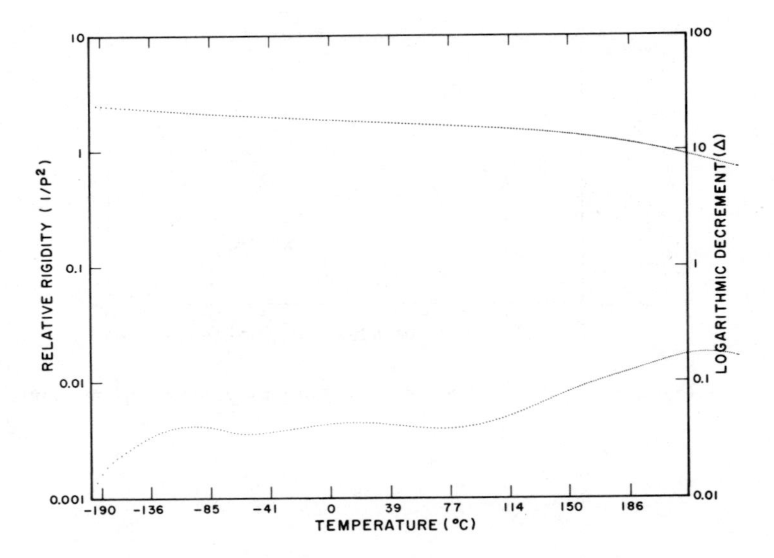

Figure 8A. Thermomechanical spectra (220° → −190°C) after 250°C/5 hr cure: $C_{6\text{-}CH_3}$

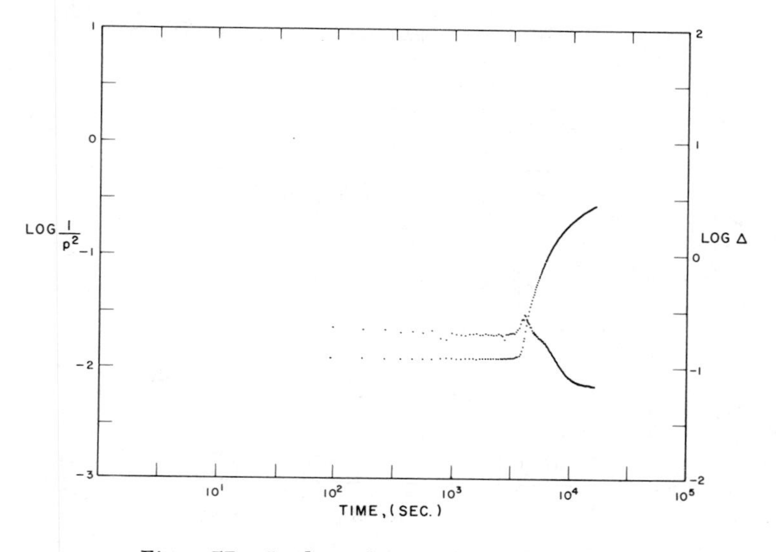

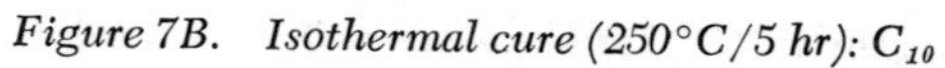
Figure 7B. Isothermal cure (250°C/5 hr): C_{10}

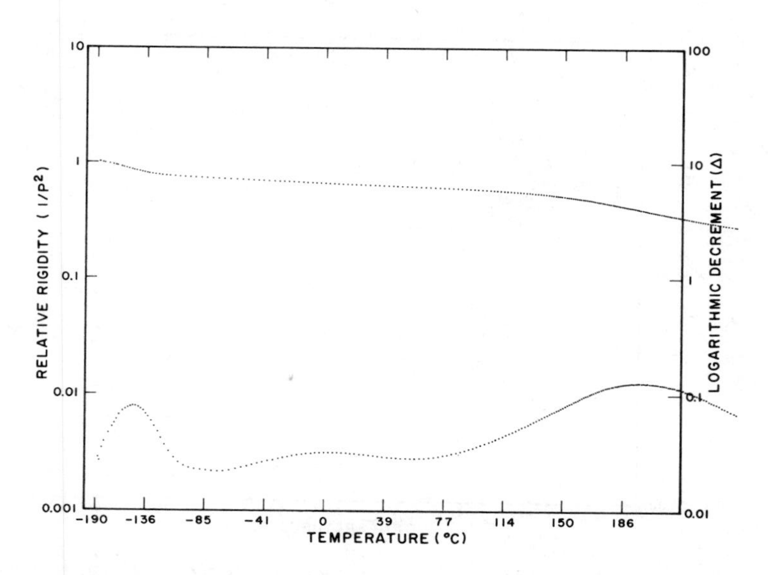

Figure 8B. Thermomechanical spectra (220° → −190°C) after 250°C/5 hr cure: C_{10}

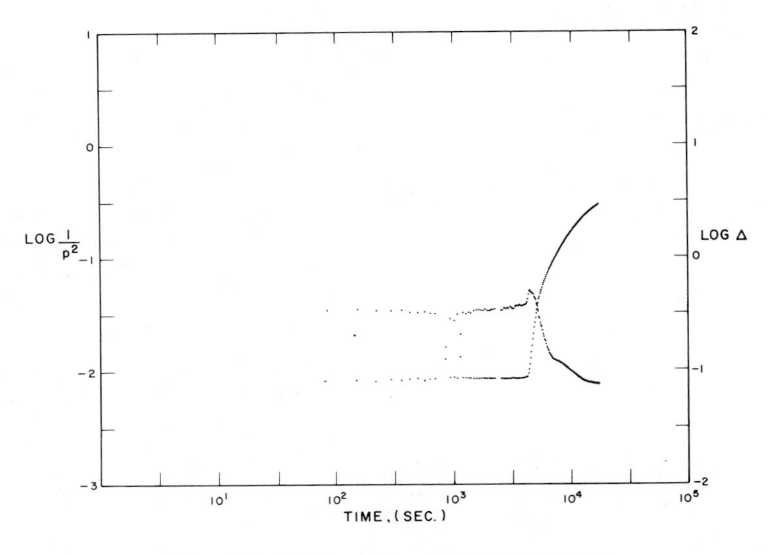

Figure 7C. Isothermal cure (250°C/5 hr): C_{14}

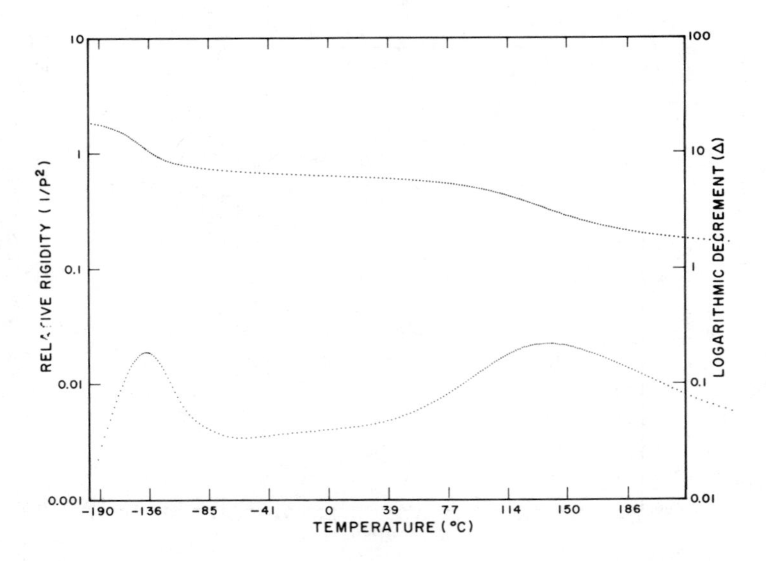

Figure 8C. Thermomechanical spectra (220° → −190°C) after 250°C/5 hr cure: C_{14}

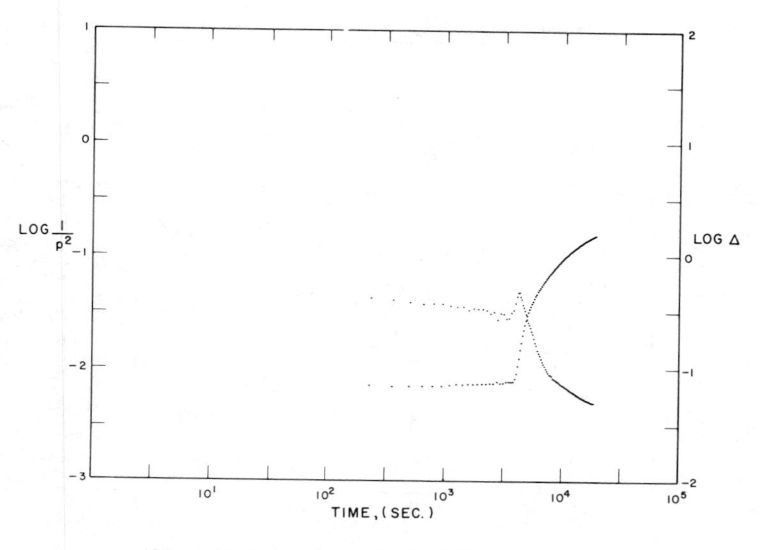

Figure 7D. Isothermal cure (250°C/5 hr): C_{22}

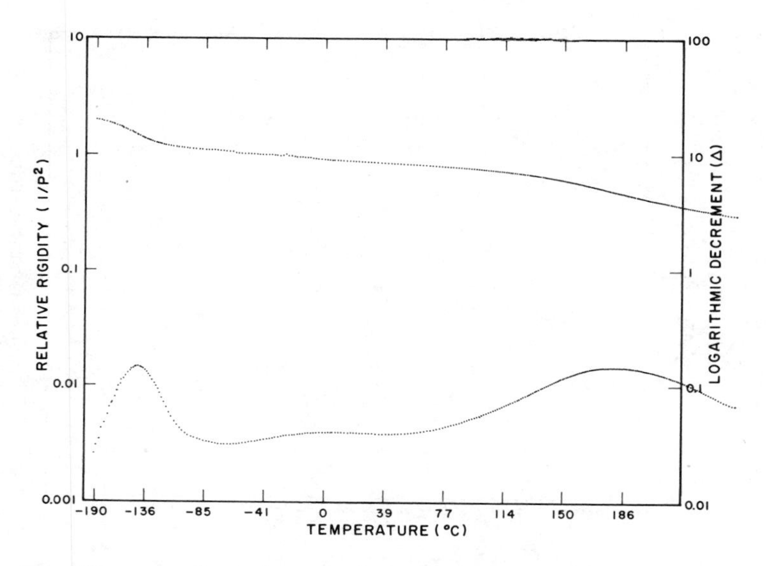

Figure 8D. Thermomechanical spectra (220° → −190°C) after 250°C/5 hr cure: C_{22}

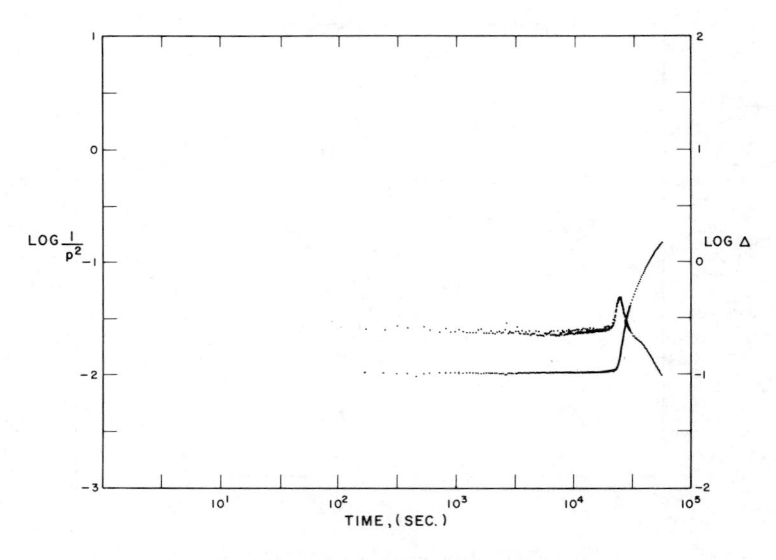

Figure 9. Isothermal cure (220°C/16.5 hr): C_{14}

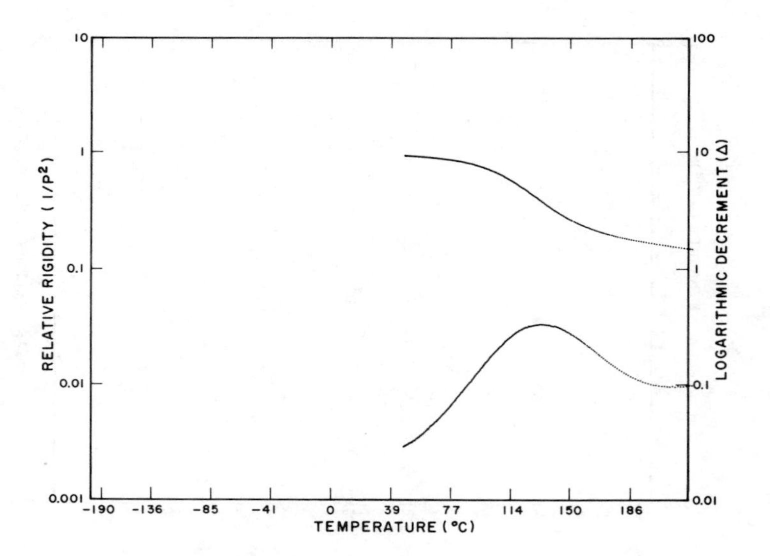

Figure 10. Thermomechanical spectra (220° → 45°C, $\Delta T/\Delta t \leq 1.5°C/min$) after 220°C/16.5 hr cure (same specimen as for Figure 9): C_{14}

vitrification) would be observed isothermally as a prolonged (because the rate of reaction would be controlled by the low concentration of reactive groups at late stages of chemical conversion) and weak process. It does appear that of the four materials the maximum glass transition temperature may be above 250°C only for the C_6-methyl material; comparison of Figures 7A and 8A shows that the maximum T_g attained (after 5 hours at 250°C) is close to 250°C.

The thermomechanical plots (Figures 8A, 8B, 8C and 8D) for the four materials obtained after curing at 250°C for 5 hours display two transitions below T_g. Numerical values are given in Table VI. The cryogenic, more intense transition, designated T_{sec}, is ascribed (1,4,7) to motion of the flexible inter-aromatic segments. As anticipated, the more flexible the segment the more intense the cryogenic mechanical loss peak: the intensity therefore increases in the order C_6-methyl $< C_{10} < C_{14} < C_{22}$. The least flexible segment, in the C_6-methyl material, is also responsible for its loss peak being at a higher temperature (-98°C) than for its higher homologues ($\sim$ -140°C). The weaker T_{sec}' transition between T_g and T_{sec} also occurs at higher temperatures the lower the segmental flexibility; however its intensity increases relative to the background and relative to that of its T_{sec} relaxation in the order C_6-methyl $> C_{10} > C_{14} > C_{22}$. The origin of the T_{sec}' relaxation is unknown.

Just as two relaxations occur prior to vitrification in isothermal cure, so two relaxations occur above T_g (designated $T' > T_g$ and $T > T_g$) in TBA thermomechanical plots of low molecular weight material (1,6). They move quickly to higher temperatures with extent of cure, as indicated in Figures 2 to 5 for $T > T_g$ and Figures 4 and 5 for $T' > T_g$, and revealed in the corresponding thermomechanical plots (not shown here, but see reference 1 for the C_{10} diamide). They should be of importance to processing since they occur above T_g.

In the course of this work high resolution thermomechanical spectra of low molecular weight material were obtained, including the following example.

Melting of the C_{14} monomer (RT $\rightarrow$ 250°C, $\Delta T/\Delta t$ 5°C/min; 250°C/5 min) and subsequent cooling ($\Delta T/\Delta t \leq$ 0.5°C/min) resulted in a material the thermomechanical behavior (RT $\rightarrow$ 200°C, $\Delta T/\Delta t$ 0.5°C/min) of which is shown in Figure 11 (see also footnote to Table III). Interpretation in terms of the sequence of transitions (designated by the loss peaks) follows:

The rigidity of the glass decreases through T_g (50.5°C), increases as a consequence of crystallization (68°C) and decreases through a series of well-defined melting steps (132°C, 143°C, 152°C and 163°C). The reported melting range was 163-165°C (5). The thermomechanical spectra on heating were reproducible for a fixed cooling rate from the melt. By changing the cooling conditions from the melt in the prehistory, the relative intensities of the transitions could be changed systematically. Multiple melting peaks have been observed on phthalocyanine prepreg material using DSC (8).

TABLE VI. Polyphthalocyanines: Transitions (°C) After 5 Hour Cure at 250°C.

Polymer	T_g (Δ)		T_{sec}' (Δ)		T_{sec} (Δ)	
C_6-CH_3	234	(0.180)	19	(0.0443)	-98	(0.0415)
C_{10}	194	(0.136)	3	(0.0342)	-145	(0.0858)
C_{14}	178	(0.160)	8	(0.0427)	-140	(0.155)
C_{22}	137	(0.234)	1	(0.0459)	-139	(0.229)

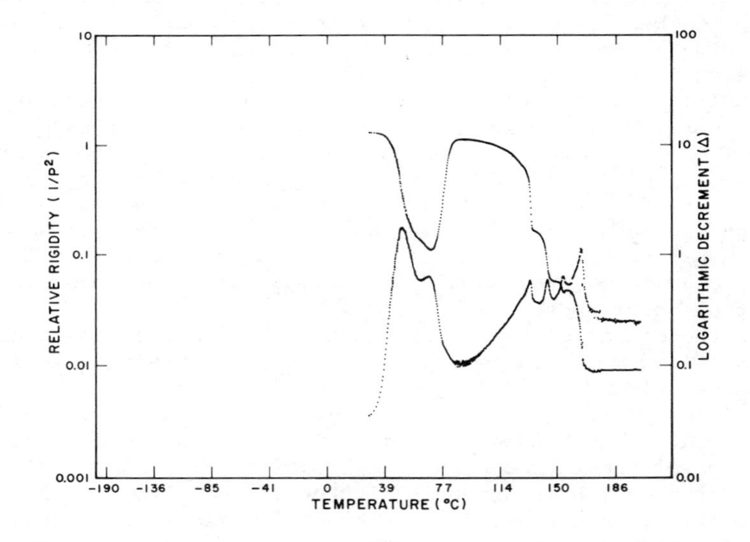

Figure 11. C_{14} monomer: thermomechanical spectra (RT → 200°C, $\Delta T/\Delta t \leq$ 0.5°C/min) after melting (250°C/5 min) and subsequent cooling ($\Delta T/\Delta t \leq$ 0.5°C/min)

Summary

An automated torsion pendulum has been used to study structure/property relationships of four phthalocyanine resins designated C_6-CH_3, C_{10}, C_{14} and C_{22}. The designation reflects the different segments between the aromatic nuclei of the monomers and of the resulting molecular networks. Cure at 250°C resulted in limiting glass transition temperatures, ${}_{250}T_{g\infty}$, which decreased as the segmental length increased: the values were C_6-methyl > 244°C, C_{10} $\sim$ 202°C, C_{14} $\sim$ 185°C and C_{22} $\sim$ 157°C. After a standard cure of 250°C/5 hr. the thermomechanical spectra revealed two transitions below T_g. That at cryogenic temperatures (designated T_{sec}) was attributed to motions of the inter-aromatic molecular segments; the greater the flexibility of the segments, the lower the temperature and the more intense the T_{sec} relaxation. The relaxation between T_g and T_{sec} (designated T_{sec}') also occurred at lower temperature the greater the segmental flexibility but with decreasing intensity (relative to the background). Two other relaxations occurring above T_g moved to higher temperatures more rapidly than T_g on cure: these bear on the processibility of the materials in the early stages of cure.

Acknowledgment

Appreciation is extended to Dr. W. D. Bascom and Dr. J. R. Griffith of the Naval Research Laboratory for providing samples of the four monomers and for their technical interactions.

Literature Cited

1. Gillham, J. K., Polym. Eng. and Sci., 19(4), 319 (1979).
2. Gillham, J. K., A.I.Ch.E. J., 20(6), 1066 (1974).
3. Gillham, J. K., Polym. Eng. and Sci., 19(10), 676 (1979).
4. Gillham, J. K., Amer. Chem. Soc., Preprints, Div. Organic Coatings and Plastics Chem., 40, 866 (1979).
5. Griffith, J. R. and O'Rear, J. G., in High Performance Composites and Adhesives for V/STOL Aircraft, NRL Memorandum Report 3721, pp 15-20 (Ed. Bascom, W. D., and Lockhart, L. B., Jr.), Naval Research Laboratory, Washington, D.C., Feb. 1978.
6. Gillham, J. K., Benci, J. A., and Boyer, R. F., Polym. Eng. and Sci., 16(5), 357 (1976).
7. Bascom, W. D., Cottington, R. L., Bitner, J. L., Hunston, D. L., and Oroshnik, J., in High Performance Composites and Adhesives for V/STOL Aircraft, NRL Memorandum Report 3721, pp 23-35 (Ed. Bascom, W. D., and Lockhart, L.B., Jr.), Naval Research Laboratory, Washington, D.C., Feb. 1978.
8. Griffith, J. R., private communication (1978).

RECEIVED February 15, 1980.

27

Dynamic Fracture in Aerospace High Polymers

AKIRA KOBAYASHI and NOBUO OHTANI

Institute of Space and Aeronautical Science, University of Tokyo, Komaba, Meguro-ku, Tokyo 153, Japan

High polymers are widely used in aerospace vehicles nowadays. One widely used high polymer for aircraft windshields is polymethyl methacrylate (PMMA). Suppose the aircraft encounters violent air turbulence resulting in heavy gust load during flight. The aircraft windshield material is then subjected to loads leading toward possible dynamic fracture. Perhaps the windshield is one of the weakest portions throughout the entire aircraft structure against such dynamic loading.

In the present paper, both macroscopic and microscopic aspects of dynamic fracture in PMMA are reported. First, the running crack propagation velocity profiles at room temperatures subjected to various strain rates were investigated from the macroscopic point of view, and the strain rate dependency essential to viscoelastic solids were studied. Then, another topic related to microscopic basis is treated. That is, the released energy rates during dynamic fracture were measured, and the correlations between fracture surface patterns observed microscopically and corresponding characteristic fracture mechanics parameter such as energy release rates during dynamic crack propagation were surveyed at different test temperatures under stable crack propagation with the aid of linear fracture mechanics and microscopic tools.

Running Crack Propagation Velocity Profiles

Measurement of Crack Propagation Velocity. For years the present authors have been engaged in the experimental study on dynamic crack propagation in high polymers. There are several techniques to measure the running crack propagation velocity (1), i.e., the velocity gage method, the electric potential method, the ultrasonics method and the high speed photography method. Among these methods the velocity gage method is most favorable especially for plastics because of their electric insulation properties.

0-8412-0567-1/80/47-132-367$05.00/0

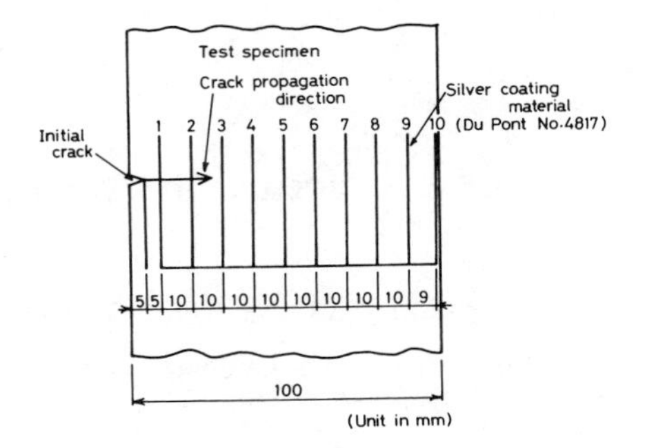

Figure 1. Velocity gauge arrangement

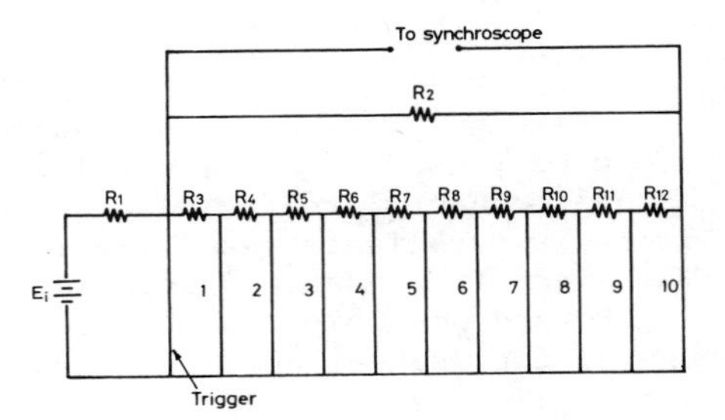

Figure 2. Electronic circuit

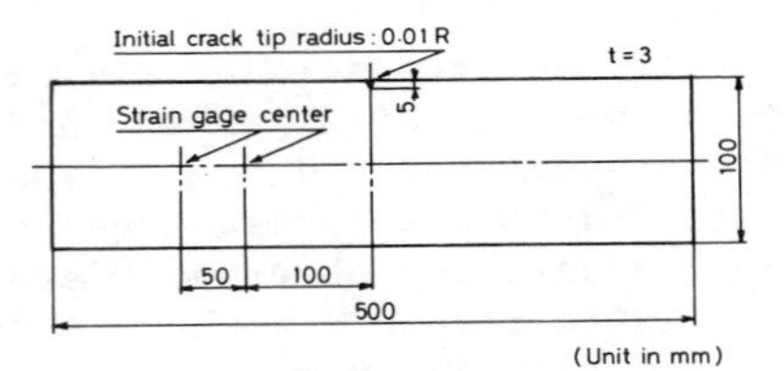

Figure 3. Test specimen for macroscopic study

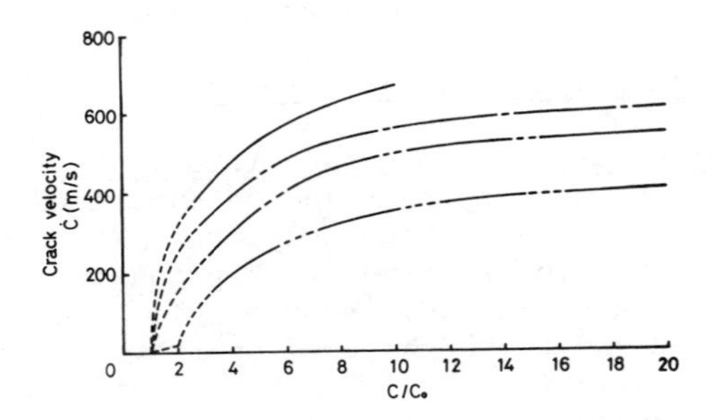

Figure 4. Strain-rate-dependent crack propagation velocity profiles: (——) $\epsilon = 48\ sec^{-1}$, (— — —) $\epsilon = 2.6 \times 10^{-2}\ sec^{-1}$, (— — — —) $\epsilon = 2.6 \times 10^{-5}\ sec^{-1}$ (— — — — —) $\epsilon = 2.6 \times 10^{-7}\ sec^{-1}$

In the present study, velocity gages were employed, of which details are as shown in Figure 1. In Figure 1 the conductive silver coating material, DuPont No. 4817, forming a series of conducting wires, is painted perpendicular to the expected crack propagation passage so as to be broken as the running crack front passes across the conducting wires, resulting in the electric resistance change in the electronic circuit shown in Figure 2. Thus the running crack propagation velocities between painted conductive wires are measured combined with the introduction of time scale from the trace on a synchroscope.

Specimens. PMMA, manufactured by Sumitomo Chemical Co. Ltd., Japan, was used and the specimen configuration is as shown in Figure 3. Foil gages of electric resistance type are placed on the specimen to check the applied strain rates.

Experimental Results of Crack Velocity Profiles. In performing the experiment a conventional Instron-type tensile tester was used to realize the mode I constant strain-rate tension loading on a specimen for the strain rates between 2.6×10^{-2}/s to 2.6×10^{-7}/s, while the specially designed high speed tensile tester, UTM-5, made by Toyo-Baldwin, Japan, was employed for the strain rate of 48/s. All the tests were done at room temperature. These test results are shown in Figure 4, where c = the arbitrary crack length and c_0 = the initial crack length of 5 mm.

Discussions. It is quite evident that there exists the strain rate dependency in dynamic crack propagation velocity profiles as seen from Figure 4. This is very interesting, because PMMA has long been treated rather elastic at least at room temperatures so far. However, the obtained strain rate dependent crack propagation velocity profiles tell us that PMMA should be treated as viscoelastic rather than elastic, since the strain rate dependency emerges more or less from the viscous component inherent to such viscoelastic solids corresponding to the applied strain rates.

Fracture Surfaces and Energy Release Rates

Next, fracture surfaces were observed microscopically and their correlations with energy release rates were studied.

Energy Release Rates. As is well-known, energy is released as a crack advances when the external load is applied on a specimen. There are various methods of characterizing resistance to cracking or released energy in materials. Among methods, one can measure the specific work of fracture of a quasi-statically propagating crack. Conventional techniques to obtain toughness in metals are rather complicated in case applied to viscoelastic solids, since the compliance calibrations vary with different testing crosshead speeds and different crack propagation velocities during

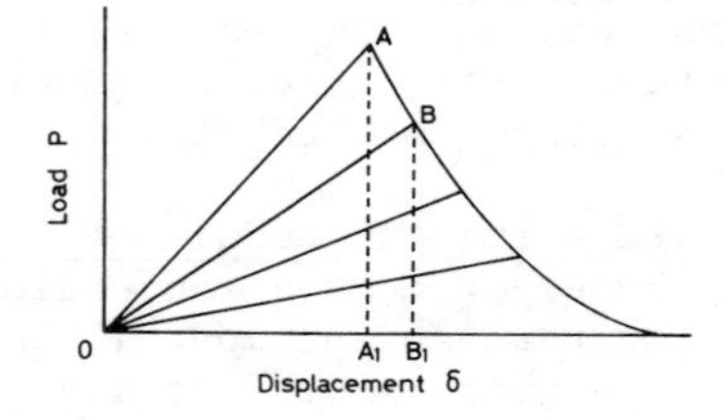

Figure 5. Load-displacement relation employed for the sector area method

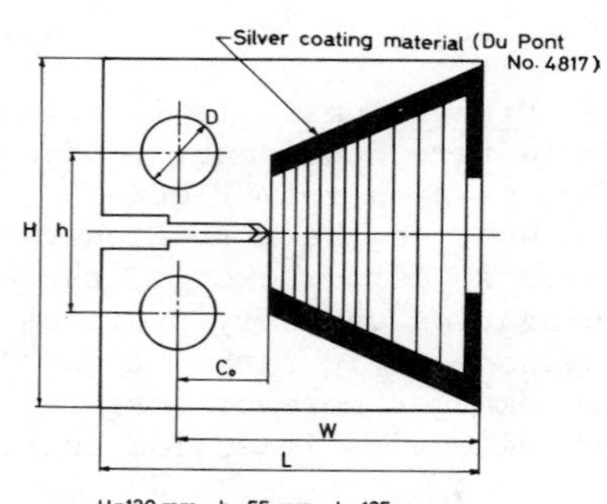

Figure 6. Compact tension specimen for microscopic study

fracture because of strain rate dependent material moduli. The sector area method proposed by Gurney and Hunt (2) is a powerful technique under these circumstances, as the very method eliminates tedious and troublesome compliance calibrations. Furthermore, it offers many incremental released energy values from a single test. If the specimen is prepared such that the crack propagation velocity varies during quasi-static crack propagation, this might reflect strain rate dependency in the work of fracture.

Via sector method, the energy release rate G prescribed by two quantities, a load P and a displacement δ at the loading pin, can be experimentally measured as shown in Figure 5, where for example, ΔOAB denotes the very energy release as the crack front advances corresponding to the displacement increment A_1B_1. P is measured by a load cell and δ by a clip gage.

Specimen. To realize the quasi-static crack propagation, a compact tension type specimen as shown in Figure 6 was used, where velocity gages are also employed to measure the crack velocity.

Experimental Procedures. In the present experiment, these energy release rates were measured during quasi-static crack propagation produced by mode I tension with an Instron type tensile tester at 0.5mm/minute crosshead speed. Three test temperatures (243°K, 285°K and 328°K) were obtained with the aid of liquid nitrogen, room temperatures and hot air blower, respectively. As described before, the running crack propagation velocity was measured by velocity gages, load by a load cell and the displacement at the loading pin by a clip gage during fracture. After fracture, fracture surfaces were observed by a microscope with some 50 to 400 times magnifications.

Experimental Results and Discussions. Thus obtained energy release rates in terms of running crack front position during stable crack propagation are shown in Figures 7 to 9. In these figures, the energy release rate G is expressed by the energy release divided by the fracture surface area corresponding to individual crack front advance. In Figures 7 to 9, each chain line shows the weighted average curve, while the shaded region denotes the scatter values. It is recognized that G is at its maximum when the stable crack propagation initiates and then the monotonously decreasing tendency follows hereafter for all cases. The maximum G value decreases with the increase of test temperature. Further, in view of Figure 10, dG/dc becomes large as the decrease in test temperature.

Crack propagation velocities in terms of a running crack front at individual test temperature are shown in Figure 11. In Figure 11, crack velocity profiles for both 328°K and 285°K are rather similar, but that for 243°K is distinctive and always showing lower values compared to the previous two higher temperature cases. All the curves are of decreasing function of a running

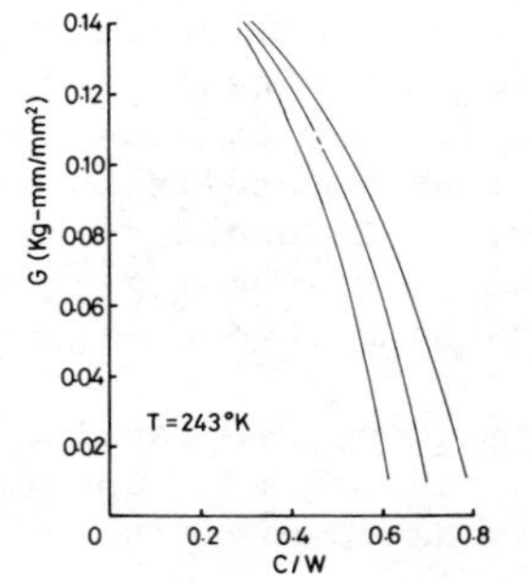

Figure 7. Variation of energy release rate caused by crack propagation (243 K)

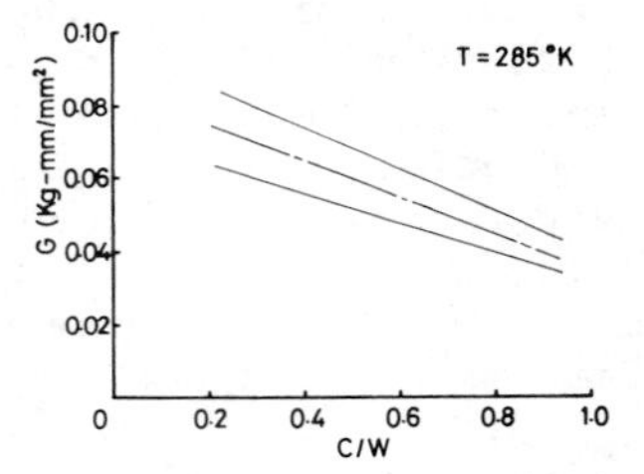

Figure 8. Variation of energy release rate caused by crack propagation (285 K)

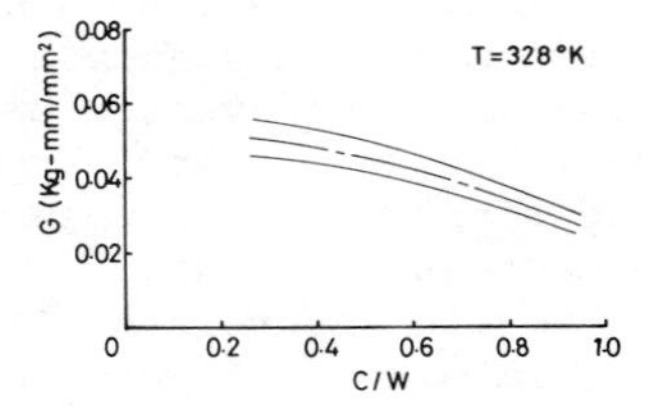

Figure 9. Variation of energy release rate caused by crack propagation (328 K)

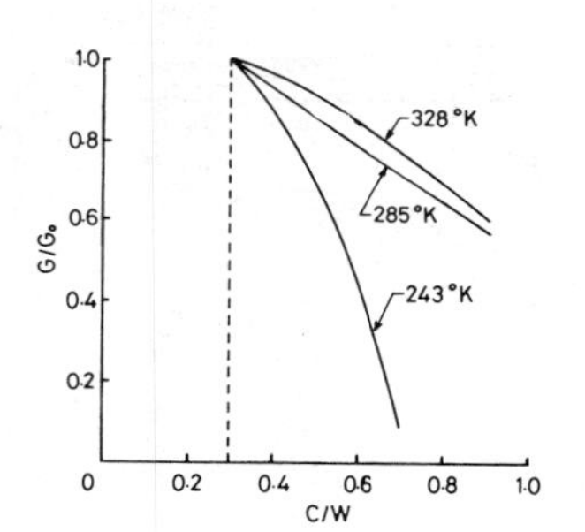

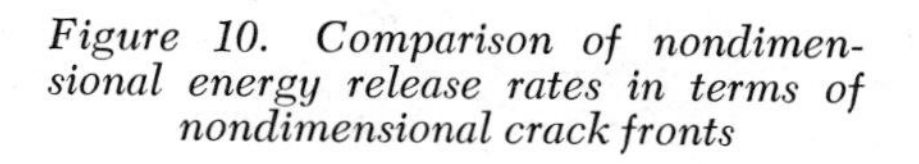

Figure 10. Comparison of nondimensional energy release rates in terms of nondimensional crack fronts

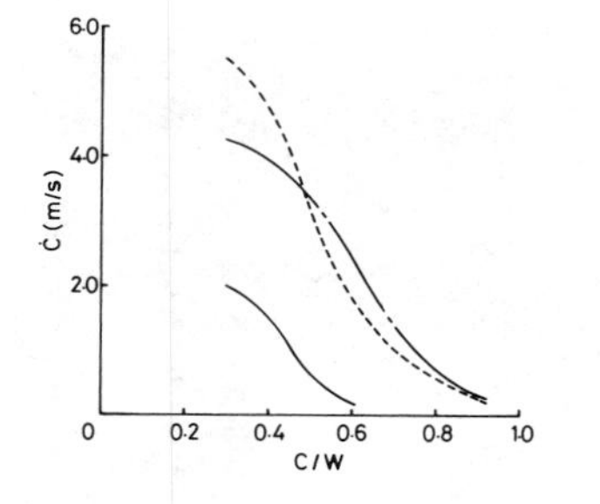

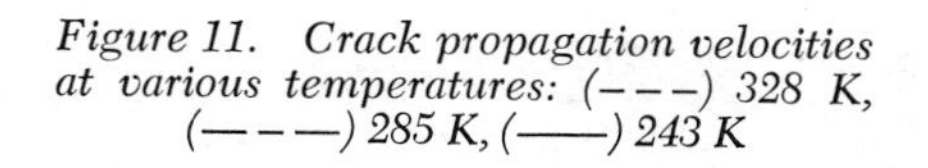

Figure 11. Crack propagation velocities at various temperatures: (- - -) 328 K, (— - —) 285 K, (——) 243 K

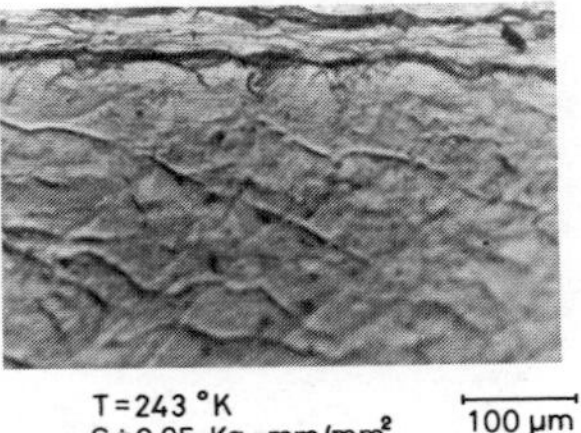

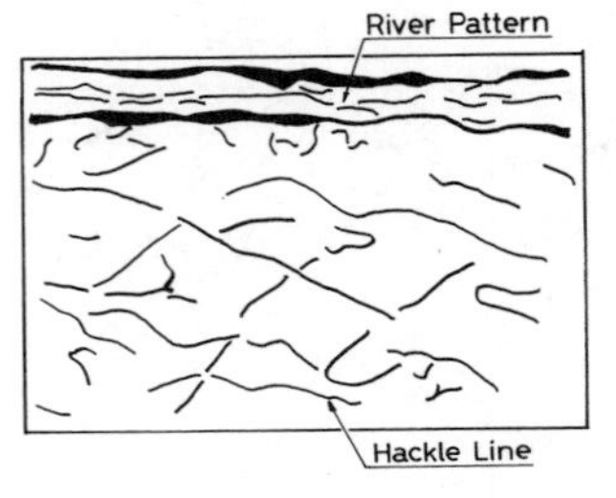

Figure 12. Microscopically observed fracture surface at 243 K (G = 0.05 Kg · mm/mm²)

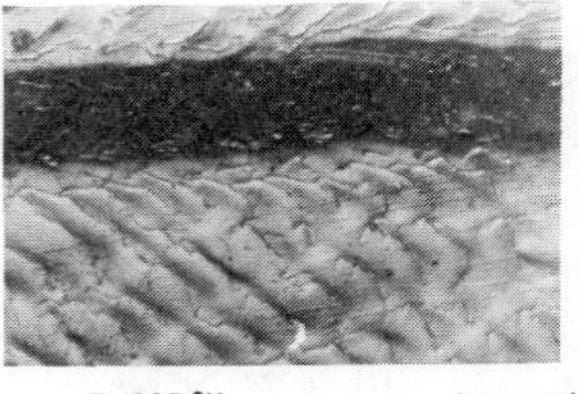

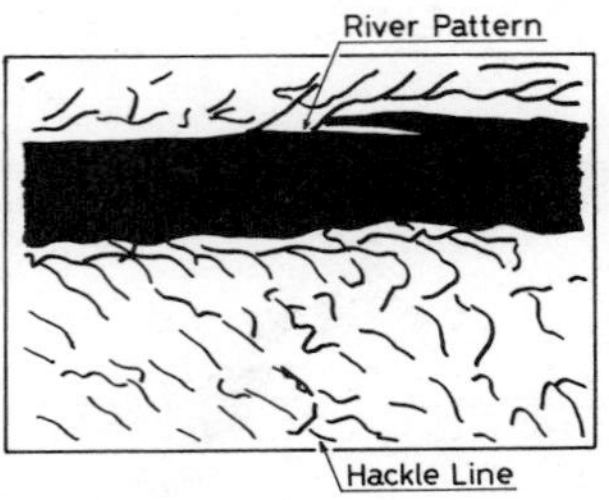

Figure 13. Microscopically observed fracture surface at 285 K (G = 0.05 Kg · mm/mm²)

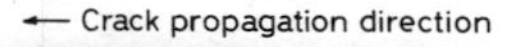

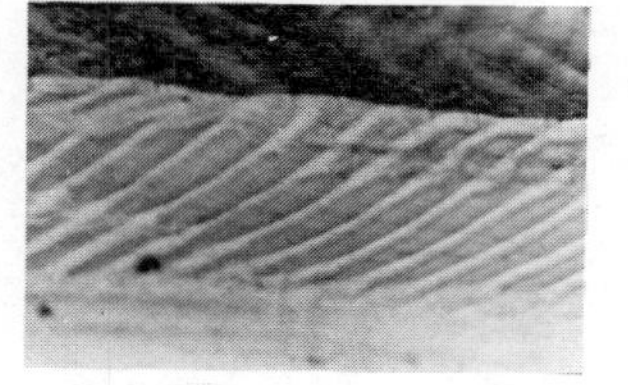

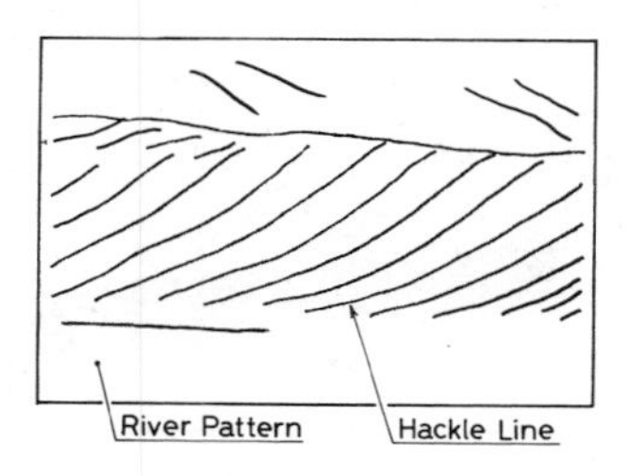

Figure 14. Microscopically observed fracture surface at 328 K (G = 0.05 Kg · mm/mm²)

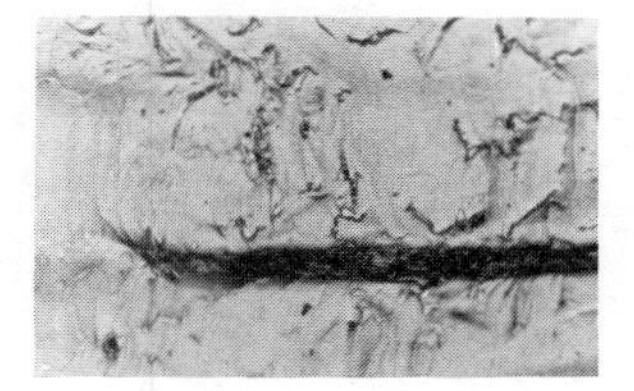

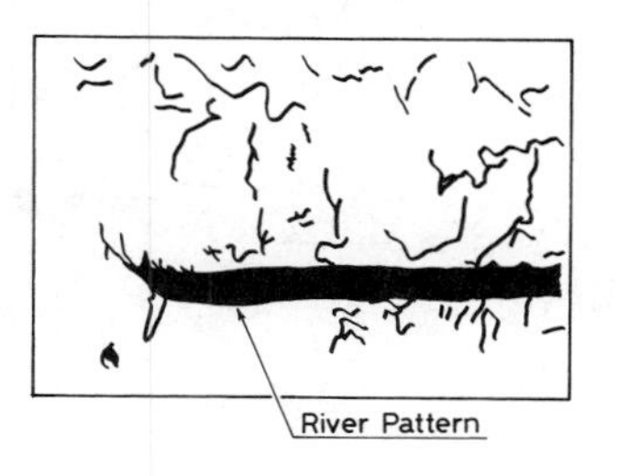

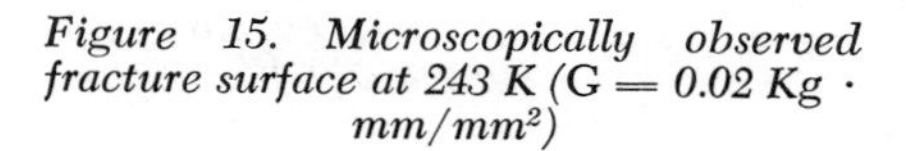

Figure 15. Microscopically observed fracture surface at 243 K (G = 0.02 Kg · mm/mm²)

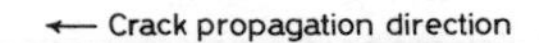

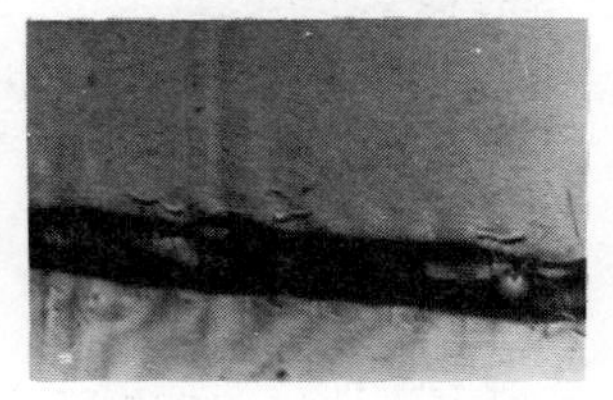

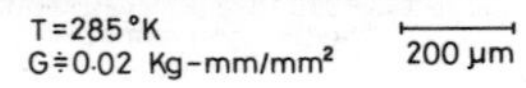

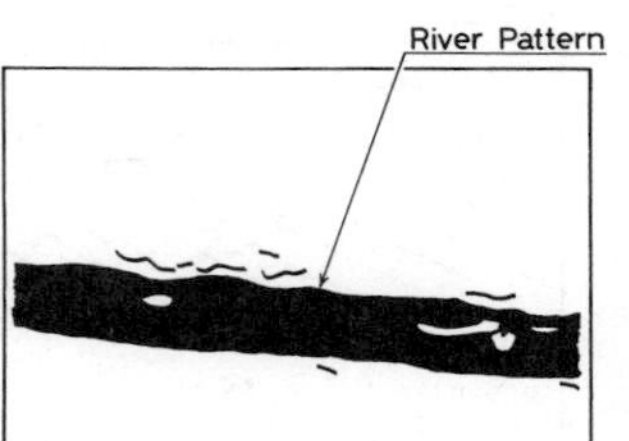

Figure 16. Microscopically observed fracture surface at 285 K (G = 0.02 Kg · mm/mm²)

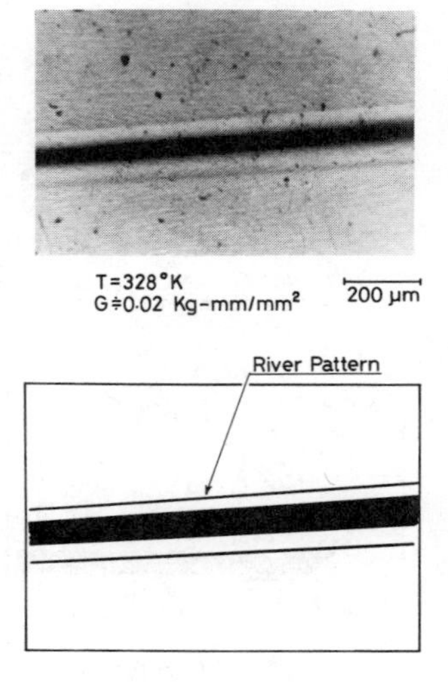

Figure 17. Microscopically observed fracture surface at 328 K (G = 0.02 Kg · mm/mm²)

crack front as the crack advances.

Microscopic inspection of fracture surfaces reveals that there exists two characteristic patterns irrespective of test temperatures. One is a combined hackle and river pattern and the other is represented by river pattern only. Furthermore, it was found that the energy release rates measured were almost identical at those characteristic pattern regions though the test temperatures were different. In Figures 12 to 14 it is observed that there exist hackle lines and river patterns concurrently, and G is equal to about o.05 $Kg\text{-}mm/mm^2$ irrespective of test temperature. Next, as the running crack front advances further, once appeared hackle lines vanish and the fracture surface is to be covered with river patterns and smooth surfaces, having G = 0.02 $Kg\text{-}mm/mm^2$ or so, for all test temperature cases as shown in Figures 15 to 17. It is interesting to find the above-mentioned, for which further effort should be tried.

Conclusions

Two aspects of dynamic fracture behavior, macroscopic and microscopic, were studied for PMMA. Macroscopically the strain rate dependent crack propagation velocity profiles were phenomenologically observed at room temperature, calling attention to the fact that PMMA should be treated as viscoelastic rather than elastic even at room temperature level. In microscopic aspect two characteristic fracture surface patterns were found, one with a combined hackle and river pattern, and the other with river pattern only, while the former has G of 0.05 $Kg\text{-}mm/mm^2$ and the latter 0.02 $Kg\text{-}mm/mm^2$, all irrespective of test temperature.

Acknowledgements

The authors are grateful to Mr. Masayuki Munemura for his effort devoted to the present study. Mr. Jun Nagashima and Mr. Hideo Hananoi are also acknowledged for their assistance in experiment.

Literature Cited

1. Liebowitz, H. ed., "Fracture", Academic Press New York, 1968, Vol. II, p. 545.

2. Gurney, C., Hunt, J., Proc. Roy. Soc. London (1967) A229, 508

RECEIVED February 15, 1980.

28

Stress Rate Dependency of the Tensile Strength of Fiber-Reinforced Plastics

TAICHI FUJII

Department of Mechanical Engineering, Osaka City University, Sugimoto-cho, Sumiyoshi-ku, Osaka 558, Japan

MITSUNORI MIKI

Department of Mechanical Engineering, Osaka Municipal Research Institute, Ogimachi, Kita-ku, Osaka 530, Japan

Several researchers have reported that the tensile strength of glassfiber reinforced plastics (GRP) increases with increasing strain rate (1,2,3,4,5). Though many materials become more brittle under impact conditions, the impact strength and fracture strain of GRP composites increase. Under impact conditions GRP composite show increased ductile fracture behavior. In addition, the impact energy is considerably larger than the static fracture energy. This advantage of GRP composites will become increasingly important in future applications. Even though this phenomena is well recognized, it remains to be analyzed.

With increasing stress rate, the constituent materials react mechanically in a different way than they do under static conditions. The resin-fiber interface may also react differently. Accordingly, we propose two models: a time dependent probabilistic failure model and a mechanical model which recognizes the nonlinear behavior at the interface. These two models can be used to analyze the strength dependency of GRP on the stress rate.

Time Dependent Probabilistic Failure Model

Damage Function. Damage function is defined as the integral of the probability of fracture per unit time or the transition probability of fracture with respect to time. Let P be the probability of fracture of a material and D be the damage function. Then

$$P=1-\exp(-D)$$

The fracture process is regarded as a 2-state, 1-step stochastic process (see Ref. 6 for stochastic fracture process). The transition probability m has been expressed so far as fol-

0-8412-0567-1/80/47-132-379$05.00/0

lows (7).

$$m=\gamma S^{\delta} \qquad (\gamma,\delta;\text{ constants, } S;\text{ stress}) \tag{2}$$

In this case, the fracture stress S_c is given by

$$S_c=\left[\frac{(1+\delta)\dot{S}}{\gamma}\right]^{\frac{1}{1+\delta}} \Gamma\left(\frac{2+\delta}{1+\delta}\right) \tag{3}$$

where $\dot{S}$ is stress rate and Γ denotes gamma function. It should be noted that the fracture stress changes by 47% when the stress rate changes by 10 times if $\delta=5.0$ (8). However, the increase in the strength of GRP composite is about 10% when the stress rate changes by 10 times.

To remove this discrepancy, the following form of the transition probability is proposed.

$$m(t;S)=\gamma S^{\delta} f(t) \tag{4}$$

where f is an arbitrary function which is defined for $t \geq 0$ and can be integrated with respect to time. When the constant stress S is applied to a specimen for time t, the damage function D is defined as

$$D(t;S)=\int_0^t m(\tau;S)d\tau=\gamma S^{\delta} F(t) \tag{5}$$

where

$$F(t)=\int_0^t f(\tau)d\tau \tag{6}$$

Reduced Time Method When the applied stress changes, the evaluation of the damage function is performed by "the reduced time method". Figure 1 shows the simplest stress history. Let m_0 be the transition probability for $S=S_0$ and m_1 for $S=S_1$.

$$m_0=\gamma S_0^{\delta} f(t) \tag{7a}$$

$$m_1=\gamma S_1^{\delta} f(t) \tag{7b}$$

These are shown in Fig. 2.

The value of the damage function at $t=t_1$, D_1, is equal to the area OABC.

$$D_1=\gamma S_0^{\delta} F(t_1) \tag{8}$$

The damage for S_0, D_1, should be shifted to an equivalent damage in order to evaluate the damage occurred from $t=t_1$ to $t=t_2$. This

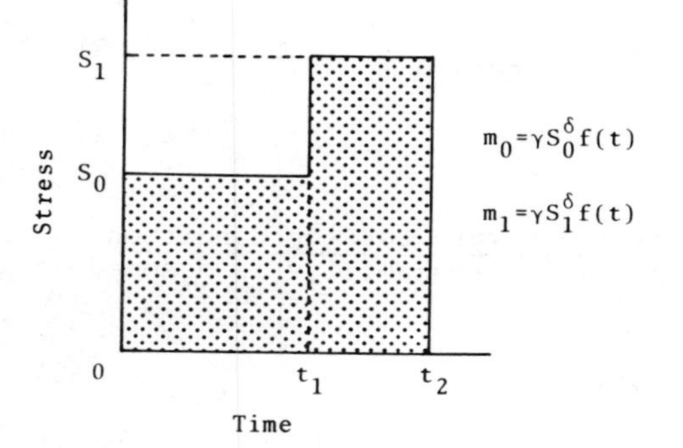

Figure 1. Stress history

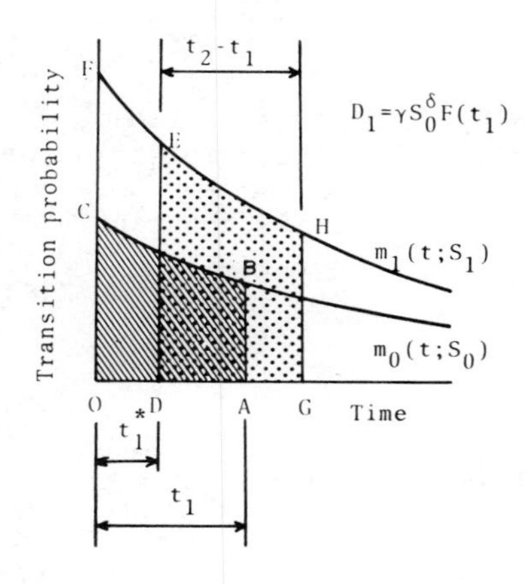

Figure 2. Transition probability

can be done by putting

$$\gamma S_1^{\delta} F(t_1^*) = D_1 \tag{9}$$

where t_1^* is the reduced time of t_1. From Eqs. (8) and (9), t_1^* can be obtained.

$$F(t^*) = \left(\frac{S_0}{S_1}\right)^{\delta} F(t_1) \tag{10}$$

When $t=t_1$, the damage function D is expressed as

$$D = \gamma S_1^{\delta} F(t_1^* + t_2 - t_1) \tag{11}$$

Fracture Stress under a Constant-stress-rate Loading Condition

In this paper, f(t) in a transition probability is assumed as the following form

$$f(t) = t^{\lambda} \qquad (\lambda\text{: constant}) \tag{12}$$

Then

$$F(t) = \int_0^t f(\tau)d\tau = \frac{1}{1+\lambda} t^{1+\lambda} \tag{13}$$

A constant-stress-rate loading condition is expressed as

$$S = at \qquad (a\text{; constant}) \tag{14}$$

Since the damage function is evaluated by "the reduced time method", the increasing stress is approximated to the combination of constant stresses as shown in Fig. 3.

Let Δt be the time interval where the stress is constant, then

$$t_r = r\Delta t \qquad (r=1,2,\cdots,n) \tag{15}$$

and

$$S_r = at_r = ar\Delta t \tag{16}$$

Using Eqs. (13) and (16), Eq. (10) can be written as

$$t_r^* = \left(\frac{r-1}{r}\right)^{\xi} (t_{r-1}^* + \Delta t) \tag{17}$$

where

$$\xi = \delta/(1+\lambda) \tag{18}$$

Therefore,

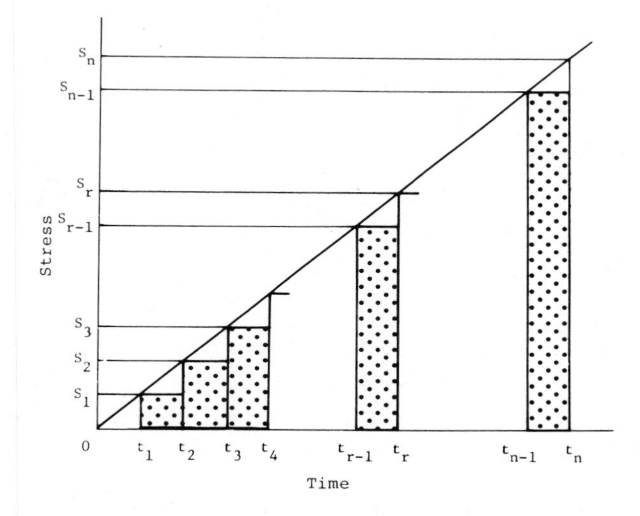

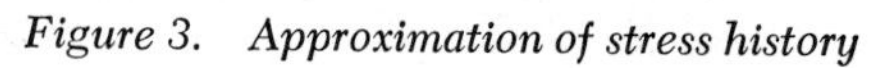
Figure 3. Approximation of stress history

$$t_n^* = \left(\frac{n-1}{n}\right)^{\xi} (t_{r-1}^* + \Delta t)$$

$$= \left(\frac{n-1}{n}\right)^{\xi} \Delta t + \left(\frac{n-1}{n}\right)^{\xi} \left(\frac{n-2}{n-1}\right)^{\xi} (t_{n-2}^* + \Delta t)$$

$$= \cdots\cdots$$

$$= \left[\left(\frac{n-1}{n}\right)^{\xi} + \left(\frac{n-2}{n}\right)^{\xi} + \left(\frac{n-3}{n}\right)^{\xi} + \cdots + \left(\frac{2}{n}\right)^{\xi} + \left(\frac{1}{n}\right)^{\xi}\right] \Delta t$$

$$= \sum_{r=0}^{n-1} \Delta t \left(\frac{r}{n}\right)^{\xi} \quad (19)$$

From Eq. (15), Δt is given by

$$\Delta t = t_n / n \quad (20)$$

Substitution of Eq. (20) into Eq. (19) yields

$$t_n^* = t_n \sum_{r=0}^{n-1} \frac{1}{n}\left(\frac{r}{n}\right)^{\xi} \quad (21)$$

When n approaches infinity, Eq. (21) becomes

$$t_n^* = \lim_{n\to\infty} t_n \sum_{r=0}^{n-1} \frac{1}{n}\left(\frac{r}{n}\right)^{\xi} = t_n \int_0^1 x^{\xi} dx = \frac{1}{1+\xi}\, t_n \quad (22)$$

Substitution of Eq. (18) into Eq. (22) yields

$$t_n^* = \frac{1+\lambda}{1+\lambda+\delta}\, t_n \quad (23)$$

When $t=t_n$, the damage function is evaluated by using Eqs. (5), (13), and (23).

$$D = \frac{\gamma S_n^{\delta}}{1+\lambda} \left(\frac{1+\lambda}{1+\lambda+\delta}\, t_n\right)^{1+\lambda} \quad (24)$$

Substituting Eq. (14) into Eq. (24) and letting $D=D_c$ (D_c: critical value of damage function. D_c=0.693 when the probability of failure is 50%), the fracture stress under constant-stress-rate loading condition can now be obtained as

$$S_c=a^{\frac{1+\lambda}{1+\lambda+\delta}}\left(\frac{(1+\lambda+\delta)^{1+\lambda}D_c}{\gamma(1+\lambda)^{\lambda}}\right)^{\frac{1}{1+\lambda+\delta}} \quad (25)$$

If $\lambda=0$, that is, the transition probability does not influenced by time, Eq. (25) is reduced to

$$S_c=a^{\frac{1}{1+\delta}}\left(\frac{(1+\delta)D_c}{\gamma}\right)^{\frac{1}{1+\delta}} \quad (26)$$

which is same as Eq. (3) essentially.

Mechanical Model with Non-linear Interfacial Behavior

When the relation between shear deformation and force in the fiber matrix interface in fibrous composite materials is linear or perfectly plastic, the analytical treatment is completed(9). But the shear deformation in the interface becomes non-linear according to the increase of load, loading rate and the change of environmental condition at aerospace. It can be thought that many patterns exist as the non-linear shearing deformation and force relation. For the simplicity of the analytical consideration, we adopted the relation of the m-th power of interfacial shear force is proportional to shear deformation. Then the relation becomes soft or hard according to $m>1$ or $m<1$, respectively.

The dispersed state of fiber into the matrix is considered with various cases. That is, the fiber is continuous or discontinuous and the orientation is regular or random etc. Figure 4 is the fundamental case of the discontinuous fiber dispersing into the matrix regularly. Fig. 5 shows the deformated state of unit area of fiber and matrix.

Considering the equilibrium of displacement and force, the following equation is introduced.

$$k\frac{d}{dx}(dP/dx)^m-AP=-B \quad (27)$$

where $A=(1/E_mA_m)+(1/E_fA_f)$

$B= P_{om}/E_mA_m+P_{of}$

E_m, A_m: Young's modulus and sectional area of matrix
E_f, A_f: Young's modulus and sectional area of fiber

Figure 6(a) shows the shear deformation s at the interface. dP/dx, h and d_f are shear force, thickness of interface and fiber diameter, respectively. Fig. 6(b) is the relation between shear force and shear deformation. s_0 is any constant adopted for normalization of shear deformation.

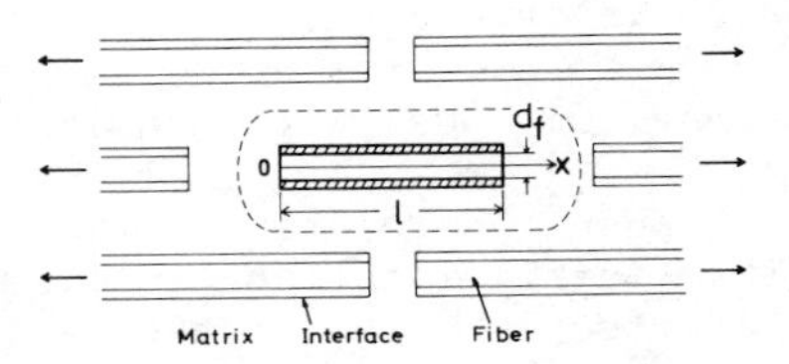

Figure 4. Schematic of short-fiber-reinforced composites

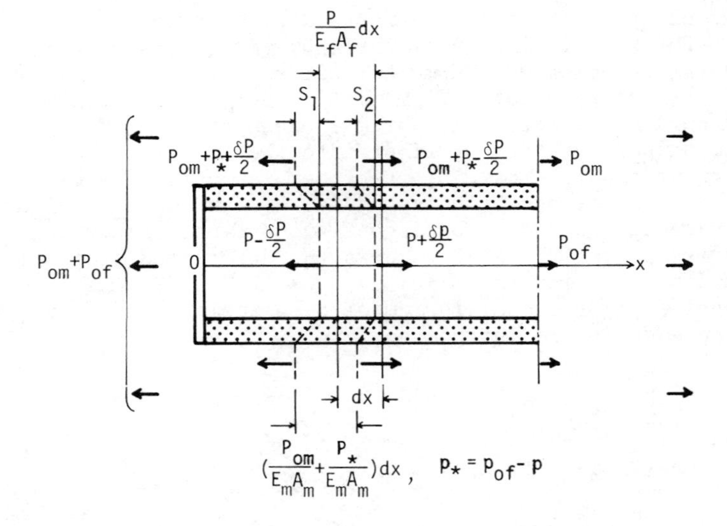

Figure 5. Schematic of deformed state of fiber, interface, and matrix

Figure 7 shows the examples of interfacial shear force and deformation. As is shown in this figure, the relation becomes soft curve when m>1 and hard when m<1. The solution of Eq. (27) becomes different types according to the value of m.

(a) m>1 Let $P=P_0+\sum_{n=1}^{\infty}\lambda^n P_n$ and substitute it into Eq. (27), then the solution is obtained from the coefficients of the same order term of λ.

$$P=-(B/A)+Cx^{\frac{m+1}{m-1}}+C_{1,1}x^{\alpha_{1,1}}+C_{1,2}x^{\alpha_{1,2}}+\cdots\cdots \tag{28}$$

$\alpha_{1,1}$ and $\alpha_{1,2}$ are the roots of the following equation.

$$\mu^2+\left(\frac{2A_*^{(m-2)}}{m-1}\binom{m-1}{1}-1\right)\mu-\frac{1}{\bar{A}A_*\left((m+1)/(m-1)\right)^{m-2}}=0 \tag{29}$$

where $A_*=\bar{A}\frac{2(m+1)^m}{(m-1)^{m+1}}$, $\bar{A}=km/A$

$C_{1,1}$ and $C_{1,2}$ are the constants determined by the boundary conditions. That is, P=0 at the end of fiber and $P=P_{max}$ at the center of fiber.

(b) m<1 The solution is obtained in the case of m=1/3 as follows.

$$P=-(B/A)-Cx^{-2}+C_{1,1}x^{-1}+C_{1,2} \tag{30}$$

(c) m=1 In this case, the behavior of interface is linear. The solution is obtained as follows

$$P=(B/A)\left\{1-\frac{\cosh\sqrt{A/k}\ (\ell/2-x)}{\cosh\sqrt{A/k}\ \ell/2}\right\} \tag{31}$$

where

$$A/k=\left(\frac{1}{E_mA_m}+\frac{1}{E_fA_f}\right)\frac{\pi(d_f+h)}{h}G$$

G: tangential modulus of interface
ℓ: length of fiber

According to the "Law of Mixture", the strength of fibrous composites S_c is expressed as follows.

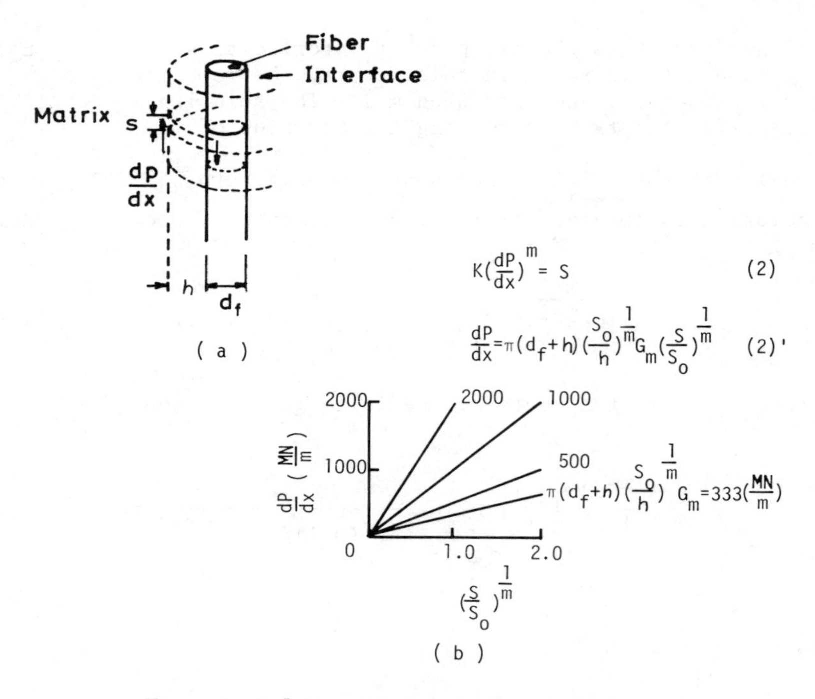

Figure 6. Relations of interfacial force and deformation

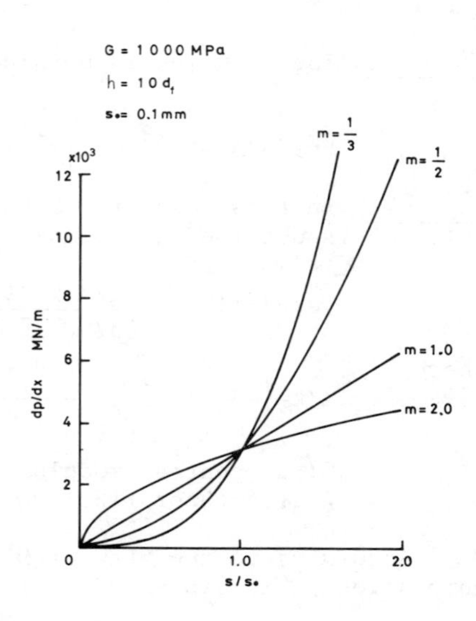

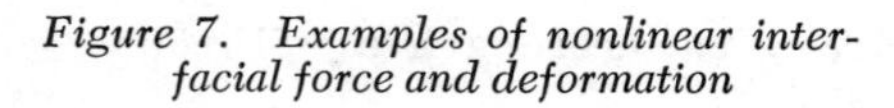

Figure 7. Examples of nonlinear interfacial force and deformation

$$S_c=(\sigma_{fu}/\alpha)V_f+(\sigma_m)_{\varepsilon_{fu}}(1-V_f) \tag{32}$$

where α is the ratio of σ_{fmax} to σ_{fmean}. V_f is the volumetric content of fiber. σ_{fu} and ε_{fu} are ultimate strength and strain of fiber respectively. Fig. 8 shows the relation between α and the exponent of non-linearlity. As is shown in this figure, the increase in α is corresponding to the increase in m.

When the behavior of interface is expressed by the Maxwell type model, linear spring and linear viscous damper are connected seriesly. Hence the following equation is introduced.

$$\frac{1}{G_*}\frac{d}{dt}(dP/dx)+(1/\eta)(dP/dx)=ds/dt \tag{33}$$

where G_*: modulus of rigidity of linear spring element of interface mechanical model
η : coefficient of viscousity

Let $dP/dx=at$ (a; constant) (34)

Substituting Eq. (34) into Eq. (33), Eq. (33) becomes the following equation.

$$s=(1/G_*)(dP/dx)+(1/2\eta)\frac{(dP/dx)^2}{a} \tag{35}$$

Hence $(s)_{a\to\infty}=(1/G_*)(dP/dx)$

and $(s)_{a\to 0}=(1/2\eta a)(dP/dx)^2$

In the latter case, it coincides with the case of m=2. This means the strength decrease according to the decrease of loading rate. Similarly the Voigt type model is applied to the behavior of interface, that is, linear spring and linear viscous damper are connected into parallel. Hence the following equation is introduced.

$$dP/dx=G_*s+\eta ds/dt \tag{36}$$

Let $dP/dx=at$ (a; constant) (37)

Substituting Eq. (37) into Eq. (36), Eq.(36) becomes the following equation.

$$s=(1/G_*)(dP/dx)-(\eta a/G_*)\frac{ds}{d(dP/dx)} \tag{38}$$

we assume that the modulus G_* increases according to the increase in s. Hence

$$(s)_{a\to\infty}\doteqdot(dP/dx)^{1/3}$$

In this case, the exponent of non-linearity decreases according to the increase of loading rate.

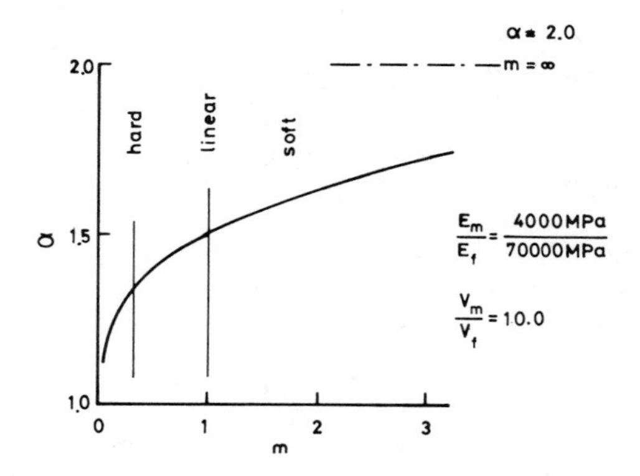

Figure 8. Ratio of fiber stress concentration and the exponent of nonlinear interfacial deformation property

Table I Material

Designation	Reinforcement	Matrix	Glass Content (vol. %)	Longitudinal Modulus (GPa)
CD	Plain woven glass cloth, MG253 (Asahi Fiber Glass Co.)	Unsaturated Polyester resin, Polymal 8279 (Takeda Chemical Industries Ltd.)	36.9	16.0
CH			25.8	
MD	Chopped strand glass mat (Asahi Fiber Glass Co.)		24.6	
MF			15.3	
R			0	3.66

Experimental

Materials Table I shows the material description and Fig. 9 shows the dimensions of the specimen used.

Testing Arrangement Static tensile tests were carried out on an Instron type universal testing machine at crosshead speed of 0.5, 5.0, 50, and 500 mm/min. The impact tensile tests were performed on a drop weight impact tester(4). It gives an impulsive tensile load by using a free fall of the weight. Lead cone buffers were applied to the impact tester to produce constant loading rate.

Results

Figure 11 shows typical load-time and strain-time curves for impact tensile tests. Figure 12 shows the relation between the tensile strength and the stress rate. The straight lines are the linear regression lines.

Discussion

As shown in Fig. 12, the tensile strength of GRP increases according to the increase of loading rate. Then the following equation can be obtained.

$$(S_c)_{exp}=S_{co}(\dot{S}/\dot{S}_o)^c \tag{39}$$

where S_c; tensile strength, S_{co}; tensile strength under unit stress rate, $\dot{S}$; stress rate, $\dot{S}_0$; unit stress rate (1MPa/s), c; constant.

the slopes of the straight lines obtained by least square analysis in Fig. 12, which are equal to the constant c, are approximately equal to each other and the mean value is 0.0453. By setting this value equal to $(1+\lambda)/(1+\lambda+\delta)$ in Eq. (25), we have

$$\lambda=0.0475\delta-1 \tag{40}$$

Consequently, we may conclude that a constant non-zero value of λ exists for GRP composites when δ is constant ($\lambda\simeq-0.7$ if $\delta=6$(8), for example), and therefore the time dependent probabilistic failure model is an analytical approach for investigating the effect of stress rate on the fracture stresses of some materials.

When the reinforcing fiber is discontinuous, not only the fiber strength but also the interfacial behavior becomes important. The analytical strength of the mechanical model with non-linear interfacial behavior is expressed by

$$(S_c)_{analytical}\simeq\sigma_{fu}V_f/\alpha \tag{41}$$

with the contribution of the matrix neglected. In this case,

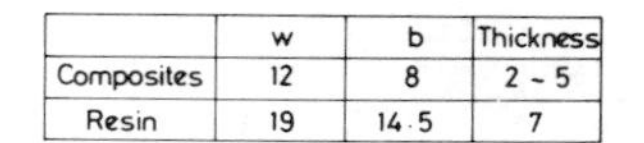

	w	b	Thickness
Composites	12	8	2 ~ 5
Resin	19	14.5	7

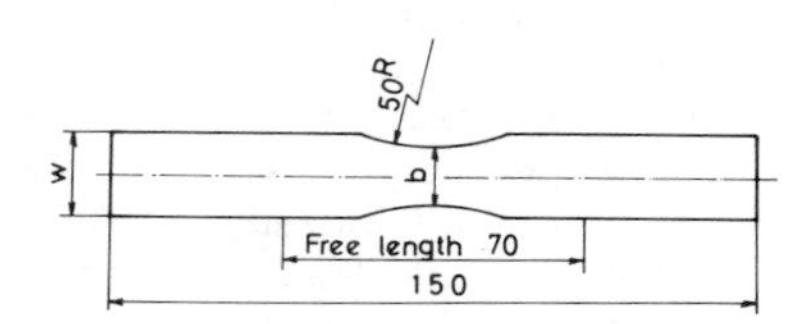

Figure 9. Dimension of the specimen

Figure 10. Drop weight tensile testing machine

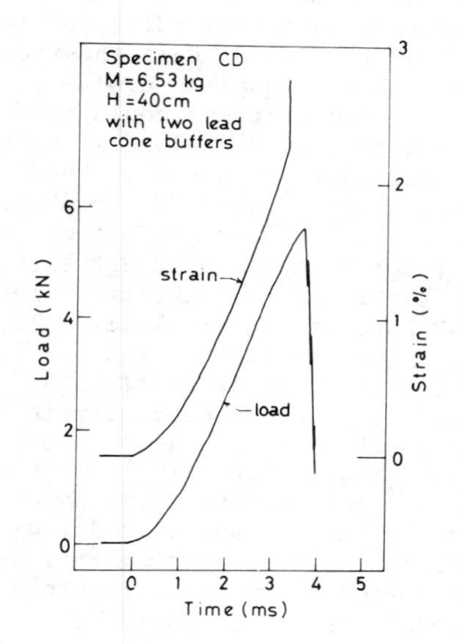

Figure 11. Typical load–time and strain–time curves for impact tensile tests

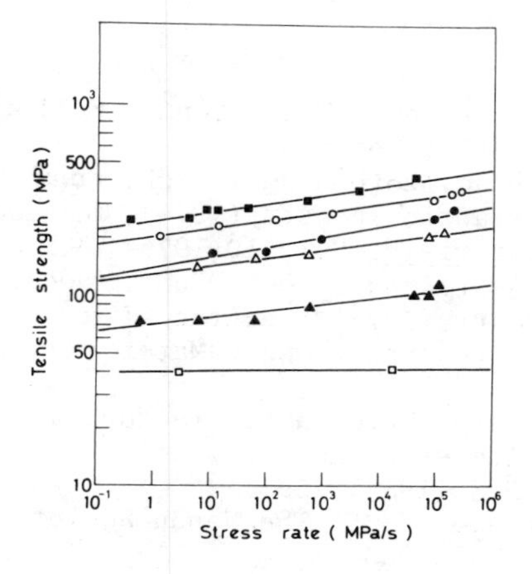

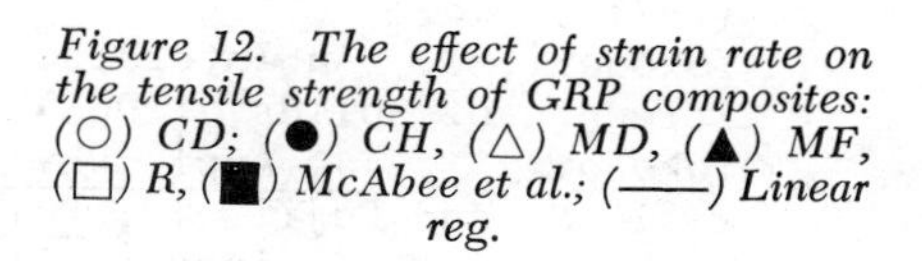
Figure 12. The effect of strain rate on the tensile strength of GRP composites: (○) CD; (●) CH, (△) MD, (▲) MF, (□) R, (■) McAbee et al.; (——) Linear reg.

$\alpha/\alpha_{m=1}$ is corresponding to $(\dot{S}_o/\dot{S})^c$.

Conclusions

We found that the linear relationships are established between the tensile strength of various GRP composites and the stress rate on log-log plot under a constant-stress-rate loading condition. The stress rate ranged from 0.57 to 270000 MPa/s and the slopes of the straight lines were all nearly equal to 0.0453. The effect of the stress rate on the fracture stress could be analyzed by using a time dependent probabilistic failure model and a mechanical model with non-linear interfacial behavior. The transition probability of fracture in the time dependent probabilistic failure model was assumed to be $\gamma S^{\delta} t^{\lambda}$ (S;stress t,time, γ,δ,λ;constants). The theoretical results showed good agreement with the experimental results. It was found that the fracture stress is proportional to the stress rate to the $(1+\lambda)/(1+\lambda+\delta)$. For a glassfiber reinforced plastics, the constants λ and δ were found to be about -0.7 and 6, respectively. Consequently, the traditional time independent probabilistic theory overestimates the effect of stress rate on the strength of materials.

Additionally, the decrease in the ratio of the maximum stress to the mean stress in the discontinuous reinforcing fiber under impact conditions could be explained by the mechanical model with non-linear interfacial behavior. Since the load is transferred only through the fiber-resin interface for short fiber reinforced plastics, this type of non-linear model is considered to be an analytical approach to investigate the stress rate dependency of the strength of the composites with discontinuous fibers.

Literature Cited

1. McAbee, E. and Chumura, M., Proc. Annual Tech. Conf., Society of Plastic Engineers, Vol. XI(1965), Section VII.
2. Rotem, A. and Lifshitz, J.M., Proc. 26th Annual Tech. Conf., RP/C Div., The Society of the Plastic Industry(1971), Sec.10G.
3. Lifshitz, J.M., J. Composite Materials, Vol. 10(1976), 92.
4. Fujii, T. and Miki, M., Proc. 31st Annual Tech. Conf., RP/C., The Society of the Plastic Industry(1976), Section 21-D.
5. Fujii, T. and Miki, M., Proc. 20th Japan Congr. Mater. Res., (1977), 246.
6. Yokobori, T., An Interdisciplinary Approach to Fracture and Strength of Solids, (1968), Wolters-Nordhoff Ltd.
7. Yokobori, T., J. Appl. Phys. 25-5 (1954), 593.
8. Miki, M. and Fujii, T., Bulletin of JSME, The Japan Society of Mechanical Engineers, 21 (1978), 963.
9. Rosen, B.W., Fiber Composite Materials , ASM 72, 75(1965).

RECEIVED January 30, 1980.

29

Experimental Analysis of Hydrothermal Aging in Fiber-Reinforced Composites

D. H. KAELBLE

Rockwell International Science Center, Thousand Oaks, CA 91360

There recently has been an increasing interest and research effort in characterizing the influence of moisture absorption on the physical response of polymer matrix composite materials. Three recent workshops have been exclusively devoted to this subject (1-3). Earlier studies indicated that moisture degrades both the interfacial bond strength and modifies the matrix bulk response by internal plasticization (4, 5). Recent studies show that the application of polyurethane protective coatings on graphite fiber in epoxy matrix (Thornell 300/NARMCO 5208) composite delayed but did not prevent the moisture degradation of composite strength (6). Quantitative chemical analysis programs for characterizing the composite prepreg for quality assurance have now been developed (7-9). These systematic approaches to chemical separation and analysis of polymer composites and adhesives provide an important new tool for analyzing the physiochemical mechanisms of enviromental degradation.

The objective of the present study is to implement an integrated chemical, physical and mechanical test program to analyze the environmental durability of a standard 177°C (350°F) service ceiling graphite/epoxy composite based on T300 graphite fiber (Union Carbide Corp.) and 5208 epoxy matrix (NARMCO Division, Celanese Corp). In order to more clearly highlight the influence of polymer matrix chemistry, comparison data for two other closely related graphite/epoxy composites are included as reference systems.

A uniaxial reinforced panel of graphite T300 in 5208 epoxy prepreg (NARMCO Lot 823 Roll 3) was layed up by using 48 plies of 0.0132 cm thick and 30 cm wide prepreg to produce a flat panel of dimensions by 100 cm length by 0.63 cm thickness. Details of the standard fabrication and cure cycle are summarized in Table I. The cure cycle described in Table I is standard for this composite. The outline of experimental methods applied for physiochemical characterization and evaluation of environmental

0-8412-0567-1/80/47-132-395$05.75/0

Table I
Fabrication and Curing Cycles for T300
Graphite Fiber in 5208 Epoxy Matrix Composite
(Volume Fraction Fiber $V_f = 0.60$)

Step No.	Procedure
1	Lay up 48 linearly aligned plies graphite epoxy prepreg between 6 bleeder plies of 120 glass cloth and 6 plies of 181 glass and vacuum bag.
2	Vacuum on part with heating rate of 1.2 to 3.3°C/min from 22 to 135°C.
3	Hold part under vacuum for 60 min at 135°C at ambient external pressure.
4	Raise external pressure to 6.0 Kg/cm^2 and release part vacuum at 2.0 Kg/cm^2 external pressure.
5	Raise temperature to 180°C at heating rate of 1.2 to 3.3°C per min. and hold 120 min at 180°C.
6	Cool from 180°C to 65°C under 6.0 Kg/cm^2 pressure with thermal gradient in part of less than 25°C.
7	Remove vacuum bag and return unconstrained part for oven post cure at 180°C for 6 hr.

durability are summarized in Table II. Details of the test methods are outlined in earlier reports (9-11).

Results and Discussion

In order to focus the discussion, the results of this study are subgrouped under specific categories.

Chemical Network Defects

Continued development and application of the physiochemical analysis system outlined in the upper portion of Table II provide further insight into the chemical network defects which limit the environmental durability of present 177°C (350°F) service ceiling graphite-epoxy composites. This system of analysis identifies the chemical network structures and curing reactions of Fig. 1 as being common to the graphite-epoxy composites studied in this program (9-11).

The original characterization program of Table II has previously been applied to uniaxial Type AS graphite fiber in 3501-5 epoxy matrix and to Union Carbide T300 graphite fiber in fiberite 934 epoxy matrix. The T300 in fiberite 934 epoxy matrix composite forms the primary skin structure of the cargo bay doors of the Space Shuttle Orbiter and thus is exposed to the environmental regions of ambient terrestial exposure, space flight, and re-entry to earth atmosphere (12). By comparing these two reference composite materials with the T300 in 5208 epoxy of this study, a much clearer picture of the role of chemical structure defects on environmental durability is established.

The lower portion of Fig. 1 shows two crosslinking reactions and chemical crosslink structures. Current analysis shows that the epoxy can homopolymerize (crosslink reaction 1) to form the crosslink shown in the lower left of Fig. 1 (13, 14). Conversely, the co-reaction of epoxide with amine (crosslink reaction 2) produces the crosslink shown in the lower right view of Fig. 1 (13, 14). The data of Table III show that the matrix glass transition temperature Tg is lowered and equilibrium moisture uptake is increased by chemical compositions that favor reaction 1 over reaction 2 in Fig. 1. The use of the BF_3 catalyst is known to favor reaction 1 and to lower the temperature range for curing for easy processability, as shown in the DSC (differential scanning calorimeter) thermograms of Fig. 2.

In the curves of Fig. 2, the rate of chemical curing correlates directly with the amplitude dH/dt of heat release rate. Thermal polymerization of pure DGMDA epoxy requires very high temperature. Addition of 2% by weight catalyst is shown to initiate curing at relatively low temperature. The co-reaction of DGMDA epoxy with DDS cirative is shown to initiate and complete during at intermediate temperatures. The curves of Fig. 2 and data of Table III show that the network dominantly formed from

Table II
Outline for Composite Durability Characterization

Part 1: Analysis of Separated Fiber and Matrix
- 1a. Obtain and separate uncured prepreg components
- 1b. Analyze fiber and matrix surface energies
- 1c. Analyze resin chemistry and curing mechanism
- 1d. Define curing kinetics and network structure
- 1e. Analyze hydrothermal aging effects on network structure

Part 2: Analysis of Composite Laminate Aging
- 2a. Obtain composite laminates for aging studies
- 2b. Measure kinetics of water diffusion into composite
- 2c. Determine interlaminar shear strength versus moisture content.

Outline for Composite Durability Characterization

- 2d. Determine fracture energy versus moisture content
- 2e. Measure dynamic mechanical (NDT) response versus moisture content

Part 3: Data Analysis and NDT Methodology
- 3a. Determine relation between strength degradation mechanisms and NDT methodology
- 3b. Design NDT experiments and statistical analysis for tracking strength degradation
- 3c. Define improved matrix and interface chemistries

Table III

Effects of Chemical Composition on Tg(dry) and Moisture Uptake of Cured Resin (cured 5 hr at 188°C)

Composition (wt%)		Tg(°C)	Equilibrium Moisture Uptake at 100°C Immersion (wt%)
71	Epoxy (TGMDA)		
29	Curative (DDS)	249	5.5 - 6.0
70	Epoxy (TGMDA)	228	5.5 - 6.0
28	Curative (DDS)		
2	BF_3 Catalyst		
98	Epoxy (TGMDA)	202	---
2	BF_3 Catalyst		
100	Epoxy	198	13

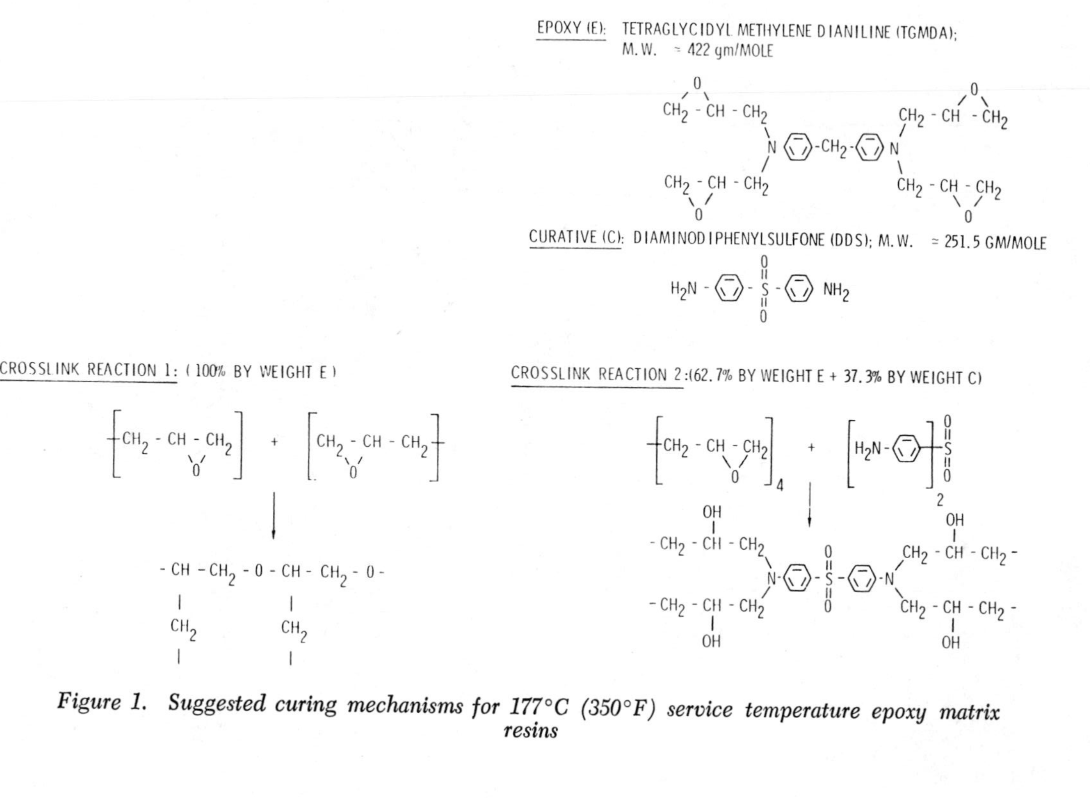

Figure 1. *Suggested curing mechanisms for 177°C (350°F) service temperature epoxy matrix resins*

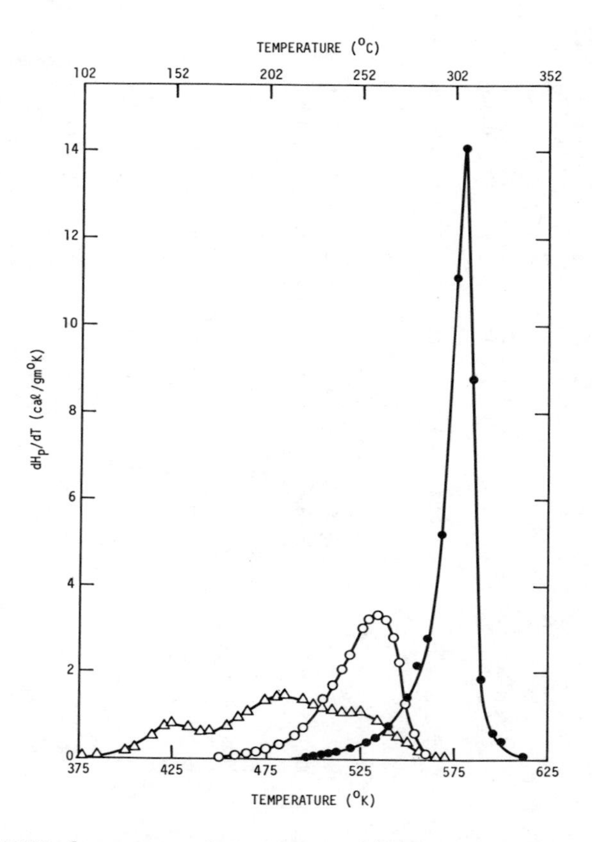

Figure 2. DSC thermograms for reactions of known compositions: (●)(pure TGMDA epoxy, (○) 71 wt % TGMDA epoxy + 29 wt % DDS curative, (△) epoxy + 2 wt % BF3 catalyst

cure reaction 1 (see Fig. 1) can be expected to display lower environmental durability than that formed from reaction 2.

Chemical Defects and Environmental Response

The DSC thermograms for the commercial epoxy matrix materials are shown in Fig. 3. The low temperature initiation of curing in Fiberite 934 and Hercules 3501-5 resins correlates with the BF_3 catalyst-assisted reaction 1, shown in Figs. 1 and 2. NARMCO 5208 shows a curing kinetics in Fig. 3 which closely correlates with reaction 2 of Figs. 1 and 2. The thermograms of Fig. 3 thus show that Fiberite 934 and Hercules 3501-5 resins are more readily cured at low temperatures than is NARMCO 5208. The catalyzed reaction of epoxy groups in Fiberite 934 and Hercules 3501-5 is shown by the curves of Fig. 4 to produce onset of the glass transition damping peak at lower temperature than in amine crosslinked NARMCO 5208 composite. The effect of exposure of these cured composites to moisture aging, followed by thermal scanning of damping properties, is shown in Fig. 5. Fiberite 934 and Hercules 3501-5 resins soften and display onset of high damping at 150°C due to higher moisture sensitivity of their crosslink networks as compared to NARMCO 5208 which retains low damping solid state response to 200°C.

The curves of Figs. 3-5 for commercial epoxy resins reflect the relation between chemical network composition, processability, and environmental durability, which are consistent with the two network reactions of Fig. 1 and the thermograms of Fig. 2 for known reaction mixtures.

Methods of Chemical Characterization

A knowledge of prepreg chemical properties is now established as important information for quality assurance and manufacturing process optimization. The summary for chemical characterization of graphite-epoxy pregreg materials is shown in Table IV. The methodologies used are based on instrumental techniques and yield quantitative information concerning the physiochemical properties of the pregreg material (7-9). For example, the higher level of free DDS curative in NARMCO 5208 correlates with the lower degree of cure and higher heat of polymerization, as shown in the data summary of Table IV. The data of Table IV shows the importance of differentiation between total DDS curative, as measured by IR spectroscopy, and free amine, as measured by quantitative molecular separation using liquid chromatography. The numeric information relating the chemical composition of the prepreg systems summarized in Table IV forms part of the materials and processes approach to chemical defects definition outlined in earlier reports (7-9).

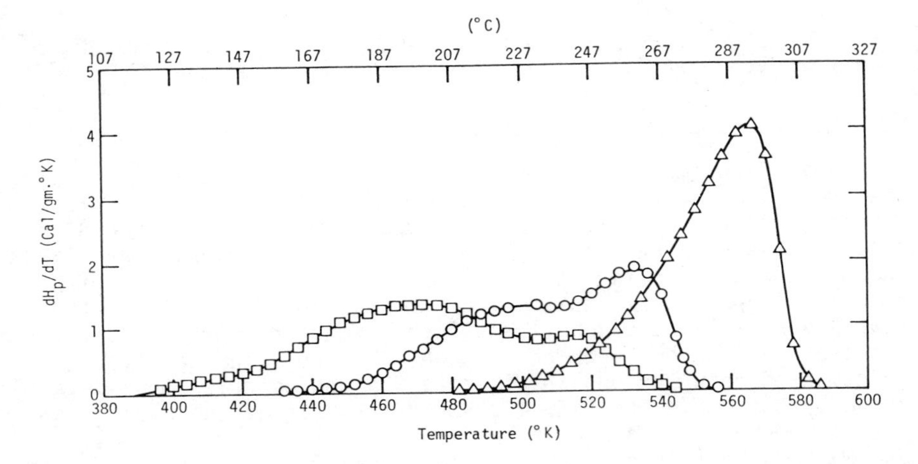

Figure 3. DSC thermograms for curing reactions of commercial epoxy matrix materials extracted from prepreg (DSC scan rate φ = 20°C/min): (□) Fiberite 934, (○) Hercules 3501-5, (△) Narmco 5208

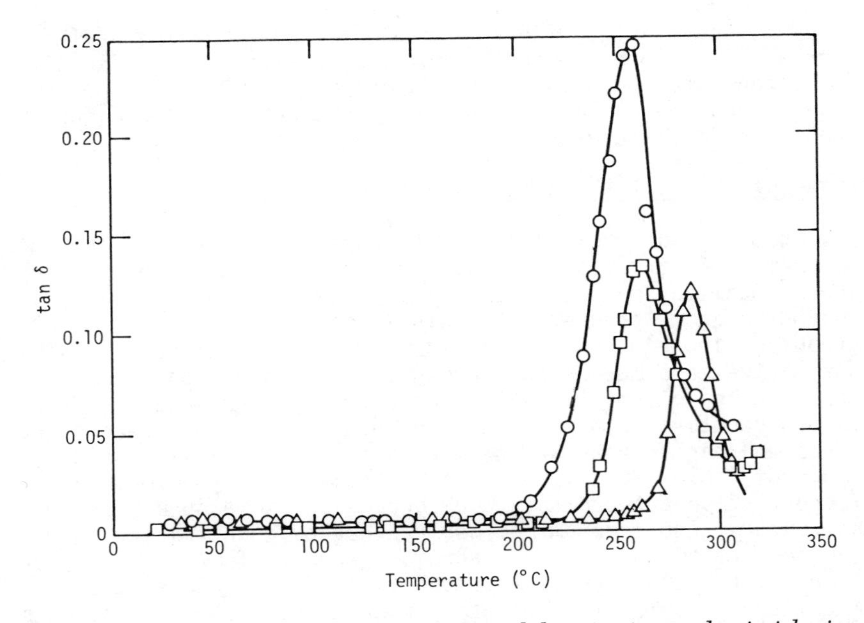

Figure 4. Rheovibion thermal scans for flexural damping in cured uniaxial reinforced graphite–epoxy composite in the dry unaged condition: (○) Fiberite 934/T300, (□) Hercules 3501-5/AS, (△) Narmco 5208/T300

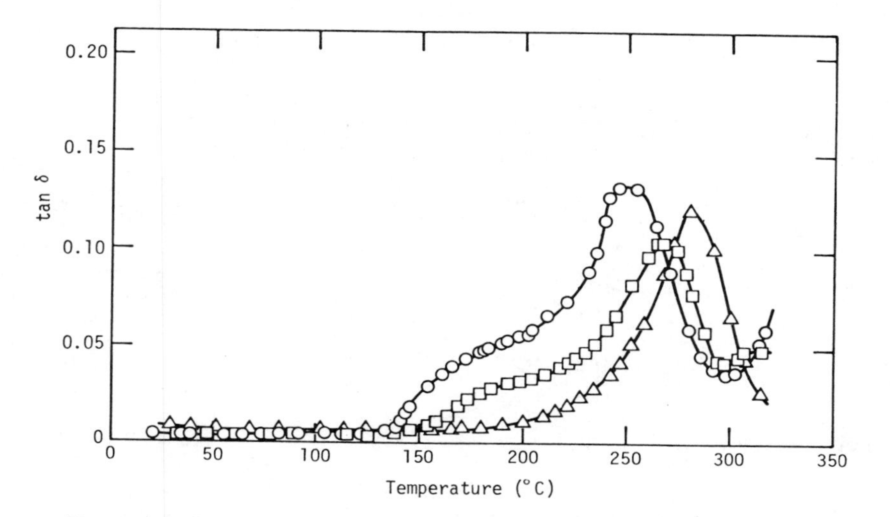

Figure 5. Rheovibion thermal scans of flexural damping in cured uniaxial reinforced graphite–epoxy composite in the wet-aged condition: (○)Fiberite 934/T300, (□) Hercules 3501-5/AS, (△) Narmco 5208/T300

Table IV
Chemical Characterization of Graphite-Epoxy Prepreg Materials

	Reference Systems		This Study
Epoxy Matrix	Hercules 3501-5	Fiberite 934	NARMCO 5208
Graphite Fiber	Hercules Type AS	U. Carbide T300	U. Carbide T300
% Total DDS Curative by IR Spectroscopy	29.2	27.8	22.1
% Free DDS Curative by Liquid Chromatography	18.1	14.5	17.8
Epoxide Equivalent	205	227	173
Wt. % BF_3 Type Boron	0.047	0.022	.0005
Relative Degree of Cure by Liquid Chromatography	22	27	6.9
Heat of Polymerization by DSC (Cal/g polymer)	107	107	140

Table V

Environmental Durability Characterization of Major 177°C (350°F) Service Temperature Graphite Epoxy Composites (Standard Cure, Uniaxial)

		Reference System		This Study
1. General				
Epoxy Matrix		Hercules 3501-5	Fiberite 934	NARMCO 5208
Graphite Fiber		Hercules Type AS	U. Carbide T300	U. Carbide T300
Fiber Finish		None	1% by Fiber wt bisphenol-A epoxy	1% by Fiber wt bisphenol-A epoxy
Post Cure Cycle		3 hr at 188°C	2 hr at 177°C	6 hr at 180°C
Fiber Volume Fraction		0.60	0.60	0.60
2. Surface Energies (dyn/cm)				
Uncured Matrix	γ^d	30.9 ± 2.1	29.9 ± 2.4	29.2 ± 2.3
	γ^p	10.1 ± 2.2	14.4 ± 2.9	13.4 ± 2.8
Cured matrix	γ^d	28.8 ±2.0	28.0 ± 2.2	28.5 ± 2.1
	γ^p	12.4 ± 2.1	14.5 ± 2.4	13.2 ± 2.3
Fiber	γ^d	26.7 ± 3.0	24.4 ± 2.6	24.9 ± 2.6
	γ^p	29.9 ± 5.8	23.4 ± 3.8	25.2 ± 3.8
3. 75°C Matrix Moisture Absorption				
Max. H_2O Uptake (wt%)		6.50	6.39	5.84
Diffusion Coefficient D_R (10^{-8} cm^2/sec)		2.08	.89	1.45
4. Dynamic Mechanical Damping (110 Hz, Flexure, Unaged)				
Alpha Transition (°C) tan δ_{max}, 110 Hz		252.0	258.0	282.0
Range (½ tan δ_{max})(°C)		43.0	37.0	24.0
tan δ_{max}		0.13	0.25	0.12

Table V (Continued)
Environmental Durability Characterization of Major 177°C (350°F) Service Temperature Graphite Epoxy Composites (Standard Cure, Uniaxial)

5. 75°C Moisture Absorption Coefficients ($D_i = 10^{-8}$ cm^2/sec)

		References	System	This Study
1st Cycle	D_1	1.30	0.53	0.76
	D_2	0.44	0.20	0.38
	D_3	0.39	0.40	0.47
2nd Cycle	D_1	3.94	2.09	1.90
	D_2	0.81	0.50	1.17
	D_3	0.85	0.80	0.74

D_1 = transfibrous surface, D_2 = translaminar surface, D_3 = interlaminar surface of uniaxial composite.

6. Ultrasonic Response 23°C, 2.25 MHz, Longitudinal Wave

Spatial Attenuation α_L (neper/cm)

	References	System	This Study
Unaged, dry	3.80	4.75	5.37
Aged, redried	4.82	6.20	5.17

Wave Velocity C_L (km/s)

	References	System	This Study
Unaged, dry	3.13	3.13	3.13
Aged, redried	3.13	3.13	3.13

7. 23°C Interlaminar Shear Strength Distributions $S=\exp(-\lambda/\lambda o)^m$

	References m	References $\ln\lambda_o$	System m	System $\ln\lambda_o$	This Study m	This Study $\ln\lambda_o$
Unaged	7.60	6.73	11.2	6.37	15.0	6.61
Aged, wet	3.90	6.46	10.0	6.13	16.2	6.52
Aged, redried	3.90	6.46	10.2	6.52	12.2	6.63

8. 23°C Transfibrous Fracture Energy Distributions $S=\exp(-W_b/W_{bo})^m$

	References		System		This Study m	This Study $\ln W_{bo}$
Imaged	-	-	-	-	10.8	4.18
Aged, wet	-	-	-	-	17.0	4.23
Aged, redried	-	-	-	-	14.1	4.33

Environmental Durability Characterization

A comprehensive environmental durability characterization for AS/3501-5 and T300/934 composites had been carried out prior to this study. These reference results are compared in Table V with data on T300/5208, which is shown to display lower hydrothermal sensitivity. The total analysis of composite response depends in an important way on characterization of bulk phases, bonded interfaces, and composite system response.

Eight categories of information are introduced in Table V. Category (1) describes general properties of three composites. Category (2) describes the results of surface energy analysis of both matrix and reinforcing fibers. Categories (3) through (8) summarize the results of composite characterization by both nondestructive testing and fracture studies.

Surface energy analysis has now been extensively applied (11, 13) to predict the moisture degradation of the fiber-matrix bond. A modified Griffith fracture mechanics relation which describes the ratio of critical fracture stress in air σ_c (air) and water σ_c (water) is given as follows (9, 15).

$$\frac{\sigma_c(\text{water})}{\sigma_c(\text{air})} = \frac{(R^2_{(H_2O)} - R_o^2)^{1/2}}{(R^2_{(\text{air})} - R_2^2)^{1/2}} = 0.39 \qquad (1)$$

$$R^2 = [\alpha_2 - 0.5(\alpha_1 + \alpha_3)]^2 + [\beta_1 + \beta_3)]^2 \qquad (2)$$

$$R_o^2 = 0.25\ [(\alpha_1 - \alpha_3)^2 + (\beta_1 - \beta_3)^2] \qquad (3)$$

where

α_1, β_1 = surface properties of epoxy matrix $= (\gamma_1{}^d)^{1/2}, (\gamma_1{}^p)^{1/2}$

α_3, β_3 = surface properties of graphite fiber $= (\gamma_3{}^d)^{1/2}, (\gamma_3{}^p)^{1/2}$

α_2, β_2 = surface properties of immersion phase $= (\gamma_2{}^d)^{1/2}, (\gamma_2{}^p)^{1/2}$

when the surface properties reported in Table V are introduced into Eqs. (1) through (3) the prediction of a 61% strength reduction of the fiber-matrix interface bond is obtained for all three composites. The magnitude of internal stress σ_T normal to the fiber surface at the fiber-matrix interface at ambient temperature T can be estimated from the following relation (16):

$$\sigma_T = \alpha(T_o - T)[E_o - a(T - T_o)] = 77.9 \text{ MNm}^{-2} \text{ (11.3 } K_{SI}) \quad (4)$$

where

α = temperature coefficient of linear thermal expansion [α = 2.92 $10^{-5}(°C^{-1})$], T_o = reference temperature of zero stress, (T_o = 177°C), E_o = transverse Young's modulus (E_o = 7.88 Gnm^{-2} = 1143 KSI), and $a = dE/dT$ is the thermal coefficient of modulus change [a = 63.6 MNm^{-1} = 5.13 Ksi $(°F)^{-1}$ which is assumed constant between T_o and the final temperature T = 22°C.

The combination of moisture degradation given by Eq. (1) combined with the high internal stress predicted by Eq. (4) explain many of the irreversible property changes due to moisture cycling shown in measurement categories 5-7 of Table V.

Solution to the interfacial moisture sensitivity is graphically indicated by the surface energy diagram of Fig. 6. Both the epoxy and fiber surface properties, as shown in Fig. 6, lie close to those of water. By shifting these surface properties to a region of high $\alpha \geq 6.0(\text{dyn/cm})^{1/2}$ and low $\beta \leq 1.5\ (\text{dyn/cm})^{1/2}$ the theory stated by Eq. (3) predicts high interface bond strength in both air and water immersion. Equally important is modification of the matrix network structure so that the present high bulk water absorption ($\phi(H_2O)$ = 5.8 to 6.5 wt %, see Table V.3) in the matrix is substantially decreased, to less than $\phi(H_2O) \leq 1.0$ wt %. This second modification relates back to basic changes in the chemical structures shown in Fig. 1 to minimize or eliminate polar groups which produce the moisture interactions. Current analysis dealing with structure-property relations in structural polymers are being developed to explore this critical issue of optimized chemistry for hydrothermal resistant polymers.

Nondestructive Evaluation

Dynamic mechanical response of graphite-epoxy composites as measured by the damping properties in the alpha (or glass) transition are quite sensitive to matrix chemical structure, as shown in Table V.4 and in the curves of Fig. 4. These properties are irreversibly modified by moisture exposure, as shown by intercomparing the curves of Fig. 4 and Fig. 5. Moisture diffusion analysis is shown to be a sensitive method for recognizing irreversible effects of moisture degradation, as shown in the data of Table V.5. A new methodology for moisture diffusion analysis has been recently developed (17), which provides a measure of the three principal diffusion coefficients for successive cycles of moisture absorption and desorption. As shown in Table V.5, the moisture absorption coefficients for diffusion parallel (D_1) and transverse (D_2, D_3) to the fiber axis of the three graphite epoxy composites are substantially increased from first to second cycles of moisture absorption and desorption at 75°C. These increased values of D_1, D_2, and D_3 for second cycle absorption result from

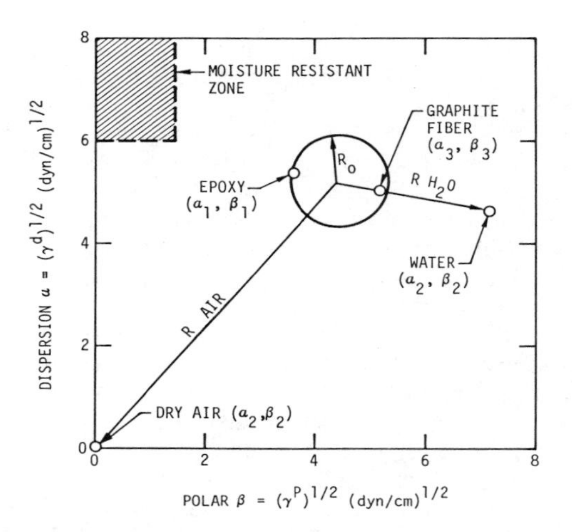

Figure 6. Griffith surface energy analysis of graphite epoxy composites (see *Equation 3)*

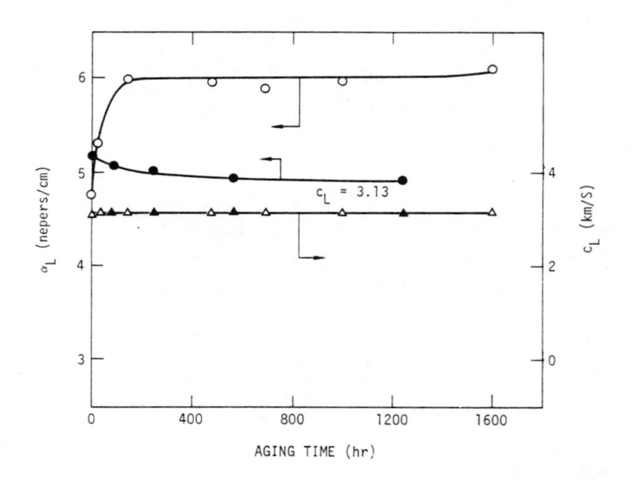

Figure 7. Effects of high-moisture aging (100°C water immersion) and redrying upon ultrasonic wave velocity c_L *and attenuation* α_L *for T300 graphite fiber in (△, ○) 934 epoxy matrix composite and in (▲, ●) 5208 epoxy matrix composite; sound propagation normal to plys*

microstructure degradation during the first cycle of moisture absorption.

Measurement of ultrasonic properties, transverse to the fiber axis in a uniaxial composite for moisture aged and dried specimens shows, as summarized in Table V.6 and Fig. 7, that ultrasonic wave velocity, C_L, is not influenced by prior moisture exposure. Extensions of these studies show that ultrasonic wave velocity changes at 23°C can be correlated with moisture concentration in the composite (11). This finding can be utilized to convert maps of ultrasonic wave velocity to maps of mositure concentration.

Conversely, the data of Table V.6 and the upper curves of Fig. 7 show that ultrasonic attenuation is irreversibly increased by prior moisture exposure in the T300/934 epoxy and remains essentially constant in T300/5208 epoxy. In this case, mapping of ultrasonic attenuation is similar to moisture diffusion analysis in recgnizing irreversible microstructure degradation. The combined ultrasonic mapping of both velocity, C_L and spatial attenuation, α_L is shown by these results to offer an important new tool for chemical defect analysis and microstructure degradation in the three composites described in Table V.

Strength Distribution Analysis

The extension of standard strength testing to measure the statistical distributions of composite strengths under varied hydrothermal aging is shown in the results summarized in Table V.7 and Fig. 8 and 9. These strengths are measured in conjunction with quantitative NDE to provide correlations between fracture and physical responses. A special statement of the Weibull (or extreme value) method of statistical evaluation for shear strength, λ, is given by the following relation (11, 18):

$$S = \exp(-\lambda/\lambda_o)^m V \qquad (5)$$

where S is the probability of survival, and V = 1.0 is the normalized volume for uniform shear loading. The Weibull parameters, λ_o, and m are evaluated by taking logarithms of Eq. (5) to provide the following linear relation (11, 18):

$$\ln(-\ell n\ S) = m[\ln\lambda - \ln\lambda_o] \qquad (6)$$

with slope m and intercept $\lambda = \lambda_o$ when $S = e^{-1} = 0.37$. The test results shown in Figs. 8 and 9 are arranged serially j = 1, 2,...N in increasing order of λ, and the survival probability is defined as follows (11, 18):

$$S = 1 - F = 1 - \frac{(j - 0.50)}{N} \qquad (7)$$

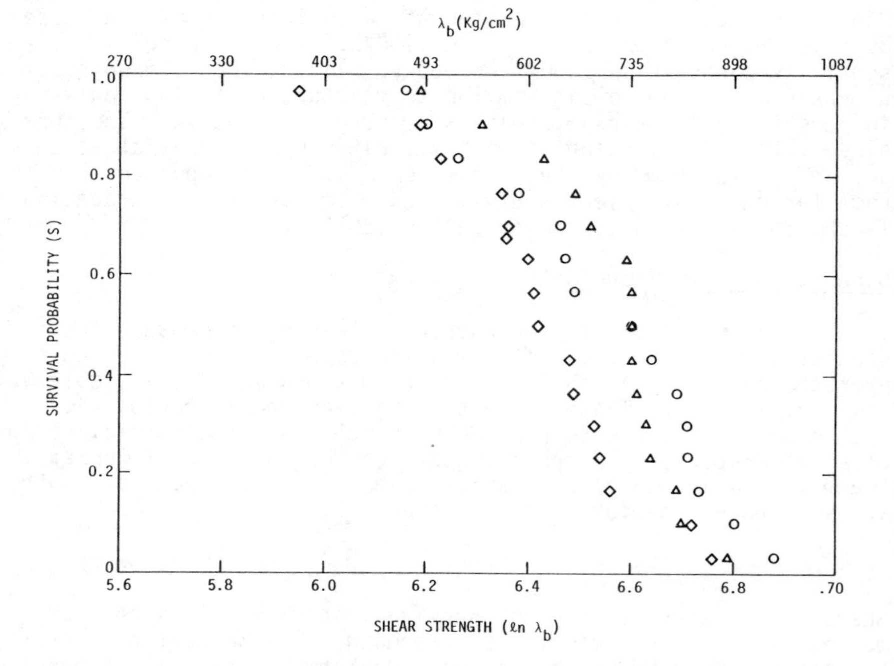

Figure 8. Survival probability (S) vs. interlaminar shear strength λ for graphite T300 fiber in 5208 epoxy matrix; interlaminar: (○) dry unaged, (◇) aged 2000 hr in 100°C H_2O, (△) aged 2000 hr in 100°C H_2O and dried at 100°C

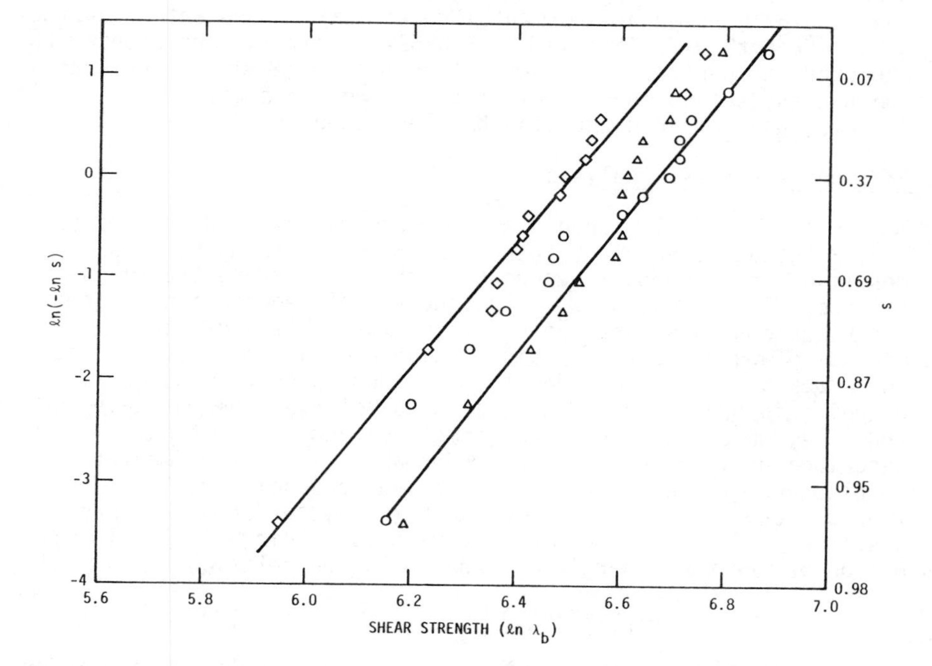

Figure 9. Weibull plots of interlaminar shear strength data for graphite T300 fiber in 5208 epoxy matrix; interlaminar: (○) dry unaged, (◇) aged 2000 hr in 100°C H_2O, (△) aged 2000 hr in 100°C H_2O and dried at 100°C

where F is failure probability and N is the number of observations. The data of Table V.7 summarizes the effects of hydrothermal aging on the distributions of interlaminar shear strengths at 23°C.

Inspection of Table V.7 shows that Hercules AS/3501-5 composite displays the largest loss in strength response on wet aging and the lowest strength recovery on redrying. The T300/508 composite characterized by this study is shown by the curves of Figs. 8 and 9 and data summary of Table V.7 to suffer the lowest strength loss on wet aging and highest strength recovery on redrying. The high environmental durability demonstrated for the shear strength distribution of T300/5208 epoxy is consistent with expectations derived from chemical analysis, network structure, and nondestructive characterization of the three composite systems, as earlier discussion has indicated.

Fracture Toughness Analysis

Analytic modeling predicts and experimental studies verify (19, 20) that lowering of interlaminar shear strength, λ, as a result of environmental degradation should increase the transfibrous fracture energy W_b. The additional degradation related contributions to W_b are the result of more extensive fiber debonding and added energy dissipation due to fiber pull out at the crack tip (9). In order to test these predictions for T300 graphite in 5208 epoxy, a special study of hydrothermal aging on fracture, which used specimen geometries and test methods described previously (20), was undertaken. The experimental results are graphed in plots of survival probability, S, versus fracture energy, W_b in Fig. 10 for three hydrothermal exposure histories. These data can be analyzed for their Weibull distribution of fracture energies by the following relation:

$$S = \exp(-W_b/W_{bo})^m V \qquad (8)$$

where W_b is fracture energy, V = 1.0 is the normalized volume for fracture damage, and W_{bo} and m are Weibull parameters. The linear least squares fit of the experimental data (see Fig. 10) to the logarithmic form of Eq. (8) provides the Weibull parameters for fracture energy tabulated in Table V.8. It is evident that hydrothermal aging, which shifts the distribution of shear strengths λ to lower values in T300/5208 composite, produces the predicted upward shift in the distribution of fracture energies, W_b, for aged wet specimens tested at 23°C. It is interesting to note, however, that redrying produces a further upward shift in the distribution of fracture energies, W_b (see Fig. 10), while the shear strength distribution, λ (see Fig. 8), is restored to values similar to unaged composite.

Summary and Conclusions

Results summarized under this study show that both chemical network (see Fig. 1) and manufacturing defects (voids, delaminations, etc.) produce concurrent and interactive degradation of composite microstructure and strength. Rational approaches to improved composite design of repair procedures can benefit by utilization of the outline for composite durability characterization summarized in Table II and detailed in the data summary of Table V. The detailed outline presented in Table I represents a specialized application of a general technical approach for polymer composite reliability studies defined by the block diagram of Fig. 11.

The systematic organization and combining of the capabilities, as shown in Fig. 11, provides a general approach that is applicable to many specific problem areas involving both polymers and composites. The technical program outlined in Fig. 11 and utilized in this study encompasses matrix chemistry and bulk properties, interfacial bonding mechanisms, composite environmental response, and fracture mechanics. The special organization of capabilities in Fig. 11 moves from macro to micro response by following the upward directed arrows and from materials to composite system response in moving from left to right. The overall output of this approach is the decomposition of the composite reliability analysis into fundamental mechanistic constituents and methodologies for detection and correction (9).

The following detailed conclusions have been derived from this study:

1. The processability and environmental durability of three state-of-the-art 177°C (350°F) service temperatue graphite epoxy composites can be directly correlated with chemical composition and crosslink network structure of the epoxy matrix.
2. Analytic methods for determining chemical composition and state of cure in graphite epoxy prepregs have been established and are suitable for production control.
3. A comprehensive environmental durability characterization of each composite type clarifies the separate roles of matrix rheology and interface bonding in composite durability.
4. The low strength loss under hydrothermal aging and high (essentially 100%) strength recovery of T300 graphite in 5208 epoxy composite is traceable to chemical structure and curing mechanism.
5. Irreversible interlaminar shear strength degradation due to hydrothermal aging can be detected nondestructively by increased values of translaminar ultrasonic signal attenuation.
6. Ultrasonic velocity perpendicular to the fiber provides a direct measure of current moisture content and is insensitive to micrstructure damage due to hydrothermal aging in the three composites studied.

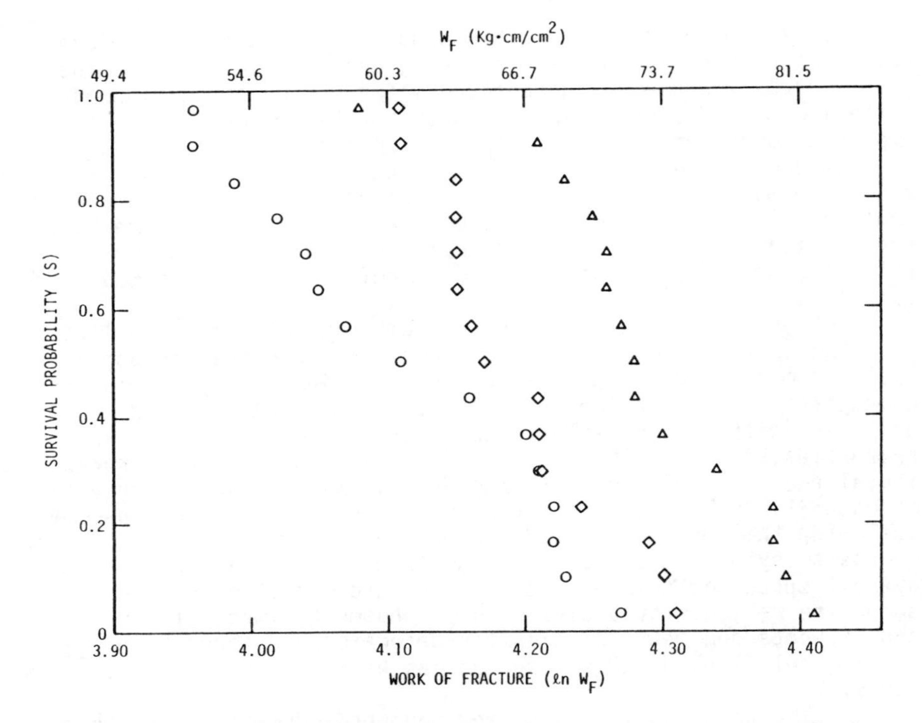

Figure 10. Survival probability (S) vs. transfibrous fracture energy W_F for graphite T300 in 5208 epoxy matrix: (○) dry unaged, (◇) aged 2000 hr in 100°C H_2O, (△) aged 2000 hr in 100°C H_2O and dried at 100°C

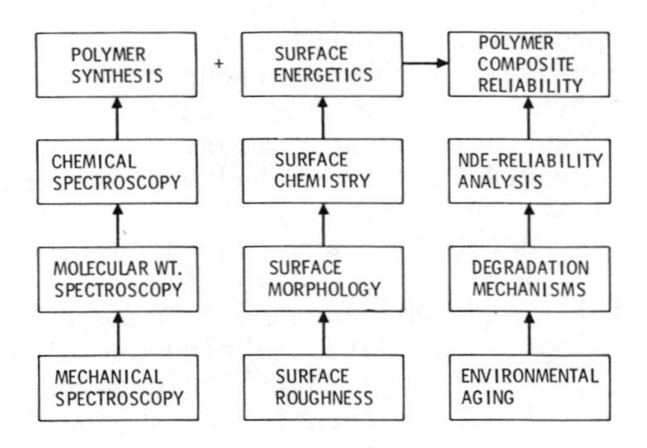

Figure 11. Technical approach for characterizing polymer composite reliability.

7. The Weibull distributions of interlaminar shear strength and transfibrous fracture energy provide quantitative analysis of environmental durability in terms useful to a design engineer concerned with high survival probability and long term durability.

Acknowledgement

This study of the T300/5208 graphite in epoxy composite was conducted under Research Grant No. DAAG 29-77-C-0005 supported by the U.S. Army Research Office. Reference data on AS/3501-5 and T300/934 graphite/epoxy composites were generated under the Rockwell International IR&D program.

Abstract

Hydrothermal exposure (combined high moisture and temperature) of graphite fiber reinforced epoxy matrix composites can produce irreversible reduction in shear strength and modify the Weibull distribution for survival probabilities. Nondestructive analysis of moisture diffusion kinetics and ultrasonic response shows that strength degradation is dominated by matrix bulk properties. Ultrasonic 2.25 MHz wave velocity is sensitive to current moisture content, while acoustic attenuation transverse to fibers correlates with microcracking as indicated by diffusion kinetics and irreversible shear strength loss. Comparative chemical analysis of two catalytically cured graphite-epoxy composites and an aromatic amine cured epoxy matrix clearly shows that curing mechanisms and epoxy network structure influence thermal response and environmental durability. The development of a systematic program for composite durability characterization which encompasses matrix chemistry and bulk properties, interfacial bonding mechanisms, composite environmental response, and fracture mechanics is demonstrated.

References

1. "Proceedings of the Air Force Workshop on Durability Characteristics of Resin Matrix Composites," Batelle-Columbus Laboratories, Columbus, Ohio, October 1975.
2. "Transactions of the Workshop on the Effects of Relative Humidity and Elevated Temperature on Composite Structures," Report AFOSR TR-77-0030, University of Delaware, Newark, Delaware, March 1976.
3. "Proceedings of the Workshop on the Role of the Polymer Substrate Interphase in Structural Adhesion," sponsored by AFOSR and AFML, University of Dayton, Dayton, Ohio, Sept. 1976.

4. Kaelble, D. H., "Adsorption and Interdiffusion Mechanisms of Bonding in Polymer Composites," Proc. 23rd International Congress on Pure and Applied Chemistry, Vol. 8, Butterworths, London, 1971, p. 265-302.
5. Browning, C. E., "The Effects of Moisture on the Properties of High Performance Structural Resins in Composites," Society of Plastics Industry, SPI Reinforced Plastics/Composites Division, 28th Annual Conference 1973, p. 15A; 1-16.
6. Lundemo, C. Y. and Thor, S-E, "The Influence of Environmental Cycling on the Mechanical Properties of Composite Materials," J. Comp. Matls., 11, 176, 1977.
7. Carpenter, J. F. and Bartels, T. T., "Characterization and Control of Composite Prepregs and Adhesives," Proc. 7th Nat. SAMPE, Vol. 7, SAMPE, Azusa, Calif., 1975, p. 43-52.
8. May, C. A., Helminiak, T. E. and Newey, H. A., "Chemical Characterization Plan for Advanced Composite Prepregs," Proc. 8th Nat. SAMPE Conference, Vol. 8, SAMPE, Azusa, Calif., 1976, p. 274-294.
9. Kaelble, D. H. and Dynes, P. J., "Preventative Nondestructive Evaluation (PNDE) of Graphite Epoxy Composites," AIAA Paper No. 77-478, AIAA Conference on Aircraft Composites, San Diego, Calif., March 1977.
10. Kaelble, D. H., Dynes, P. J., and Maus, L., "Hydrothermal Aging of Composite Materials," J. of Adhesion, 8, 1976, p. 121, p. 155.
11. Kaelble, D. H. and Dynes, P. J., "Methods for Detecting Moisture Degradation in Graphite-Epoxy Composites," Materials Evaluation 35, 4, 1977, p. 103.
12. Bergmann, H. W. and Dill, C. W., "Effect of Absorbed Moisture on Strength and Stiffness of Graphite-Epoxy Composites," Proc. 8th Nat. SAMPE Conf., Vol. 8, SAMPE, Azusa, Calif., 1976, p. 244-256.
13. Tanaka, Y. and Mika, T. F., "Epoxy Curing Reactions," Chapter 3 in Epoxy Resins; Editors: May, C. A. and Tanaka, Y.,Dekker, New York, 1973, p. 135-139.
14. Lee, H. and Neville, K., "Handbook of Epoxy Resins," McGraw-Hill New York, 1967.
15. Kaelble, D. H., "Interface Degradation Processes and Durability," Polymer Engineering and Science, 17, p. 474-477.
16. Hahn, H. T. and Pagano, N. J., "Curing Stresses in Composite Laminates," J. Composite Materials, 9, 1975, p. 91.
17. Kaelble and Dynes, P. J., "Methods for Detecting Moisture Degradation," Interdisciplinary Program for Quantitative Flaw Detection, Special Report Third Year Effort, Covering Period July 1, 1976-June 30, 1977, Contract Number F33615-74-C-5180, Prepared for Advanced Projects Research Agency and Air Force Materials Laboratory.
18. Robinson, E. Y., "Estimating Weibull Parameters for Composite Materials," Proceedings of Colloquium on Structural Reliability, Editors: Sedlow, J. L., Cruse, T. A. and Halpin,

J. C., Carnegie Mellon University, Pittsburgh, 1972, p. 463-526.

19. Kaelble, D. H., "Theory and Analysis of Fracture Energy in Fiber Reinforced Composites," J. Adhesion, 5, 245, 1973.
20. Kaelble, D. H., Dynes, P. J., Crane, L. W. and Maus, L., "Interfacial Mechanisms of Moisture Degradation in Graphite-Epoxy Composites," J. Adhesion, 7, 25, 1974.

RECEIVED January 8, 1980.

30

Moisture Diffusion and Microdamage in Composites

CHUK L. LEUNG and DAVID H. KAELBLE

Rockwell International Science Center, Thousand Oaks, CA 91360

Because of their favorable performance characteristics, advanced composites using graphite or boron fibers as reinforcement are increasingly being used in large scale primary structures in aerospace applications. There has been much concern, however, about the extent and mechanisms of strength degradation of these composites when exposed to moisture for long periods of time. Moreover, a logic is needed for damage analysis and repair of composite laminates, wherein the quality and environmental durability of the composites are to be quantified through the techniques of hydrothermal stress analysis. This analysis will also provide guidance for the mechanical requirements for composite reliability and material replacement or repair due to temperature/moisture loads.

Theory

Analytical Profilometry Modeling. A method for determining the depth/concentration profile of moisture in a composite is discussed in terms of mathematical analysis and experiment. This new methodology is termed moisture profilometry. The experiment involves heating of the composite to a desired control temperature, followed by measurement of the rate of moisture effusion as a function of time. The analytical problem is then to determine the initial distribution of water, i.e., the concentration $c_o(x) = c(x,o)$ as a function of x (where x is the distance into the thickness of the composite), from the subsequent history of effusion rate measurements. The attempt to treat this as a deterministic problem leads to difficulties. One of these difficulties is associated with the inevitable incompleteness of the experimental data, namely, any real experiments, can yield only a finite set of numbers. A particularly convenient way is to apply statistical estimation theory to the analysis.

0-8412-0567-1/80/47-132-419$05.00/0

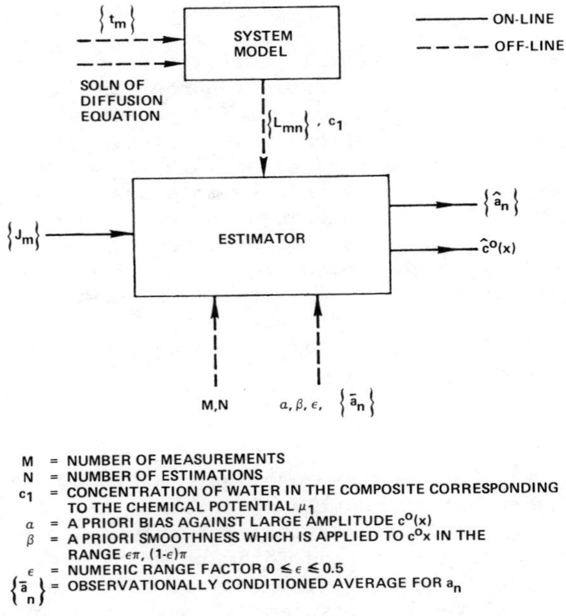

M = NUMBER OF MEASUREMENTS
N = NUMBER OF ESTIMATIONS
c_1 = CONCENTRATION OF WATER IN THE COMPOSITE CORRESPONDING TO THE CHEMICAL POTENTIAL μ_1
α = A PRIORI BIAS AGAINST LARGE AMPLITUDE $c^o(x)$
β = A PRIORI SMOOTHNESS WHICH IS APPLIED TO $c^o x$ IN THE RANGE $\epsilon\pi$, $(1-\epsilon)\pi$
ϵ = NUMERIC RANGE FACTOR $0 \leq \epsilon \leq 0.5$
$\{\bar{a}_n\}$ = OBSERVATIONALLY CONDITIONED AVERAGE FOR a_n

Figure 1. Schematic of the estimation process

The inverse-diffusion system is illustrated in more explicit form in Fig. 1. The material is assumed to be a rectangular block of polymeric material occupying a region of space defined by the inequalities

$$\begin{array}{l} 0 \leqslant x \leqslant L_x \\ 0 \leqslant x \leqslant L_y \\ 0 \leqslant x \leqslant L_z \end{array} \tag{1}$$

where x, y, z are the usual Cartesian coordinates. It is also assumed that the material is homogeneous but anisotropic, with the principal axes of the diffusion tensor parallel to the coordinate axes. The principal values of the diffusion tensor are denoted by D_x, D_y and D_z. The initial concentration is assumed to depend only on x, i.e.,

$$C(x,y,z,\ t = o) = c_o(x) \quad . \tag{2}$$

It is further assumed that the measurements involve the total diffusion rate from the entire surface at a specific set of later times. The inversion problem therefore involves a theoretical solution of the above diffusion problem with an arbitrary $c_o(x)$ and dry boundary conditions ($c_1 = o$). The solution is given by Richardson.[1] The matrix L_{mn} shown in Figure 1 is then used to obtain from the measured J_m (the total diffusion rate) at times t_m, estimates $\hat{a}_n$ of the coefficients of $\sin(n\pi x/L_x)$, i.e.,

$$c_o(x) = c_1 + \sum_{n=1}^{N} a_n \sin \frac{nx\pi}{L_x} \quad . \tag{3}$$

This process of going from the J_m obtained from experimental measurements to the estimates a_n is incorporated in the box labelled "estimator" in Figure 1. The final result of the entire estimation process is presented as a plot of the original moisture concentration at time t = o on the ordinate and the depth of moisture penetration the abscissa.

To establish the validity and accuracy of the estimator, called the Inverse Diffusion Solution, a Direct Model is developed;[2] this is based on a numerical solution of Fickian diffusion equation. The Direct Model gives data for:

(a) Moisture profile at any specified time,
(b) Moisture diffusion rate at each time interval,
(c) Moisture content at each time interval.

Theoretical moisture effusion rates obtained from the Direct Model, item (b), can therefore be analyzed by the Inverse

Diffusion Model, yielding moisture profiles which should correlate to those generated directly from Fick's Law, item (a), thus validating the Inverse Diffusion Model.

The Moisture Program. A computer program developed by Shen and Springer[3] computes, by finite difference method, the moisture distribution within each layer of a multi-layered composite material. For each layer, the following parameters are supplied to the program as input variables:

(a) density
(b) thermal conductivity
(c) boundary condition coefficients
(d) Arrhenius-type diffusion coefficients
(f) temperature and R.H. for each time interval
(g) cycling of T'/R.H. as specified in (f).

The output of the program consists of:

(a) original input data
(b) moisture distribution for each time interval; the average moisture concentration for each layer
(c) if the number of cycles or number of hydration/dehydration steps within each cycle is greater than one, the moisture concentration is of course additive, so that each line of output reflects the accumulated moisture distribution within the sample.

The Stress Analysis Program. This computer program was originally developed by the Los Angeles Division of Rockwell International.[4] It calculates:

(a) the laminate flexural and extensional constants when treated as a set of laminae or as an orthotropic homogeneous material,
(b) the stress, strain and margins of safety for prescribed layers, given the external load. These external loads being in plane, rotational, and transverse as well as temperature differential within each lamina.

The output of the program gives the layer-by-layer stress analysis of shear load (Q), extensional load (σ) both in the laminae (σ_T, σ_L, σ_{LT}) and in the laminate (σ_x, σ_y, σ_{xy}). The margins of safety for the fiber and matrix material for each layer are also calculated.

The Moisture-Stress Program. The combination of the Moisture Program in Paragraph 2 and the Stress Analysis Program in Paragraph 3 would therefore give the effect of stress-strain forces due to moisture absorption on the performance and reliability of the composite when an external load is imposed on the composite. This should be a valuable tool to a design engineer when working with composite materials.

Consider a material subjected to a stress, σ_i, at a temperature, T, and moisture concentration, H. The total strain on the material is then the sum of the mechanical (ε^M) and nonmechanical strains (ε^N):

$$\varepsilon_i = \varepsilon_i^M + \varepsilon_i^N \quad , \qquad (i = 1, 2, 6) \quad . \tag{4}$$

The mechanical strain in Eq. (4) is, according to lamination theory,

$$\varepsilon_i^M = S_{ij} \; (T,H) \; \sigma_j \tag{5}$$

where S_{ij} is the compliance. The nonmechanical strain is divided into thermal, ε^T, and swelling due to moisture, ε^H:

$$\varepsilon_i^N = \varepsilon_i^T \; (T) + \varepsilon_i^H \; (H) \tag{6}$$

Assuming transverse isotropy with the axis of symmetry in the longitudinal direction, the nonmechanical strains of the composite are:

$$\varepsilon_L^N = (V_m \, E_{mL} \, \varepsilon_{mL}^N + V_f \, E_{fL} \, \varepsilon_{fL}^N)/(V_m \, E_{mL} + V_f \, E_{fL})$$

and (7)

$$\varepsilon_T^N = V_m \, \varepsilon_{mT}^N + V_f \, \varepsilon_{fT}^N + V_m \, \nu_{mLT} \, \varepsilon_{mL}^N + V_f \, \nu_{fLT} \, \varepsilon_{fL}^N$$

$$-\varepsilon_L^N \, (V_m \, \nu_{mLT} + V_f \, \nu_{fLT}) \tag{8}$$

where
V = volume fraction
m = matrix
f = fiber
E = modulus
ν = Poisson's ratio

It has been shown that for epoxies, swelling increases proportionally to the additions of moisture content. The proportionality constant, called coefficient of hygroscopic expansion, is approximately equal to the specific gravity of the epoxies, S_m, indicating volume dependence. Thus for the epoxy matrix, due to its isotropy,

$$\varepsilon^H_{mL} = \varepsilon^H_{mT} = \frac{1}{3} S_m H \quad . \tag{9}$$

Moisture absorption in the fibers is assumed to be negligible and so is ε^H_L in Eq. (8). Therefore, Eq. (8) becomes

$$\varepsilon^H_T = \varepsilon^H_{mT} + \nu_{mLT}\, \varepsilon^H_{mL} \tag{10}$$

per unit volume. Substitution of Eq. (9) into Eq. (10) generates the transverse swelling strain of the composite,

$$\varepsilon^H_T = \frac{1 + \nu_m}{3} S_m H \quad . \tag{11}$$

The total strain is therefore the sum of the swelling strain and strain due to thermal expansion. Thus in the transverse direction, Eq. (6) becomes

$$\varepsilon_T = \frac{1 + \nu_m}{3} S_m H + \alpha_T\, \Delta T \tag{12}$$

where α_T is the coefficient of thermal expansion in the transverse direction and ΔT is the temperature differential from room temperature. The strain of the longitudinal direction, other than that due to the external load, is

$$\varepsilon_L = \alpha_L\, \Delta T \tag{13}$$

since swelling is assumed negligible in the longitudinal direction.

The addition of the Stress Analysis Program into the Moisture Program therefore calculates the hydrothermal stress in each ply of the composite. The most likely location for ply failure can be predicted when the hydrothermal stress of each ply is compared with the allowed stress designated by the designer and the margins of safety.

Procedures

Analytical modeling[1,2] was achieved and coded into computer programs to generate the moisture depth profiles within a composite by measuring the moisture effusion rates and then subjecting the data to a statistical estimation analysis; this enabled the depth concentrations of moisture at time t=0 to be calculated. The accuracy of this inversion technique was verified by a separate program which generates the moisture depth profile through an analysis of diffusion with arbitrary boundary concentrations based on Fickian diffusion kinetics.

A hydrothermal stress analysis program, based on Shen and Springer's[3] Moisture Program as well as the Point Stress Analysis Program developed by Rockwell International,[4] was developed. It calculates the effect of moisture swelling on the mechanical properties and the reliability of the multidirectional fibrous composites.

Results and Discussion

The influence of the thermal/moisture environment on the behavior and performance of composite materials is complex and important. First, the absorption of moisture has been shown to lower the T_g of the polymeric matrix, thus diminishing the matrix-influenced properties of the composite at elevated temperatures. Second, moisture absorption of the matrix resin results in the swelling of the composite. Since swelling is restrained in the fiber direction, significant internal stress is developed in the multidirectional laminate by moisture absorption.

A series of the profiles generated by the estimation process are compared against the profiles from a direct Fickian solution, as shown in Figs. 2 and 3. These specimens were assumed to be completely saturated with moisture prior to effusion measurements and the moisture profiles were calculated at times of 20 and 100 minutes after effusion had started.

A much more severe test of the estimator is for the case of varying boundary concentrations, in which case the hypothetical specimen was partially saturated, removed from moisture and the moisture penetration mapped after various lengths of drying times. Analytical solutions of moisture profiles solved for 20 and 100 minutes of absorption and various drying times are shown in Figs. 4 and 5.

Several imortant conclusions can be drawn from the results shown in Figs. 2-5. In each case, there is excellent agreement between the direct Fickian solution and the inverse estimator solution, thus validating the Estimator model. Particularly for Figs. 4 and 5, which show that after dehydration has begun, a ridge of high moisture content is present immediately beneath the dry surface. It is entirely feasible that different states of stress exist to the right (into the sample core) and to the left (to the surface) of the ridge of water beneath the surface. If a thermal spike is applied to the composite at this time, for example during a supersonic dash of an aircraft, the different stress states can cause microcracks or delaminations beneath the surface of the composite.

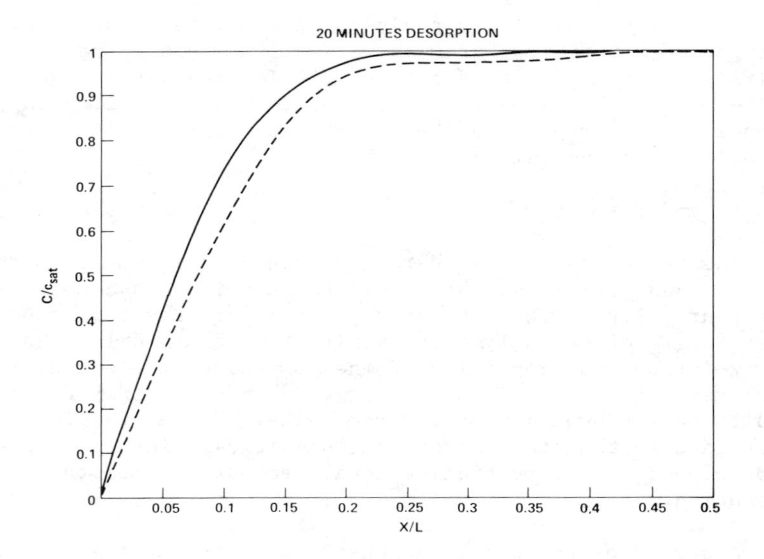

Figure 2. Moisture profile via analytical modeling of a fully saturated composite after 20 min desorption: (——) direct Fickian solution, (– – –) estimator inverse solution

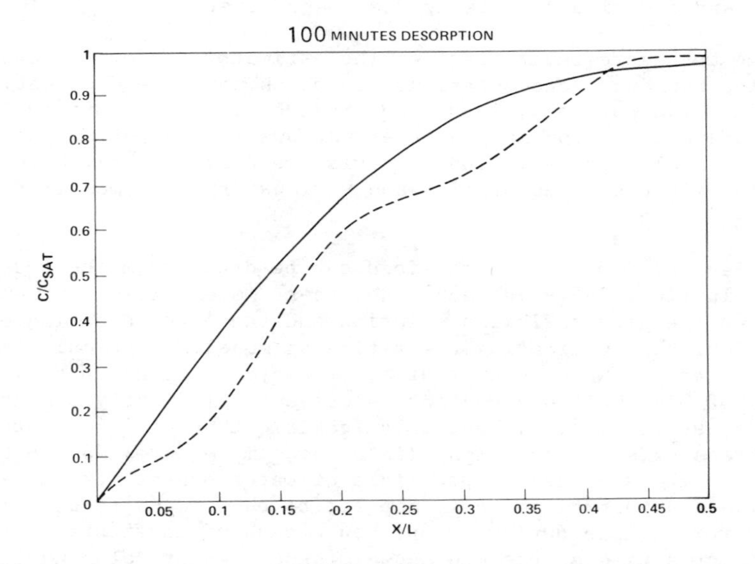

Figure 3. Moisture profile via analytical modeling of a fully saturated composite after 100 min desorption: (——) direct Fickian solution, (– – –) estimator inverse solution

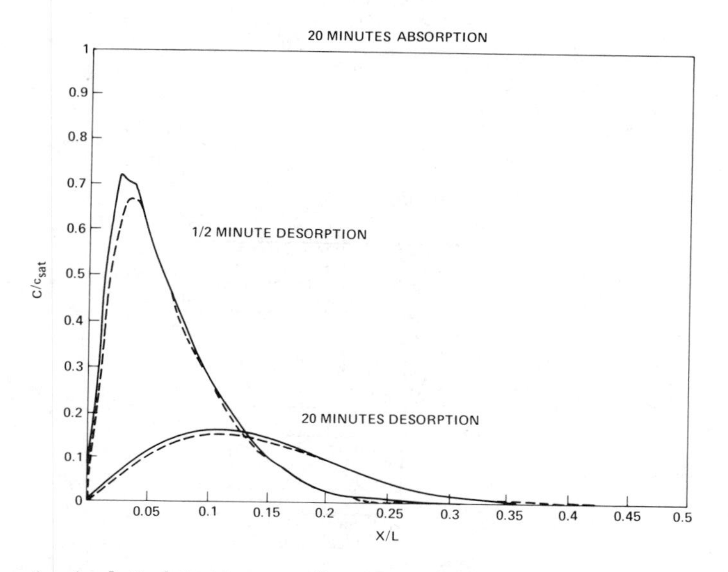

Figure 4. Analytical moisture profile of partially saturated composite after 20 min absorption: (——) direct Fickian solution, (– – –) estimator inverse solution

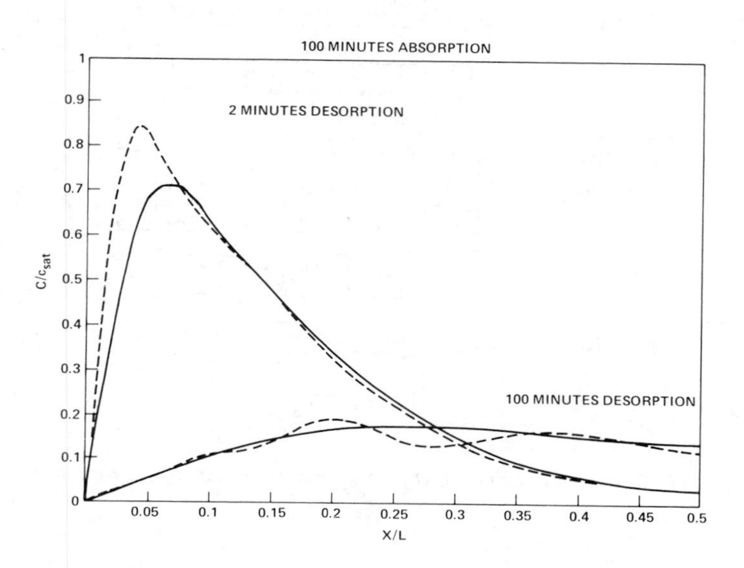

Figure 5. Analytical moisture profile of partially saturated composite after 100 min absorption: (——) direct Fickian solution, (– – –) estimator inverse solution

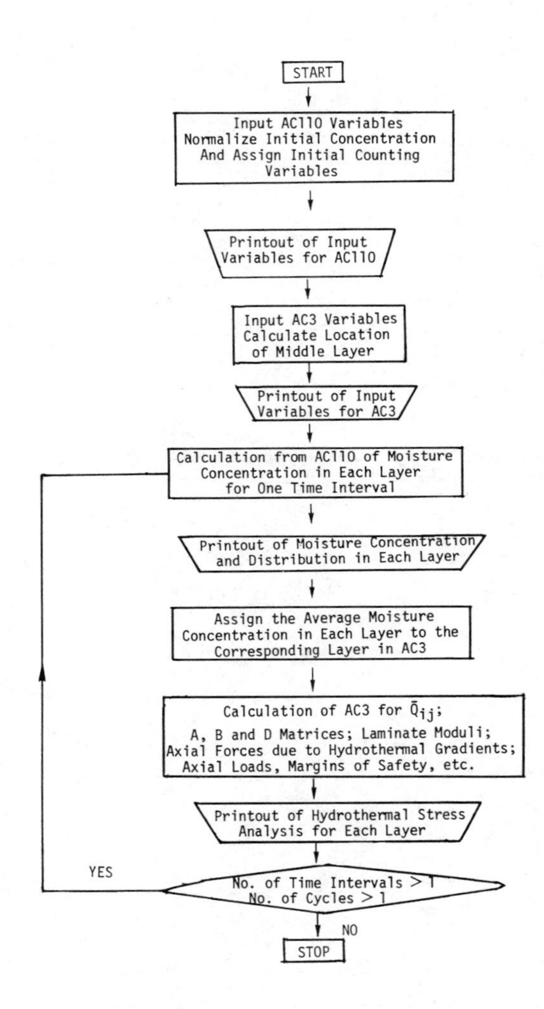

Figure 6. Flowchart of Hydrothermal Stress Analysis Program

The hydrothermal stress analysis program is therefore developed as a predictive methodology in composite design. The most likely location for ply failure can be predicted when the hydrothermal stress of each ply is compared with the maximum allowable stress/strain parameters chosen by the engineer, the margins of safety.

Figure 6 shows the flowchart for the addition of the Moisture Program (designated AC110) into the Stress Analysis Program (designated AC3). In AC110, the input variables include: the number of layers in the laminate; whether it is insulated at one or both sides; the initial moisture concentration distribution; and the thickness, thermal conductivity, density, and diffusion coefficient of each lamina in the composite. In the AC3 program, the input variables include the pertinent mechanical properties of each lamina such as the Young's modulus, Poisson's ratio, coefficients of thermal and moisture expansions, the axial and rotational strains, the axial and shear loads, temperature differential among the laminae, the applied stress, and the allowable strain and stress of each lamina.

For a hypothetical 8-ply composite of fiber orientation $(0/\pm45/90)_s$, which had undergone a complete environmental temperature/moisture cycle as shown in Fig. 7, the moisture distribution in the composite calculated by the AC110 program is shown in Fig. 8. This distribution profile bears a close resemblance to that shown in Figs. 4 and 5 in that, during dehydration, the near-surface moisture concentration is higher than that near the core of the sample.

When the composite sample is subjected to an extensional load of 2000 lb/in during the environmental cycling, different levels of stress exist within the composite. The stress distribution as calculated by the hydrothermal stress program, is shown in Figs. 9-11. The x,y and x-y components of the stress within the initially dry sample are shown in Figs. 9A, 10A and 11A, respectively. During each environmental cycle, the computer code is able to follow the moisture and stress distributions within each ply, as shown by Figs. 9B to E, 10B to E and 11B to E. The most prominent result from hydrothermal effect is shown in Figs. 10B to E, which describes the y component of the total stress in the laminate. Due to the fact that (1) the outermost ply has the highest moisture content and (2) swelling occurs only in the direction transverse to fiber orientation, the first ply and fourth ply undergo rapid stress fluctuation as moisture goes into and out of the laminate. It is because of these different types and levels of stress that microcracks and delaminations within a composite are produced.

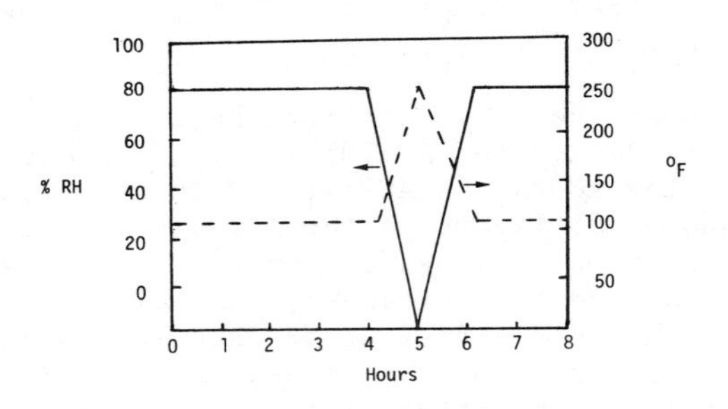

Figure 7. Environmental conditions of 1 cycle

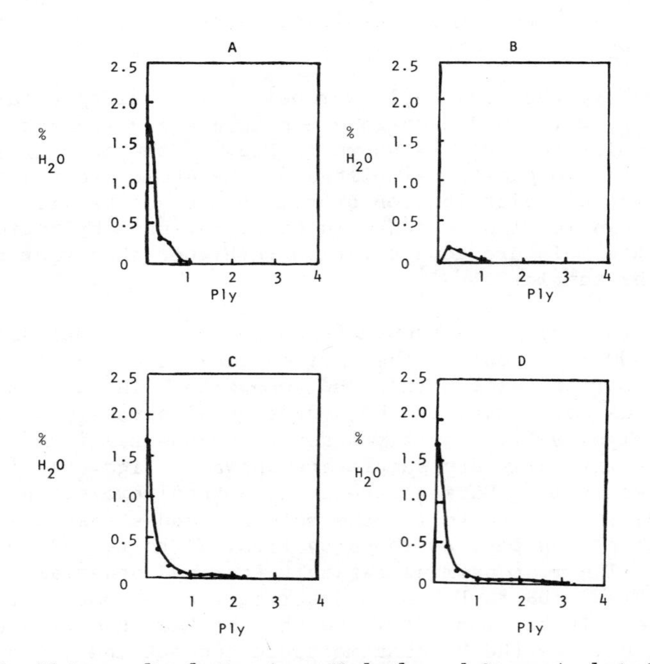

Figure 8. Moisture distribution from Hydrothermal Stress Analysis Program after 1 cycle: (A) interval 1; (B) interval 2; (C) interval 3; (D) interval 4

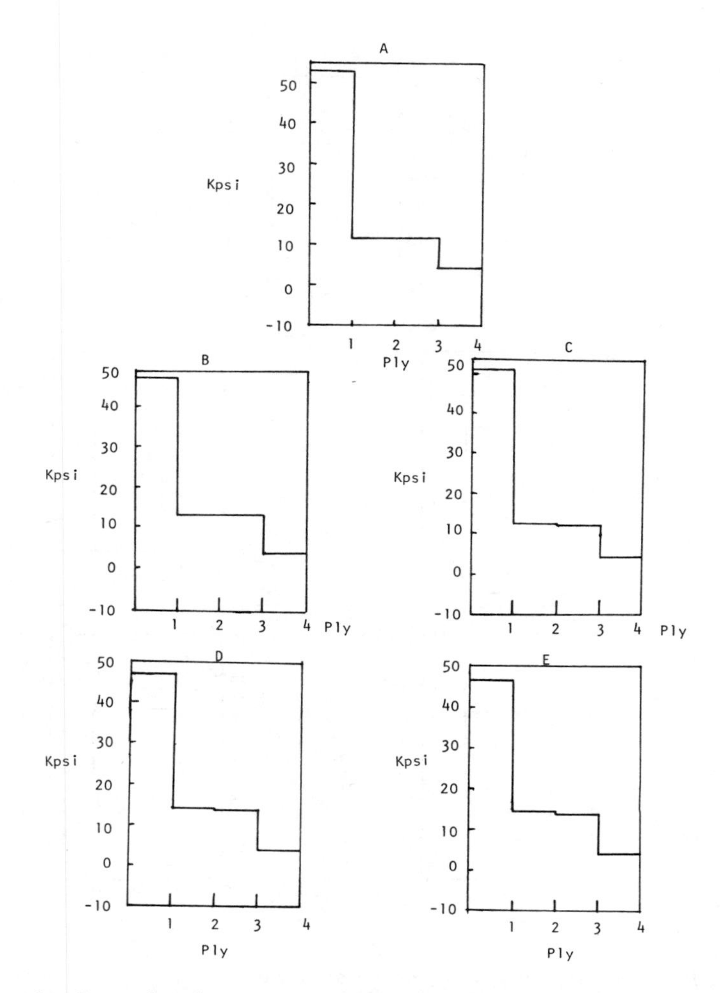

Figure 9. Stress distribution of composite in x-direction: (A) dry sample; (B) cycle 1; (C) cycle 2; (D) cycle 3; (E) cycle 4

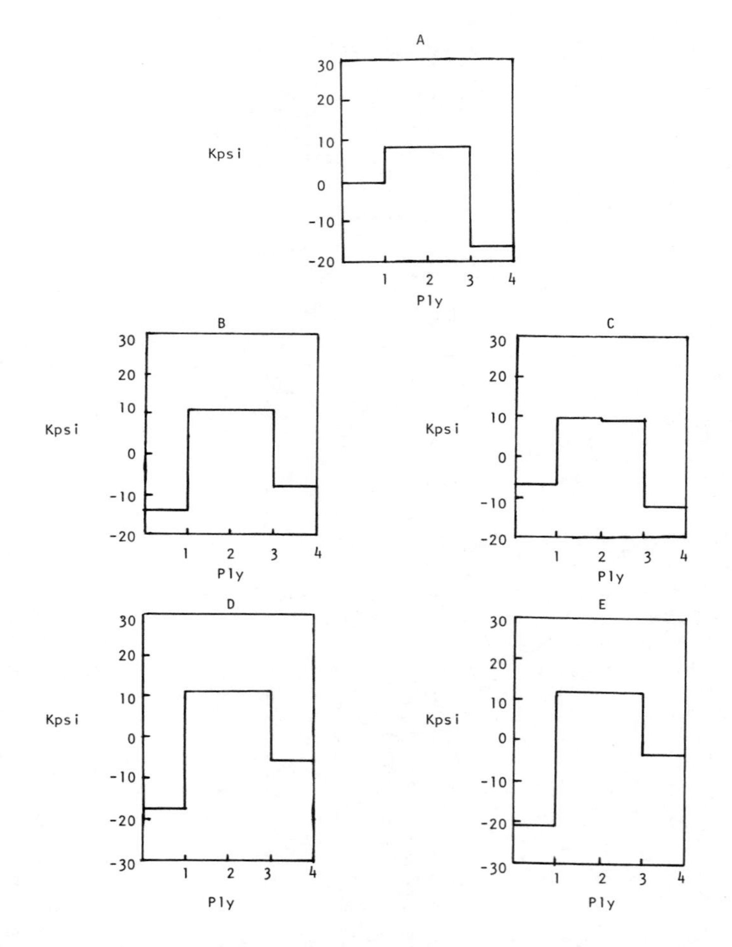

Figure 10. Stress distribution of composite in y-direction: (A) dry sample; (B) cycle 1; (C) cycle 2; (D) cycle 3; (E) cycle 4

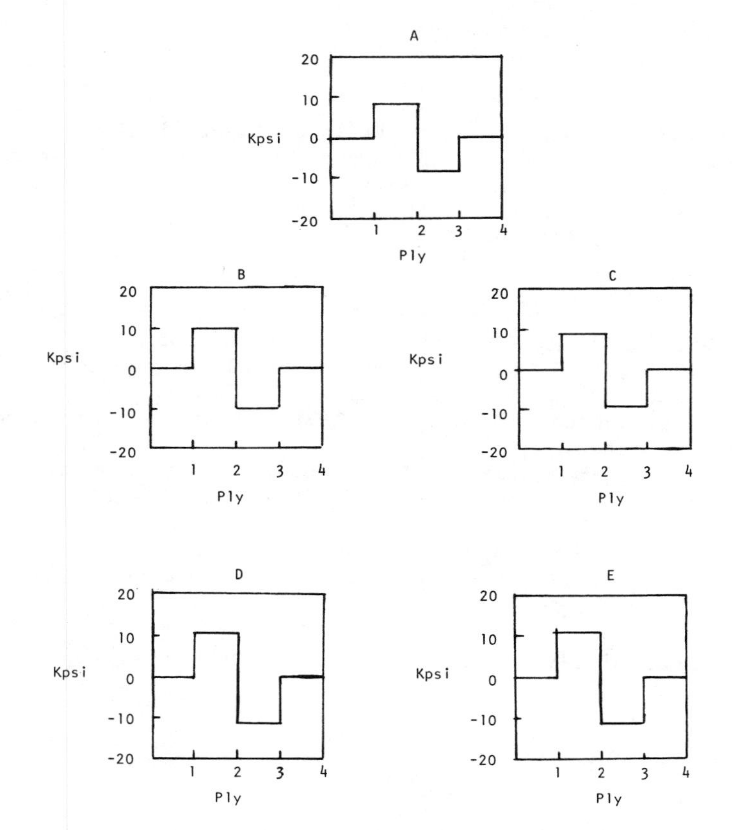

Figure 11. Stress distribution of composite in xy-*direction: (*A*) dry sample; (*B*) cycle 1; (*C*) cycle 2; (*D*) cycle 3; (*E*) cycle 4*

Conclusion

It has been shown that moisture effusion measurements can be used as an effective methodology in nondestructive evaluation of moisture penetration in fibrous composites. It has also been shown that hydrothermal stress analysis can be used to predict the performance of a composite laminate, whether in component design or repair circumstances, as a part of computer-aided-design (CAD) capabilities.

Acknowledgement

Part of this work (the Inverse and Direct Fickian Solutions) was sponsored by the Center for Advanced NDE operated by the Rockwell International Science Center for the Advanced Research Projects Agency and the Air Force Materials Laboratory under contract F33615-74-C-5180.

References

1. Richardson, J. M. "The Inverse Diffusion Problem," SCTR-77-38, 1977, Rockwell International Science Center.
2. Richardson, J. M. Internal Letter: "Diffusion with Arbitrary Boundary Concentrations," Feb. 1978, Rockwell International Science Center.
3. Shen, S. H.; Springer, G. S. "Moisture Absorption and Desorption of Composite Materials," AFML-TR-76-102, June 1976.
4. "Advanced Composites Design Guide," 3rd Ed., Vol. II, Chapter 2b.1, AFML Contract No. F33615-71-C-1362, WBAFB, January 1973.

RECEIVED January 8, 1980.

31

Self-Stress-Enhanced Water Migration in Composites

N. R. FARRAR[1] and K. H. G. ASHBEE

H. H. Willis Physics Laboratory, Royal Fort, Tyndall Avenue, Bristol BS8 1TL, England

The resin swelling that accommodates water uptake by fibre reinforced plastics is strongly inhomogeneous. A fillet of resin between three closely packed fibres (Figure 1A) would, if unconstrained by the presence of the fibres, undergo the shape change illustrated in Figure 1(B). The concentration of water reaches saturation in the resin adjacent to the external surface and a well defined region of saturated swelling begins to move inwards as indicated in Figure 1(C). However, the resin is not free to adopt these shape changes because of fibre constraint and, as a result, the swollen resin experiences compression and, correspondingly, the unswollen resin experiences tension. These mechanical constraints are exerted radially in the core and circumferentially in the flanges of the fillet. The tensile stresses so generated attract absorbed water thereby giving rise to enhanced migration rates.

In order to study this phenomenon in a specimen geometry representative of the thin flange section between adjacent fibres, samples of polyester resin and samples of epoxy resin measuring ~10 μm in thickness and containing entrapped air bubbles, were cast between glass cover slips, cured in strict accordance with recommendations by the respective manufacturers and immersed in water. Rates of water uptake at 100° C. were measured by noting the times at which water droplets appeared inside the entrapped air bubbles which were conveniently distributed at various radial distances from the edge of each resin layer. The data so obtained, indicate water diffusion coefficients of $\sim 10^{-4}$ cm^2 s^{-1} in polyesters and $\sim 10^{-7}$ cm^2 s^{-1} in epoxies, i.e. diffusion coefficients which are over an order of magnitude greater than values published for water diffusion in the respective bulk resins.

Subsequent work, described here, concerns measurements of the states of stress and strain generated in thin layers of mechanically constrained resin during water uptake.

[1]Current address: Department of Materials Science and Engineering, Cornell University, Bard Hall, Ithaca, New York 14853 USA.

0-8412-0567-1/80/47-132-435$05.00/0

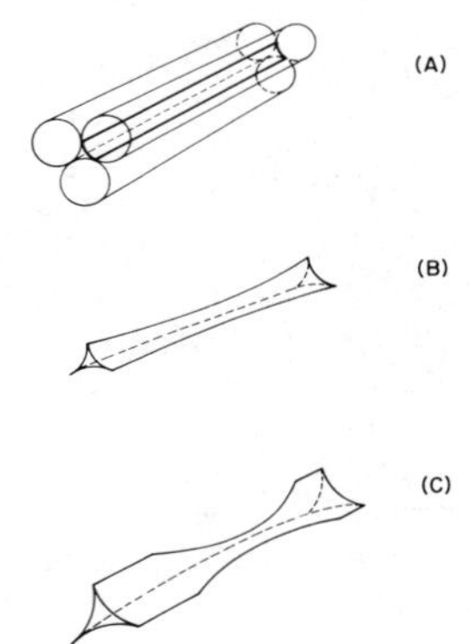

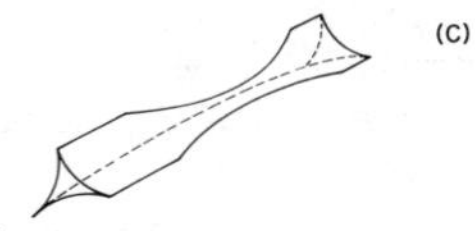

Figure 1. Swelling of resin fillet in fiber reinforced composite

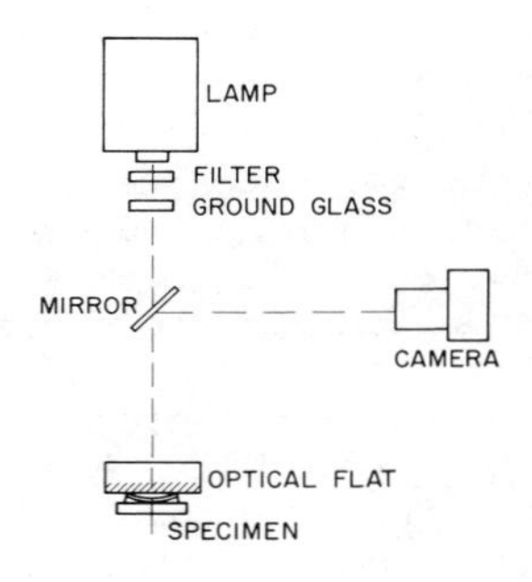

Figure 2. Experimental arrangement for Newton's rings experiment. (8)

Use of Newton's Rings to Monitor Swelling

The displacement normal to a thin layer of resin between two adherends can be monitored during swelling caused by water uptake at its edge, by observing changes in the pattern of Newton's rings created between one of the adherends, and a reference flat. A regular 150 μm thick microscope cover slip serves as an adherend which is thin and flexible such that it follows the deformation of the resin film. A slab of metal is used as the other adherend. The experimental arrangement is shown in Figure 2. Filtered light, of wavelength λ, from a mercury vapour lamp is directed through the optical flat towards the cover slip, and interference between incident and reflected beams occurs within the variable thickness gap located between the two. The Newton's rings are photographed using light reflected into a 35 mm camera by the half silvered mirror. In its simplest form, the pattern of Newton's rings is a consequence of interference between just two beams, namely the incident beam and the reflected beam. The fringe contrast is considerably enhanced by silvering the upper surface of the cover slip and half silvering the lower surface of the optical flat in order to produce multiple reflections and hence multiple beam interference. To ensure that changes in the pattern arise only from distortion of the cover slip caused by resin swelling, it is essential that each experiment be carried out without disturbing the specimen/optical flat assembly. All of the components are set up on an optical bench and, for hot water tests, it has been necessary to develop a rig which avoids condensation onto the optical components.

Adjacent rings of the same colour (black or white) are loci of points for which the optical path length, in the space between cover slip and optical flat, differs by one wavelength. By the same token, a displacement in the pattern of Newton's rings by an amount equal to one ring width corresponds to a change in path length equal to one wavelength. By observing changes in the number of rings between fixed markers, such as entrapped air bubbles, displacements normal to the joint during water uptake can be measured to an accuracy of $\lambda/4$. If required, displacements which are at least as small as $\lambda/10$ can be measured by superimposition of images in order to create Moiré patterns.

Experimental

Materials. Long term experiments have been carried out using CIBA-GEIGY MY 750 DGEBA epoxy resin mixed with the manufacturer's recommended proportions of methyltetrahydrophthalic anhydride hardener and triamyl-ammonium phenate accelerator. The mix is cast between a degreased and nitric acid cleaned thin glass cover slip and a block of 99.99% purity aluminium cleaned by successive degreasing, de-oxidising and anodising procedures. The resin film thickness is about 15 μm and is controlled by the mass of resin used for bonding, the cover slip being allowed to settle under its own weight. After the manufacturer's recommended curing schedule a reflective coating is vacuum deposited on the cover slip and the sample exposed to distilled water. Tests have been carried out at room temperature and at the boiling point of water. Good fringe contrast requires a water resistant reflective coating and minimisation of the number of highly reflecting air/glass interfaces or the use of transmitting coatings of intermediate refractive index.

Observations. Figure 3 shows a typical sequence of Newton's rings patterns observed after progressively longer exposure times in room temperature water. One characteristic feature is the inward migration of a circumferential locus of kinks in the individual fringes; the kinks are best resolved at points such as B, where the fringe orientation is such that a prominent kink is produced. The kinks delineate an abrupt change in resin film thickness. The water concentration, and hence the swelling associated with water uptake, saturates and gives rise to a shoulder separating fully saturated from less than fully saturated resin and this shoulder progressively moves inward from the rim of the specimen. This is shown schematically in Figure 4A-B.

A second characteristic feature is the occurrence and growth of an edge crack lying parallel to the interfaces and giving rise to circumferential interference fringes. To maintain contact with the outer annulus of uniformly swollen resin (Figure 4C), the adherends would need to bend with curvature opposite to that inside the shoulder. Failure to adopt such "S" wise bending manifests itself as the observed interfacial crack.

Analysis of Measurements from Newton's Rings Experiments. Displacement, normal to the specimen plane, of one quarter of a wavelength is revealed as a reversal of contrast in the pattern of Newton's rings. Thus the bright → dark change at the fringe indicated by arrow A in Figure 3(a) → (c) indicates a displacement of 136.5 nm (λ = 546.1 nm for the green light emitted by the mercury vapour lamp used in the present experiments) at a point 6 mm in from the rim after only two weeks exposure to room temperature water. Across the locus of kinks in Figure 3, the fringe deviation amounts to approximately two ring widths, i.e. a normal displacement of λ (0.5461 μm) in an annular ring width of approximately 0.8 mm. A local thickening of the 15 μm thick resin film by 0.5461 μm corresponds to a linear swelling of almost 4% which is close to the value expected for saturated swelling, as determined in previous tests. Using previously measured values of resin modulus this strain indicates normal stresses of about two kilobars in the saturated resin.

Timoshenko's(1) formula for the axial stress σ produced by an axial load W on a thin plate clamped at its rim is

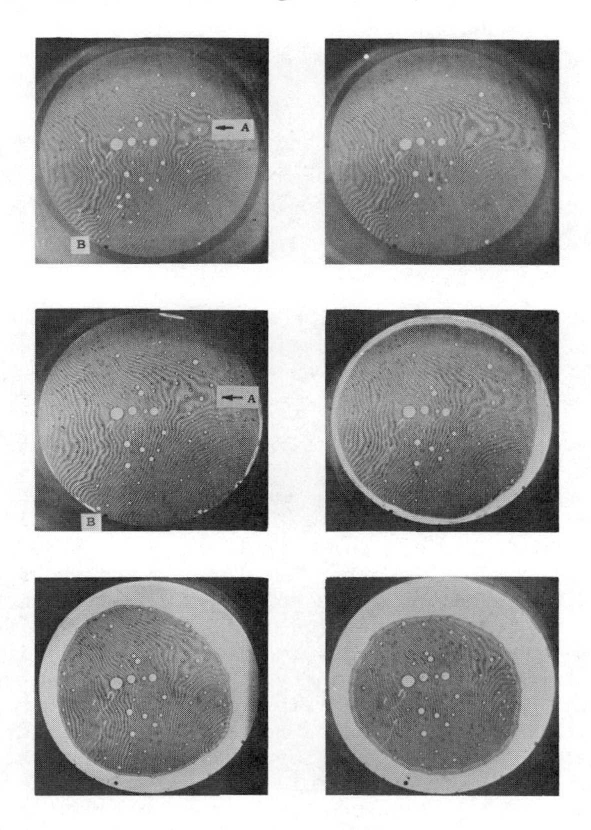

Figure 3. Swelling and debonding of an epoxy layer during water uptake at its edge (8). The changing pattern of Newton's rings is produced by the gap between an optical flat and a flexible cover slip (see text); 20°C water. Patterns recorded after (a) 0, (b) 116, (c) 356, (d) 1008, (e) 2329, and (f) 3043 hr, viewing from left to right and top to bottom.

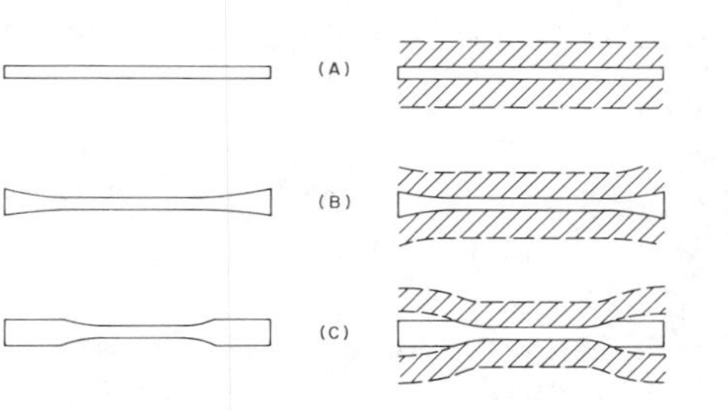

Figure 4. Inhomogeneous swelling anticipated for a layer of resin during water uptake at its edge (8). LHS without any mechanical constraint; RHS with mechanical constraint due to presence of adherends.

$$\sigma = \frac{W}{h^2} [(1 + \nu)(0.485 \ln \frac{a}{h} + 0.52) + 0.48]$$

where h is the plate thickness and 2a is its diameter. ν is Poisson's ratio. The axial displacement is given by Love (2) as

$$\omega = \frac{Wa^2}{16 \pi R}$$

where

$$R = \frac{Eh^3}{12(1 - \nu^2)} .$$

E is Young's modulus. Substituting $E = 6.3 \times 10^{11}$ dynes cm^{-2}, $\nu = 0.25$ and $h = 150$ µm, for the glass cover slip it is found that $D = \sim 10^5$ ergs, 2a = 19 mm, hence, $\omega = 136.5$ nm gives W = 144 dynes and $\sigma = 2.33$ bars.

This estimate for σ assumes that the cover slip is rigidly clamped at its rim but is otherwise free to undergo a uniform curvature deformation under the action of the axial load W. In fact, the cover slip is bonded over the whole of its area of contact with the resin and the deformation is strongly inhomogeneous, being concentrated in the vicinity of the shoulder between resin which is fully saturated and resin which is less than fully saturated with diffused water. From geometric considerations the fact that the cover slip deformation is concentrated over a 0.8 mm annulus leads to a 600 fold increase in the curvature of the slip over the value used in the stress calculation (based on plate diameter). Thus it is expected that the stress required to produce the observed bending deformation is of the same order as the normal stress estimate using resin modulus. Additional experiments using samples of resin cast between two glass adherends allowed optical stress-birefringence measurements to be made during swelling. These tests confirm that the local stresses at the deformation shoulder are indeed about two kilobars (3).

Order of Magnitude Estimate for Stress Enhanced Water Migration. An estimate of the stress required for stress induced drift of water molecules to be as important as diffusion in a concentration gradient may be made as follows. The respective fluxes are

$$j_{drift} = - \mu \; c \; \nabla \phi$$

where μ is the drift mobility, c is the water concentration and $\nabla \phi$ is the potential energy gradient arising from the tensile stress at the center of the specimen sketched in Figure 4, and

$$j_{diffusion} = - D \nabla c$$

where D is the water diffusion coefficient and ∇c is the water concentration gradient.

$$\frac{j_{drift}}{j_{diffusion}} = \frac{\mu c \nabla \phi}{D \nabla c} = \frac{c}{\nabla c} \frac{\nabla \phi}{k T}$$

using the Nernst-Einstein relationship $\mu = D/kT$. The dimensions are cm for $c/\nabla c$ and erg cm^{-1} for $\nabla \phi$. Hence,

$$\frac{c}{\nabla c} \nabla \phi$$

is a quantity of energy, say $\Delta \phi$, i.e.

$$\frac{j_{drift}}{j_{diffusion}} \sim \frac{\Delta \phi}{k T} .$$

$k = 1.4 \times 10^{-16}$ erg K^{-1} so, for the fluxes to be equal at room temperature, $\Delta \phi$ would have to be of the order of 4.2×10^{-14} erg. Assuming that water molecules migrate singly and cause swelling ΔV equal in magnitude to the natural volume of the water molecule, i.e.

$$\Delta V = \frac{18}{6 \times 10^{23}} = 30 \times 10^{-24} \text{ cm}^3,$$

then $\Delta \phi = p\Delta V = 4.2 \times 10^{-24}$ ergs and $p = 1.4 \times 10^9$ dynes cm^{-2} ~ 1 kbar. Hence, the observed stresses are certainly great enough to significantly enhance water migration.

<u>Summary of Other Newton's Rings Experiments</u>. Very similar results have been obtained from changes in the pattern of Newton's rings observed during water uptake by supported adhesive films. That is, displacements remote from the edge of the joint occur after surprisingly short exposure times, large displacements and correspondingly large stresses straddle the boundary between fully swollen and less than fully swollen adhesive, and debonding at the rim follows attainment of full water saturation. These conclusions are true for joints manufactured between a glass cover slip as one adherend and aluminium, both with and without anodising treatments, as the other adherend. Use of hot water rather than cold water as the exposure medium accelerates the incidence of all these phenomena.

Figure 5 shows a comparison of the debonding rates for epoxy resin joints tested at room temperature and 100° C, and a

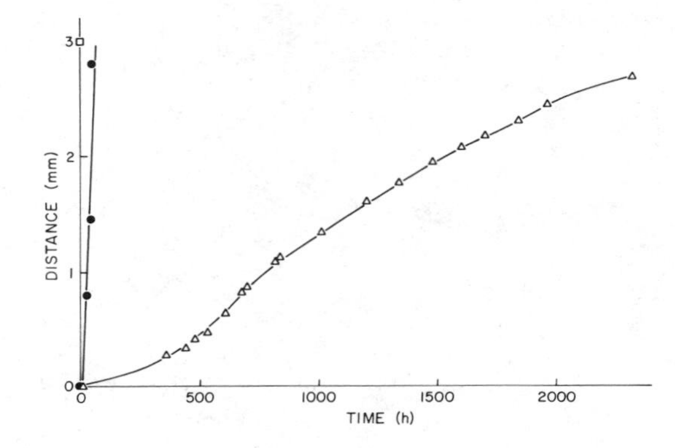

Figure 5. Debonding of epoxy resin joints at (●) room temperature and (△) 100°C and (□) polyester resin joint at room temperature (3)

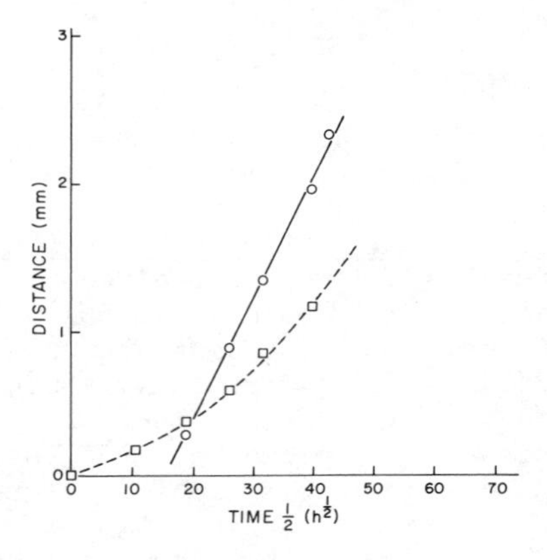

Figure 6. Migration of (□) the kink in a Newton's ring and (○) of a point on the debonding crack; 20°C water. Data taken from Figure 3 (8)

polyester resin joint tested at room temperature. The higher temperature test results in a 40 fold increase in debonding rate for epoxy resin, which yields an activation energy of about 10 kcal $mole^{-1}$ for the process, assuming it to be thermally activated. As expected, due to its more polar structure, the polyester resin is less water resistant than epoxy resin at the same temperature, by a factor of ~13.

Discussion

Kinetics of the Various Processes during Water Uptake. The spatial positions of the locus of kinks in the pattern of Newton's rings, the edge of the debonding crack and the water droplets inside entrapped air bubbles are all very well defined and have been measured as functions of time. The data obtained from the experiment reported in Figure 3 are presented in Figure 6. To test whether any of the data fit the solution to Fick's law for the case of a planar interface between infinitely long bars of solution and solvent, published by Barrer (4), for example, the measurements are plotted as functions of $(time)^{1/2}$. Migration of the shoulder defining the extent of water saturation is evidently not controlled by a $t^{1/2}$ law. Migration of the crack edge, however, does approximate reasonably well to $t^{1/2}$ behaviour. The kink monitored in order to construct Figure 6 did not fall on a radius where debonding initiated, i.e. the two sets of data in Figure 6 are for different radii and the cross-over does not indicate that the crack edge has overtaken the shoulder.

The data shown in Figure 7 are for a specimen manufactured between two glass components (a cover slip and microscope slide), using the same CIBA-GEIGY epoxy resin system and exposed to boiling water. In this experiment the appearance of penny-shaped pressure filled cavities, created by dissolution of water soluble impurities (5), was also used to detect the advance of absorbed water from the free surface. Straight lines, i.e. $t^{1/2}$ laws, may be drawn through the three data sets although both plots of absorbed water ingress change slope at $t^{1/2} = 6.5$. The time dependence shown in both Figures 6 and 7 is unexpected for an amorphous glassy polymer below its glass transition temperature, T_g. Assuming that physical sorption and activated diffusion are the permeation processes it is expected that polymer segmental motion is the rate controlling step in the migration of water molecules. The strong positive temperature dependence of the rate of ingress indicates that the process is one of diffusion rather than convection, and the lack of evidence for water flow in channels or capillaries supports this conclusion. However, there are several effects which could influence the kinetics of water uptake. In particular it is expected that the diffusion coefficient be strongly concentration dependent, although nearly Fickian behaviour is often observed at low levels of sorption. The anomalous properties of the water molecule due to its small

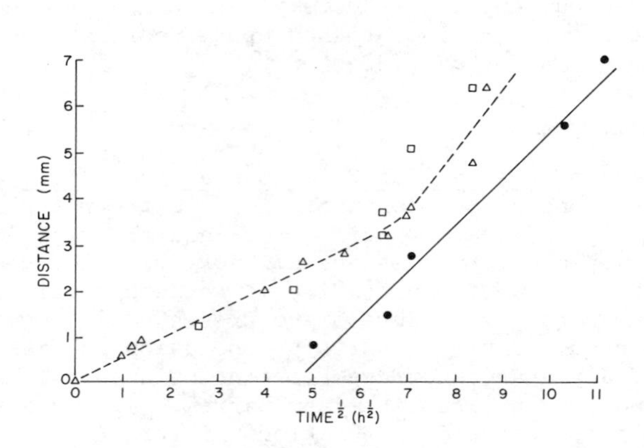

Figure 7. Migration (△) of a kink, (●) a point on the debonding crack, and (□) phase-separated water; 100°C water (8)

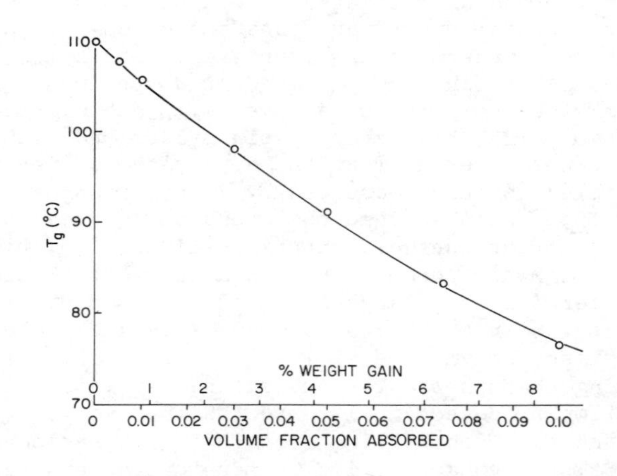

Figure 8. Theoretical prediction of the plasticization of epoxy resin by absorbed water at 100°C

size and high polarity often contribute to unexpected behaviour. Crosslinks, chemical interactions and orientation effects also lead to non-linearity. The other clear influence on water migration is swelling stresses which certainly enhance the rate of permeation, although the effect on the time dependence is not clear.

In the 100°C test, rapid resin plasticisation may have occurred and a diffusion-rate controlled process may then be expected. Figure 8 shows a theoretical prediction of the decrease in T_g with absorbed water content for epoxy resin. It is clear that only 0.02 volume fraction of water is required for T_g to be reduced below 100°C. However, absorption of this amount, equivalent to about 2 wt. %, takes about 40 hours according to previous measurements on the same resin (3). In this case the slope transition in Figure 7 may be evidence for an abrupt change in diffusion coefficient at T_g, with apparent Fickian diffusion occurring at all effective temperatures. The reason for this is not clear.

Effective diffusion coefficients have been measured from Figures 6 and 7. At 20°C, $7 \times 10^{-10} \lesssim D \lesssim 5 \times 10^{-9}$ cm^2s^{-1} and at 100°C, $7 \times 10^{-7} \lesssim D \lesssim 5 \times 10^{-6}$ cm^2s^{-1}. A constant factor of 1000 between the two temperatures would indicate an activation energy for 'diffusion' of about 20 kcal/mole. Previous measurements on bulk resin give $D \simeq 10^{-8} cm^2s^{-1}$ at 100°C, which is less than the lower limit of the present values due to the fact that constrained swelling has not created the stresses responsible for enhanced migration.

The anomalous behaviour shown in Figure 6 is indicative of a rate increase over a process dominated by polymer relaxations. In general, a time exponent of 0.5 indicates Fickian diffusion, characterised by a more diffuse advancing front than relaxation controlled migration, which is characterised by a time exponent of 1.0 and a sharp swollen/unswollen resin transition zone. The 0.7 time exponent may indicate that water ingress follows a Fickian law but with a continuously increasing value of diffusion coefficient, i.e. the curve in Figure 7 is a succession of straight lines. The sharpness of the observed boundary between swollen and unswollen resin requires further study in order to determine the rate controlling step in the permeation process. Experiments such as those of Thomas and Windle (6), who have used iodine doped solvents to allow measurements of concentration profile and diffusion front position during sorption in bulk PMMA, could be used to study this problem in epoxy resin flims. Kwei and Zupko (7) have measured the time dependence of the sorption of different solvents in epoxy samples and observe time exponents between 0.5 and 1.0 corresponding to diffuse and sharp fronts respectively. However, no explanation of the reason for anomalous time dependence and no conclusions on sorption mechanisms are made in either work.

Further work using the methods outlined here will provide data on epoxy properties particular to thin film sections typical of the resin fillets in composites. However, measurements on bulk resin will also play an important role in understanding sorption phenomena relevant to composites. For this reason an interference technique using double exposure holography has also been developed (9), which yields, simultaneously, accurate deformation and sorption data on epoxy adhesive interfaces.

General Discussion. It is apparent that enhanced water migration occurs in constrained thin resin films. The attraction afforded by tensile stress is the most likely explanation of this effect. Three mechanisms could give rise to such a stress. Curing shrinkage is assumed to be negligible, as viscoelastic flow at elevated temperatures will relieve any stresses, and differential contraction between resin and adherends on cooling will be partially relieved by self-adjustment of the film thickness. The normal stresses will be insignificant although lateral tensions may be present. Hence, it is thought that swelling stresses during water absorption are the most important source of high local stresses. If water migration within the resin layer is significantly enhanced by the self-stressing that results from the inhomogeneous nature of swelling, it follows that the effective diffusion coefficient is a function of water concentration and processes occurring at rates proportional to the square root of time cannot be expected. The self generation of high local stresses during swelling is a consequence of the sharp boundary between fully saturated and less than fully saturated resin and appears to be unaffected by changing the materials used as adherends in order to resist mechanical failure when external forces are applied to the composite itself. Measurements of the swelling stresses and calculations of enhanced migration in a stress-field show that the observed increases in diffusion coefficients may be explained, although the time dependence and kinetics of the overall process and the rate controlling mechanisms remain incompletely understood. Complete understanding of the problem will require investigation of the nature of swelling and areas of research fundamental to this include identification of the various physical and chemical states of diffused water, the mechanism(s) by which diffused water promotes so-called plasticisation, and the precise origin(s) of dimensional changes attributable to curing, water uptake and water expulsion.

Abstract

Swelling during water uptake by fibre reinforced resin composites is strongly inhomogeneous. In particular, there exists a well defined shoulder between fully saturated and less than fully saturated resin. The principal stresses introduced as a consequence of the mechanical constraint exerted by the fibres include a normal compressive stress near the flanges and a normal tensile stress near the core of a resin fillet. The magnitude of the stress has been measured by analysing the Newton's rings patterns produced between a flexible thin glass cover slip, bonded to a sample thin film of resin supported on a thick metal adherend, and an optical flat. The observed local stresses of about two kilobars in the sample are shown to be great enough to account for enhanced water mobility in the resin and non-Fickian diffusion effects.

List of Symbols

λ:	wavelength of light	ν:	Poisson's ratio
σ:	stress	2a:	plate diameter
W:	load	ω:	displacement
h:	plate thickness	E:	Young's modulus
R:	flexural rigidity	j:	flux
μ:	drift mobility	c:	concentration
ϕ:	potential energy	k:	Boltzmann's constant
T:	absolute temperature	V:	volume
p:	pressure	t:	time
D:	diffusion coefficient		

Acknowledgements

This research was supported in part by the US Army (grant No. DA-ERO-76-G-068) and in part by the US Air Force (grant No. AFOSR-77-3448).

This article was prepared with the cooperation of the Cornell Material Science Center which is funded by the National Science Foundation (DMR 76 81083).

Literature Cited

1. Timoshenko, S.; "Strength of Materials", 2nd Ed., Van Nostrand, New York, 1947.
2. Love, A. E. H.; "Treatise on the Mathematical Theory of Elasticity", 4th Ed., Cambridge University Press, Cambridge, 1959.
3. Farrar, N. R., "Ph.D. Thesis", University of Bristol, 1977.
4. Barrer, R. M.; "Diffusion in and through Solids", Cambridge University Press, Cambridge, 1941.
5. Farrar, N. R., and Ashbee, K. H. G.; J. Physics D, 1978, 11, 1009.
6. Thomas, N. L., and Windle, A. H.; Polymer, 1978, 19, 255.
7. Kwei, T. K., and Zupko, N. M.; J. Pol. Sci., 1969, A-2, 7, 867.
8. Farrar, N. R., and Ashbee, K. H. G.; U.S.A.F. Interim Scientific Report on Grant Number AFOSR-77-3448, 1978.
9. Farrar, N. R., Conners, A. M., and Kramer, E. J.; Unpublished results.

RECEIVED January 8, 1980.

32

Measurement of the Distribution of Water in a Graphite Epoxy by Precision Abrasion Mass Spectrometry

M. A. GRAYSON

McDonnell Douglass Research Laboratories, McDonnell Douglas Corporation, St. Louis, MO 63166

The study of volatile compounds evolved from polymeric samples under mechanical load by mass spectrometry has a 15-year history which was recently reviewed (1). In this technique, which we call stress mass spectrometry or stress MS, the polymeric sample is mechanically deformed in the ion-source housing of a time-of-flight mass spectrometer (TOFMS). Compounds evolved from the sample expand into the ion source where they are ionized by electron bombardment and subsequently mass analyzed.

The compounds evolved from the sample can originate from either or both of two sources: 1) they can be indigenous volatile compounds, such as water vapor, solvents, plasticizers or unreacted starting materials which diffuse from the sample as a result of mechanical loading and/or failure, or 2) they can be volatile products formed as a result of stress-induced chemical reactions in the polymer chain.

Prior to 1976, all results of stress MS experiments of polymers were interpreted as support for the latter origin, i.e., volatile products from stress-induced chemistry (1). At that time, we performed stress MS experiments on polystyrene (2) and concluded that indigenous styrene monomer was a more likely source of monomer than stress-induced chemistry in the main chain.

Consequently, a proof-of-principle device that sawed the polymer in the ion-source housing was used to demonstrate that indigenous volatile compounds can be desorbed from the polymer (3). A more sophisticated device was constructed to determine the quantitative distribution profile of indigenous volatile compounds by precise milling of the polymer (4).

This latter technique, which we call precision abrasion mass spectrometry (PAMS) depends upon abrasion of the polymeric material at a constant rate. Thus, any volatile compounds trapped in the sample are desorbed, and the amount of a compound detected by the mass spectrometer reflects its concentration in the polymeric material. If a hole is abraded in the sample, it is possible to obtain a concentration vs depth profile of the volatile compound. Abrasion as used here includes any method of pulverizing the bulk of the polymeric material into small

0-8412-0567-1/80/47-132-449$05.00/0

particles from which the indigenous volatile compounds can evolve. Thus, sawing, milling, drilling, or abrading with a diamond-coated drill can be used in PAMS experiments as various techniques to abrade the sample.

The distribution of water in composite materials is of interest because moisture can cause localized regions of increased stress (5). Further, the plasticizing effects of moisture sorbed by the epoxy matrix can reduce its ability to stabilize high modulus fibers (6). Despite the importance of sorbed moisture, little has been done to experimentally measure the moisture profile in composites.

Sandorff and Tajima (7) report a method of sectioning thin slabs of a composite with a microtome and drying the specimens to constant weight to determine the water distribution as a function of depth in the sample. The data from such work are of sufficient quality to calculate the solubility and diffusion coefficients. Their results, as expected on the basis of Fickian diffusion, indicate a smoothly varying moisture profile. Preliminary results of PAMS experiments are reported here as a technique to measure the quantitative distribution of water as a function of depth in the sample.

Experimental

Precision Abrasion Mass Spectrometry. A block diagram of the PAMS instrument is shown in Fig. 1. A time-of-flight mass spectrometer (TOFMS) was used for these studies for several reasons: a) the TOFMS has a large, open ion-source and ion-source housing, providing ample room to install the sample abrasion system; b) the gases evolved during abrasion directly enter the ionizing region of the source; and c) the TOFMS generates 20 000 mass spectra per second. Thus, complete mass spectra of compounds from short-lived transient events (typically 10 ms) can be obtained by Z-axis modulated oscilloscope display and photography techniques described by Lincoln (8).

The TOFMS used in these studies is a Bendix model 12-101 instrument refitted with a solid-state electronic chassis (CVC Products Mark V). The mass spectrometer was operated with a conventional electron bombardment ion source at a trap current of 1 μA and an ionizing potential of 70 eV. The sensitivity of the instrument was measured with n-hexane. At a mass flow rate of 10^{-10} g/s of n-hexane into the ion-source housing, an interpretable mass spectrum with a signal-to-noise ratio of 10 to 1 was obtained.

The instrument is equipped with one analog scanner (MA 010) for conventional mass spectral recording and two four-channel monitors (MA-006). Eight ions can be continuously recorded during the abrasion process. A total ion monitor is available for monitoring the sum of ion currents above a predetermined mass. A Z-axis modulated oscilloscope display data acquisition system is also available (1,9). The ion-source housing was modified by the addition of two opposing arms terminated by vacuum flanges at 90° to the axis of the flight tube. One arm is fitted with a viewing port, and the other accommodates a rotary vacuum feedthrough and the cutting tool drive system.

The sample stage (Fig. 2) is attached to the shaft of a three-axis micropositioner (Perkin Elmer-Ultek) which is integral with the bottom ion-source housing flange of the TOFMS. Each axis of the micropositioner is controlled by a stepping motor and controller. This arrangement permits precise control of the position and movement of the sample in each of the three principal axes.

A cutting tool, whose axis of rotation is along one of the three principal axes, is used to abrade the sample. A motor generator and controller (Electro-Craft) provides rotary power to the tool chuck. The output of the motor generator is connected to the shaft of a rotary high-vacuum feedthrough (Ferrofluidics) by a flexible coupling. A precision chuck (Albrecht) is press-fit to a taper on the vacuum-side of the rotary feedthrough shaft. The entire assembly is mounted to a vacuum flange which attaches to a side arm in the ion-source housing of the TOFMS.

The tool drive and three-axis micropositioner can be mounted in a bench jig to mount the sample and determine the mechanical operating characteristics of the system (see Fig. 3). A detailed description of the instrument along with a discussion of its operating characteristics will appear elsewhere (9).

Sample. The composite studied in these experiments is a graphite epoxy test panel 3.9 ± 0.05 mm thick. A photomicrograph of the cross section of the test panel is shown in Fig. 4. There are 26 graphite plies with a fiberglass scrim on each surface. The test panel was maintained at room temperature and ambient humidity for 11 months. Test specimens 20 by 40 mm were cut from the panel by a dry cutting process. Three test specimens were used to determine the total water content by drying to constant weight. The samples were dried in air at 110 ± 3°C. PAMS was used to determine the moisture profile in undried samples.

Moisture Profile Measurements. Tungsten carbide end mills (1.6 mm diam) were used to drill holes in the specimens. Tool speeds from 100 to 600 rpm and sample feed rates of 6 μm/s to 16 μm/s were used during the PAMS experiments.

Since the surface of the sample must be exposed to the vacuum of the mass spectrometer for 20 to 30 min before experiments can commence, the samples are sandwiched between two aluminum plates 0.8 mm thick to prevent loss of water from the surface. Some experiments were also performed with one surface of the sample exposed to determine the changes in the moisture profile which occur during pumpdown of the mass spectrometer.

The ion current at m/e 17, a fragmentation peak from water, was continuously monitored during the abrasion process to measure the evolution of water from the sample. Mass 17 was used since mass 18, the molecular ion of water, was saturated at the mass spectrometer operating conditions used in these experiments.

In addition, the ion current at m/e 44 was monitored to measure the evolution of carbon dioxide from the sample. Empirical evidence indicates that the amount of carbon dioxide evolved from a graphite epoxy sample during abrasion provides information concerning the sharpness of the cutting tool. Since abrasion of the sample occurs *in vacuo* without the benefit of cooling fluids or cutting lubricants, the tool wears rapidly, typically after drilling 15 to 20 mm, i.e., four to six holes in

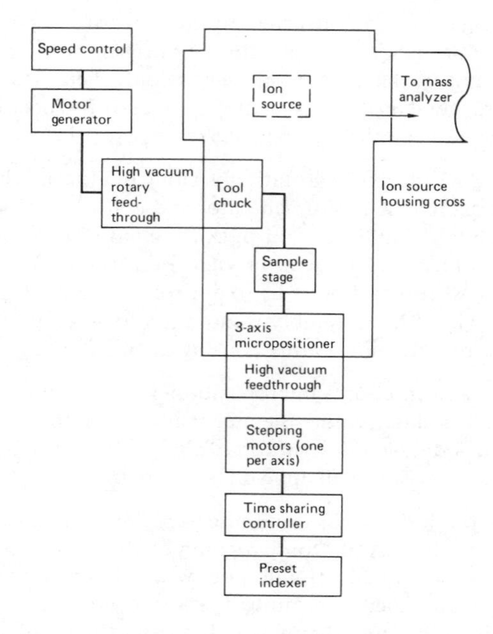

Figure 1. Block diagram of precision abrasion mass spectrometer (PAMS) (4)

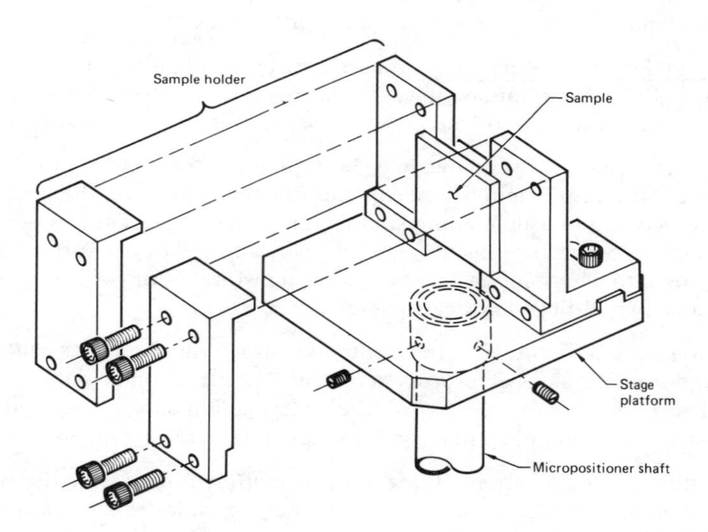

Figure 2. Sample stage for precision abrasion mass spectrometry (4)

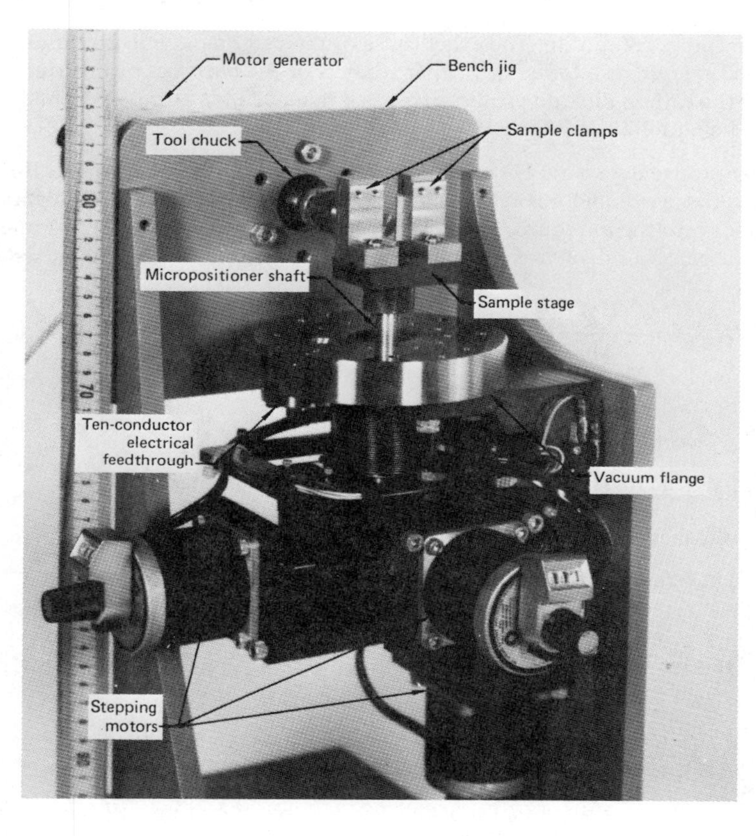

Figure 3. Precision abrasion device in bench jig

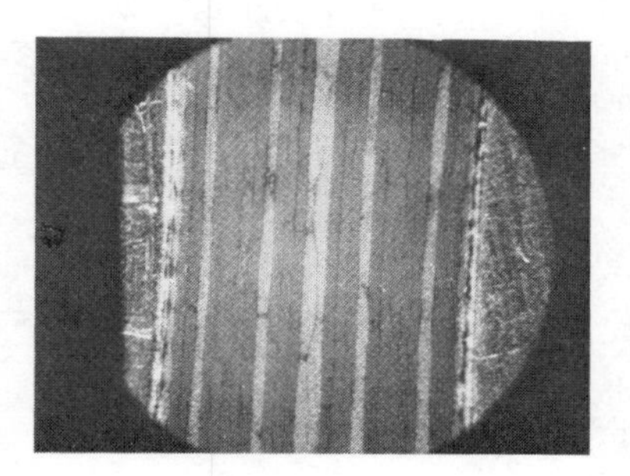

Figure 4. Photomicrograph of cross-section of graphite epoxy composite

a graphite epoxy composite. The exact cause of the carbon dioxide evolution is not known and represents a topic for further study. For the purposes of this study, however, the carbon dioxide profile serves as a valuable indicator of the condition of the cutting tool.

Quantitative data were obtained by calibrating the mass spectrometer for water vapor. A gas/liquid inlet system was used to leak a known mass flow rate of water vapor into the ion-source housing while measuring the ion current at m/e 17. The weight percent of water is the most convenient form of presenting the data:

$$W_{H_2O}\% = \frac{Q_{H_2O}}{M} \times 100, \tag{1}$$

where

$W_{H_2O}\%$ is the weight percent of water in the sample,

Q_{H_2O} is the mass flow rate of water from the sample, and

M is the mass abrasion rate of the composite.

Expressed in terms of measurable experimental quantities, Eq. (1) becomes

$$W_{H_2O}\% = \frac{I_{H_2O}\ S_{H_2O}}{\pi \rho r^2 v},$$

where

I_{H_2O} is the ion current of m/e 17 (μA),

S_{H_2O} is the sensitivity factor of the TOFMS to water at m/e 17 (g/s · μA),

ρ is the bulk density of the composite (g/cm^3),
r is the radius of the cutting tool (cm), and
v is the velocity of the sample stage (cm/s).

Results and Discussion

Results of these preliminary PAMS experiments were used to achieve two objectives. 1) the total water content of the composite measured by PAMS was compared with that measured by drying to constant weight. 2) changes in the moisture profile resulting from exposure of one surface to the vacuum of the mass spectrometer were determined.

A typical moisture profile for an undried graphite epoxy composite is shown in Fig. 5. This profile was obtained from a sample which had been clamped between two thin (0.8 mm) sheets of aluminum to prevent moisture loss from the surfaces during evacuation of the vacuum envelope of the TOFMS. It is apparent that the use of an impermeable cover is sufficient to prevent the loss of moisture from the surface of the sample when it is exposed to vacuum. The shape of the moisture profile is comprised of two features: the gross shape of the curve which corresponds to a concentration gradient as expected from Fickian diffusion, and fine structure which corresponds to short-term increases in the evolution of water from the sample. The gross shape of the water profile is similar to that obtained by Sandorff and Tajima (7). The tail in the moisture profile after the sample has been completely drilled through is due to the exponential decay of water vapor in the ion source housing as it is pumped away.

The origin of the fine structure in the moisture profile is uncertain. The variation in stage velocity during these experiments is on the order of 10%. Thus local variations in the moisture profile could be due to short-term variation in the stage velocity or to randomly distributed moisture-rich regions in the composite. To determine the actual source of the fine structure, a position transducer will be connected to the sample stage so that stage velocity data can be obtained during abrasion of the sample in the TOFMS. At the present time, stage velocity data can be obtained only when the abrasion system is mounted in the bench jig.

The integral of the area under the moisture profile is a measure of the total weight percent of water contained in the sample. The average total moisture content for three holes by PAMS was 0.1 wt %. The average total moisture content from drying to constant weight was 0.35 wt % (dried in air at 110°C for 26 days). Thus, the moisture content determined by PAMS is biased towards a lower value. The exact reason for this discrepancy is under investigation.

Several PAMS experiments were performed on a sample with one surface exposed to the vacuum of the mass spectrometer. A series of holes was drilled over a period of several days. Fig. 6 is a typical moisture profile obtained after 1 h exposure of the sample surface to vacuum. The high moisture concentration near the surface of the sample has diffused from the exposed surface. Of particular interest is the observation that the moisture profile after several days exposure to vacuum is essentially identical to that after an hour exposure. Thus, there is little, if any, change in the moisture profile despite the prolonged exposure of the sample. The cause of this phenomenon is not certain; however, it is apparent that a short-term exposure to vacuum is sufficient to permit the loss of moisture from a depth of several tenths of a millimeter below the surface.

Conclusions

The technique of precision abrasion mass spectrometry is a useful method to obtain moisture profile data from composite samples. These data are of sufficient quality that PAMS experiments on samples subjected to controlled environmental exposure can be used to calculate solubility and diffusion coefficients. In addition, this application indicates the utility of PAMS to determine distribution profiles for indigenous volatile compounds other than water.

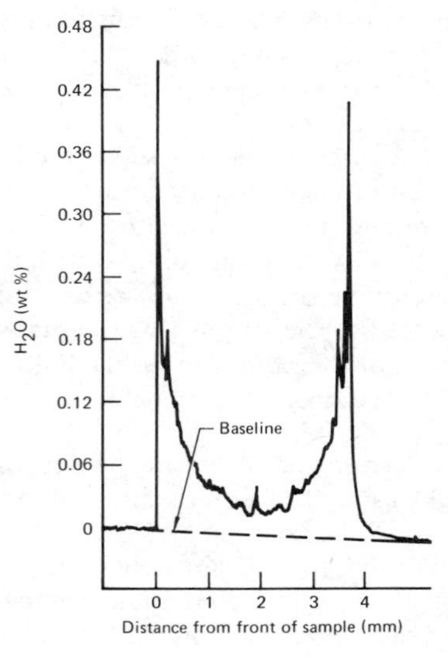

Figure 5. Typical moisture profile from a graphite epoxy composite

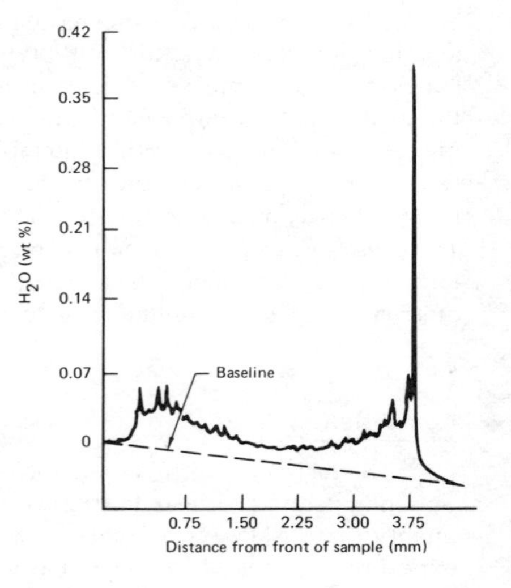

Figure 6. Typical moisture profile from a graphite epoxy composite with one surface exposed to vacuum

Several areas of future study are suggested by these results. Investigations should be conducted to determine

- the cause of the fine structure in the moisture distribution profile,
- the moisture profile in the resin system and compare it to that in the composite,
- the moisture profile normal to and parallel to the fiber axis, and
- the distribution profiles of other gases, such as oxygen in the composite.

Despite the early emphasis by other workers on stress-induced chemical reactions in polymers by stress MS, the use of this technique for the study of indigenous volatile compounds in polymers is an equally important study area which warrants further development.

Acknowledgements

This research was conducted under the McDonnell Douglas Independent Research and Development Program.

Literature Cited

1. Grayson, M.A.; Wolf, C.J.; Advances in Chemistry Series, J. Koenig, Ed., American Chemical Society Press, Washington, DC, 1979, Vol. 174, p. 53.
2. Grayson, M.A.; Wolf, C.J.; Levy, R.L; Miller, D.B.; J. Polym. Sci. (Phys. Ed.) 1976, 14, 1601.
3. Grayson, M.A.; Wolf, C.J.; Proc. Twenty-Fifth Annual Conference on Mass Spectrometry and Allied Topics, 1977, p. 677.
4. Grayson, M.A.; Theby, E.A.; Lippold, K.O.; Org. Coatings and Plastics Chem., 1979, 40, 883.
5. Farrar, N.R.; Ashbee, K.H.G.; Org. Coatings and Plastics Chem., 1979, 40, 947.
6. Stifel, P.M.; Proc. 20th AIAA Structures, Structural Dynamics and Materials Conf., 1979, p. 273.
7. Sandorff, P.E.; Tajima, Y.A.; Composites, 1979, 10, 38.
8. Lincoln, K.A.; Int. J. Mass Spectrom. Ion Phys., 1969, 2, 75.
9. Grayson, M.A.; Theby, E.A.; Lippold, K.O.; Rev. Sci. Inst. (Submitted).

RECEIVED January 8, 1980.

33

Quantitative Analysis of Resin Matrix Aging by Gel Permeation Chromatography and Differential Scanning Calorimetry

DAVID J. CRABTREE

Northrop Corporation, Hawthorne, CA 90250

The matrix resins widely used with graphite fiber for the fabrication of composite aircraft structures are epoxy formulations which, while they have at least six months storage life at -18C, have appreciable reaction rates at normal ambient temperatures. These prepregs can therefore be ruined by inadvertent exposure to ambient temperature. It is crucially important for the fabricator of composite aircraft structures to know how much aging has occurred in these prepreg matrix resins prior to fabrication. If significant aging has occurred, the fabricator may have to change the cure cycle used to cure the structure in order to meet specifications for resin content and structural thickness. Because the amount of resin flow that occurs during the normal autoclave cure process decreases with increasing age of the resin, structures cured from aged prepreg will be resin rich and overly thick unless an adjustment is made.

Ideally the prepreg will be received from the supplier frozen in dry ice and will be kept refrigerated in cold storage (-18C) prior to its use. In the optimum fabrication sequence the prepreg will be layed up on the mold and cured within two days after removal from cold storage. However, interruptions in the normal cycle of events can occur which will cause the prepreg to be exposed to ambient temperatures for some period of time. For example: (1) errors can occur in the shipment of the prepreg from the supplier and the prepreg may be exposed on a loading dock for several days, (2) power failures may occur in the refrigeration system at the fabricator's plant, and (3) the fabrication sequence may require more than 2-days lapse time for completion. It is therefore necessary to make a quantitative measurement of the resin matrix's age prior to the time it is used in the fabrication of a structure.

Northrop undertook this program to develop analytical methods for quantatively determining the amount of aging which has occurred in prepreg matrix resins. Our intention was to have a procedure whereby the amount of aging which has occurred in each batch of prepreg can be tracked from the day of its manufacture to the day it is

0-8412-0567-1/80/47-132-459$05.00/0

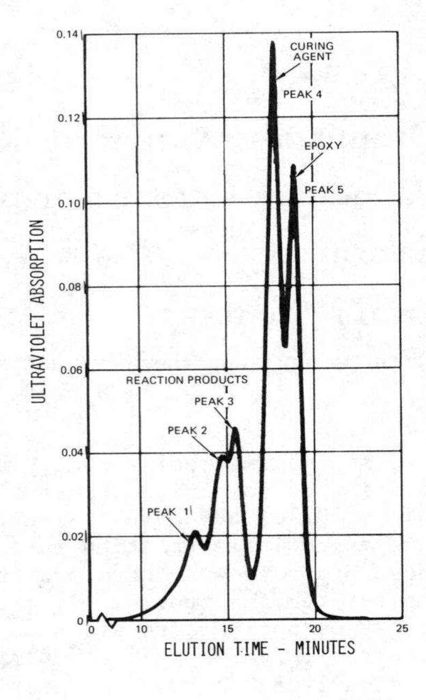

Figure 1. GPC analysis of 3501-6 resin (Lot 009) at 313 nm

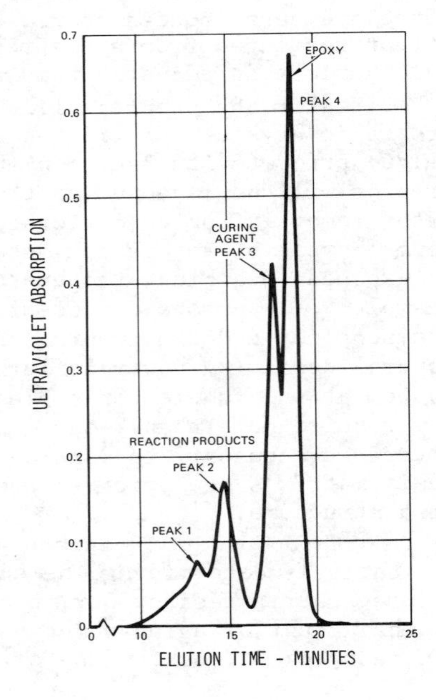

Figure 2. GPC analysis of 3501-6 resin (Lot 009) at 254 nm

used. Gel permeation chromatography (GPC) and differential scanning calorimetry (DSC) were selected as analytical methods for this work.

Materials Analyzed

Methods for quantitatively measuring how much aging occurs during ambient exposure of two epoxy resin formulations important to the aerospace industry were developed. These materials were: (1) 3501-6 (manufactured by Hercules), and (2) 5208 (manufactured by Narmco). Both of these resins are widely used for the manufacture of graphite fiber composites.

Results and Discussion

GPC and DSC methods have proved very useful in measuring changes that occur in resin matrices due to ambient exposure. In GPC analysis the increase in higher molecular weight components as well as the decrease in the curing agent content can be monitored. DSC analysis can also be used for some resins. DSC quantitatively measures the exothermic heat of reaction that is evolved by a resin during cure. Due to aging, changes occur in the amount of heat evolved. DSC was found useful in monitoring aging of 3501-6 resin.

A. 3501-6 Epoxy Resin. The GPC analyses of unaged 3501-6 (lot 009) using the ultraviolet detector at 313 and 254 nanometers are shown in Figures 1 and 2. The peaks are labeled and identified. These measurements can be made quantitative by: (1) the percentages of curing agent and/or high molecular weight reaction products can be determined using peak heights and/or peak areas; and (2) the changes in the ratios of selected peaks can be used. These ratios are useful because as the resin ages there is either a progressive increase or decrease in the ratios. Ratios that are useful and the effects aging has upon them are summarized in Table I.

TABLE I
EFFECT OF AMBIENT AGING UPON GPC PEAK RATIOS
FOR 3501-6 AND 5208 EPOXY RESINS

Peak Ratio	3501-6 Resin Wavelength 254 NM	3501-6 Resin Wavelength 313 NM	5208 Resin Wavelength 254 NM	5208 Resin Wavelength 313 NM
2/3	Increase	Increase	Increase	Increase
2/4	Increase	Increase	Increase	Increase
4/5	-	Decrease	-	-
3/4	Decrease	Increase	Decrease	Decrease
1/3	Increase	Increase	Increase	Increase

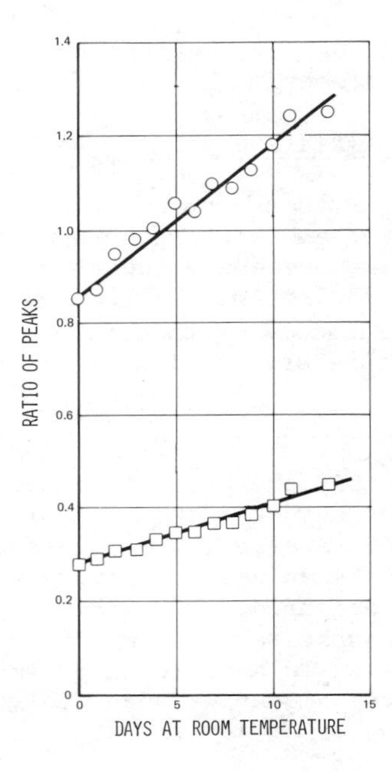

Figure 3. GPC peak ratios at 313 nm of 3501-6 resin (Lot 009) after ambient exposure: (○) 2/3 peak ratio, (□) 2/4 peak ratio

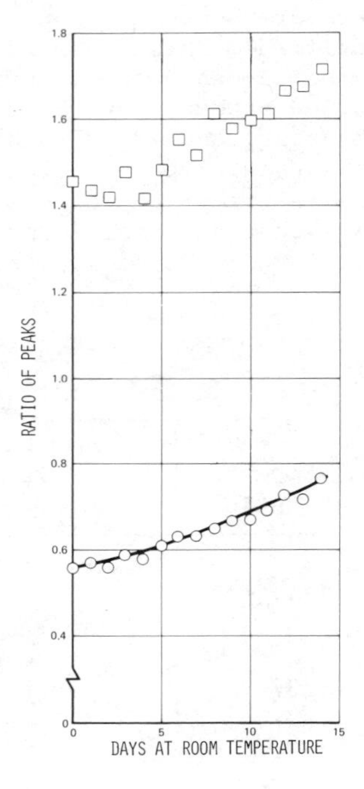

Figure 4. GPC peak ratios at 313 nm of 3501-6 resin (Lot 039) after ambient exposure: (□) 2/3 peak ratio, (○) 2/4 peak ratio

We have found the use of peak ratios an effective method of monitoring changes in resin advancement. In Figure 3 graphs of the changes that occurred in the ratio of peaks 2/3 and 2/4 at 313 nanometers are shown. Progressive increases occurred for both these peak ratios. In Figure 4 the peak ratios for 3501-6 lot 039 are shown. This lot of resin was in a much more advanced state when analyzed initially. Note that the initial values of the peak ratios are much higher and they continue to increase. An error occurred in the shipment of batch 039. It was lost somewhere in transit and arrived at the plant in an advanced state of aging.

The peak ratios from the analysis at 254 nanometers for both batches 009 and 039 are shown in Figures 5 and 6. Here again it is possible both to detect an overage condition in batch 039 and to detect changes that occur during ambient temperature aging.

The DSC analysis (Figure 7) shows three exotherm peaks which occur at approximately 135C, 170C, and 230C. Peaks 1 and 2, which are not fully resolved, appear to be generated by the reaction of an accelerator in the resin. The third peak is generated by the reaction of the curing agent. During aging the heights of peaks 1 and 2 decrease with time. Peak 3 is not sensitive to aging. The ratio of peaks 2/3 decreases as the age of the resin increases. Figure 8 shows the changes in this peak ratio as lots 009 and 039 are aged at ambient temperature. It is readily apparent that lot 039 is more advanced than lot 009.

B. 5208 Epoxy Resin. The GPC analyses of unaged 5208 made using the ultraviolet detector at 313 and 254 nanometers are shown in Figures 9 and 10. The peaks are labeled and identified. Changes that occurred in selected peak ratios at 313 and 254 nanometers (see Table I) are shown in Figures 11 and 12. These peak ratios delineate changes in the age of the resin matrix. DSC measurements were not used for 5208 because the resin only shows a single exotherm peak. There is no accelerator peak to monitor for aging analysis.

Experimental Equipment and Conditions

A. GPC Analysis. GPC analyses were made using a Waters Associates Liquid Chromatographi Model ALC/GPC-244. Five Microstragel columns were used in series. The pore sizes of the columns were 100Å (2 columns), 500Å (2 columns), and 1,000Å (1 column). Ultraviolet grade tetrahydrofuran solvent was used. The instrument was operated at room temperature with a flow rate of 1 ml/min. Concentrations of the samples used for injections were typically in the range 2 to 3 mg/ml. An ultraviolet detector operated at 254 and 313 nanometers was used for all analyses.

B. DSC Analysis. DSC analyses were made on a Du Pont Model 990 Thermal Analyzer using the Model 910 DSC cell. Analyses were

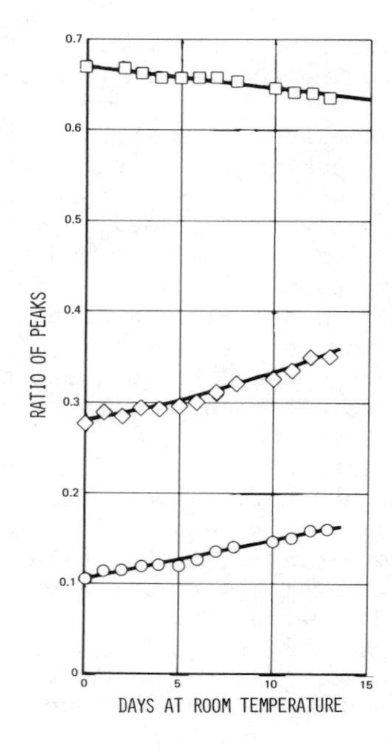

Figure 5. GPC peak ratios at 254 nm of 3501-6 resin (Lot 009) after ambient exposure: (□) 3/4 peak ratio, (◇) 2/3 peak ratio, (○) 1/3 peak ratio

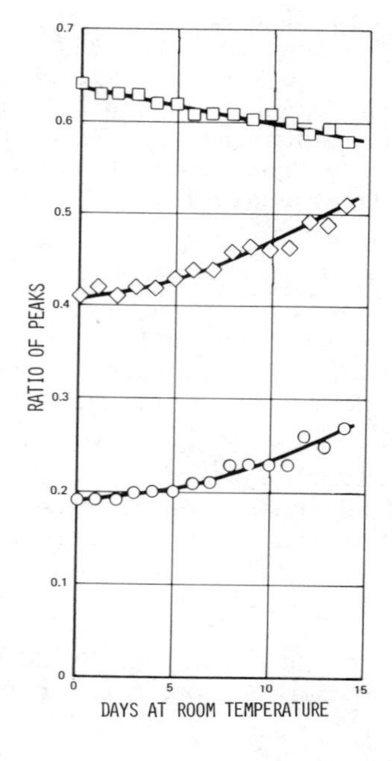

Figure 6. GPC peak ratios at 254 nm of 3501-6 resin (Lot 039) after ambient exposure: (□) 3/4 peak ratio, (◇) 2/3 peak ratio, (○) 1/3 peak ratio

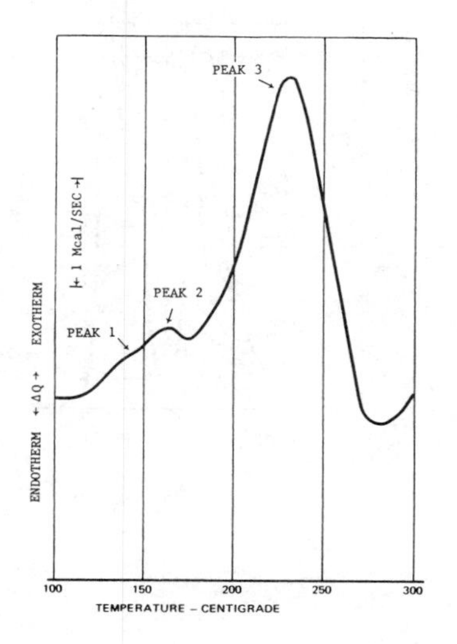

Figure 7. Differential scanning calorimetric analysis of 3501-6 composite resin

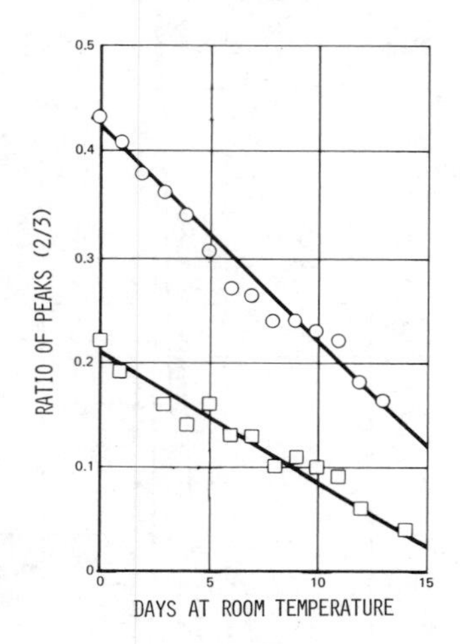

Figure 8. DSC peak ratios of 3501-6 resin (Lots 009 and 039) after ambient exposure: (○) Lot 009, (□) Lot 039

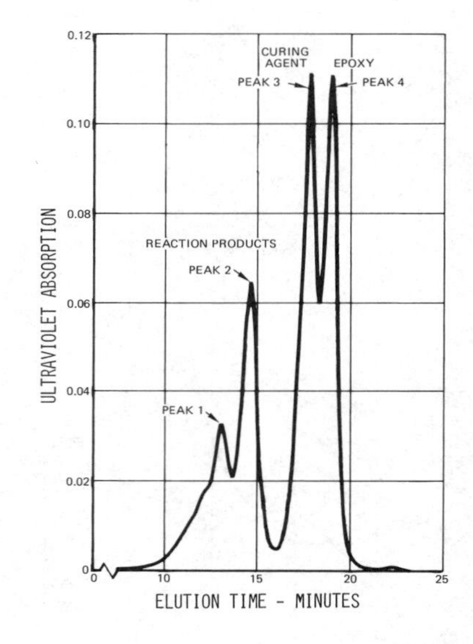

Figure 9. GPC analysis of 5208 resin at 313 nm

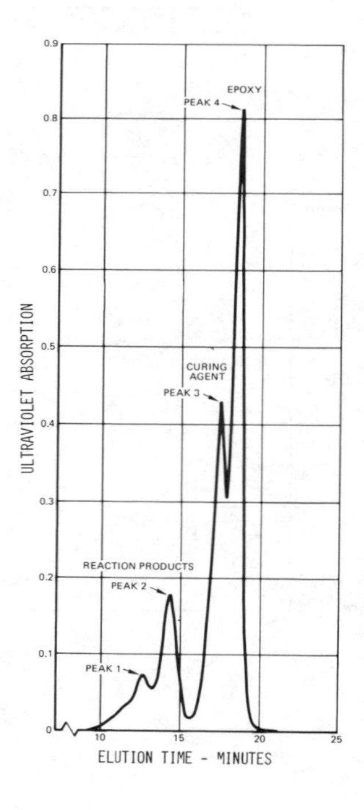

Figure 10. GPC analysis of 5208 resin at 254 nm

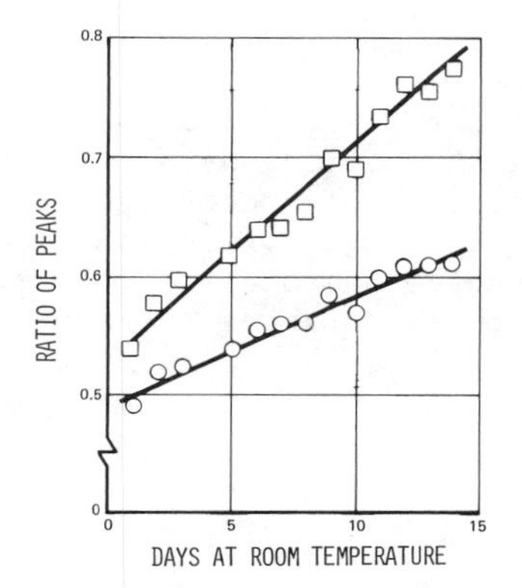

Figure 11. GPC peak ratios at 313 nm of 5208 resin after ambient exposure: (□) 2/3 peak ratio, (○) 1/2 peak ratio

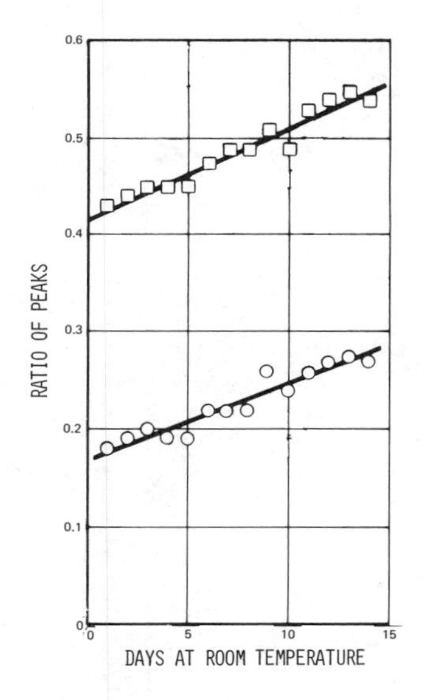

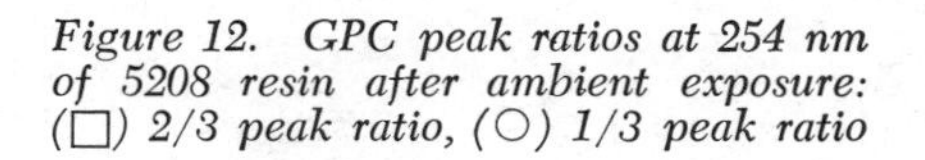
Figure 12. GPC peak ratios at 254 nm of 5208 resin after ambient exposure: (□) 2/3 peak ratio, (○) 1/3 peak ratio

made at a heat up rate of 10C/min. An atmosphere of flowing nitrogen gas was maintained in the cell during the analyses. Sample sizes were in the range 10 to 30 milligram. Analyses were made with instrument sensitivities in the range 0.5 to 2.0 mcal/second.

Conclusions

GPC and DSC have been used to monitor aging that occurs in matrix resins during storage. It is therefore possible to track whatever changes occur during the lifetime of a prepreg. The fabricator can determine how much aging has occurred in a material before using it, and if significant aging has occurred, he may alter the cure cycle as required to ensure cure of a structure which meets specifications.

RECEIVED February 28, 1980.

Effect of Room-Temperature Aging on Graphite/Polyimide Prepreg Materials

H. C. NASH, C. F. PORANSKI, JR., and R. Y. TING

Chemistry Division, Naval Research Laboratory, Washington, DC 20375

Fiber-reinforced composite materials are of increasing importance in many aerospace applications. The driving force behind this trend is the requirement for lighter-weight structural materials for advanced vehicles. Fiber composites seem to satisfy this need by offering a very high strength to weight ratio when compared with conventional metallic structural materials. This weight consideration is especially important to the successful development of vertical and short-take-off and landing (V/STOL) aircrafts in order to compensate for the large and heavy engines required for vertical lift. Since many epoxy resins only have a maximum use temperature around 135°C (275°F), a number of high temperature resin systems are being evaluated as possible candidates for the resin matrix material in graphite reinforced composites on future V/STOL aircraft. A resin is being sought which will meet operational requirements at temperatures in excess of 204°C (400°F), but will be relatively insensitive to moisture and will have the ease of fabrication of state-of-the-art epoxies. One of the promising systems under consideration is an addition polyimide, Hexcel's F-178 resin (1).

A major difficulty in handling many prepregs is that the material must be stored at very low temperatures (i.e. -20°C). The shelf life of these prepregs is generally only a few weeks at room temperature and only a few months at freezer temperatures. Presumably, continual B-staging occurs at room temperature and, at the same time, the resin may degrade by absorbing moisture from the atmosphere. Also, since before lay-up for fabrication the prepreg must warm up to room temperature, it could absorb moisture through condensation.

We therefore carried out an aging study to examine the effects of out-time and humidity at room temperature on the properties of F-178/graphite prepreg. We fabricated laminates from material which had been exposed to various humidities at room temperature and evaluated the mechanical properties as a function of prepreg out-time. We also analyzed extracts of the aged prepreg using proton nmr spectroscopy to see if this technique could reveal anything about the chemistry of the aging process.

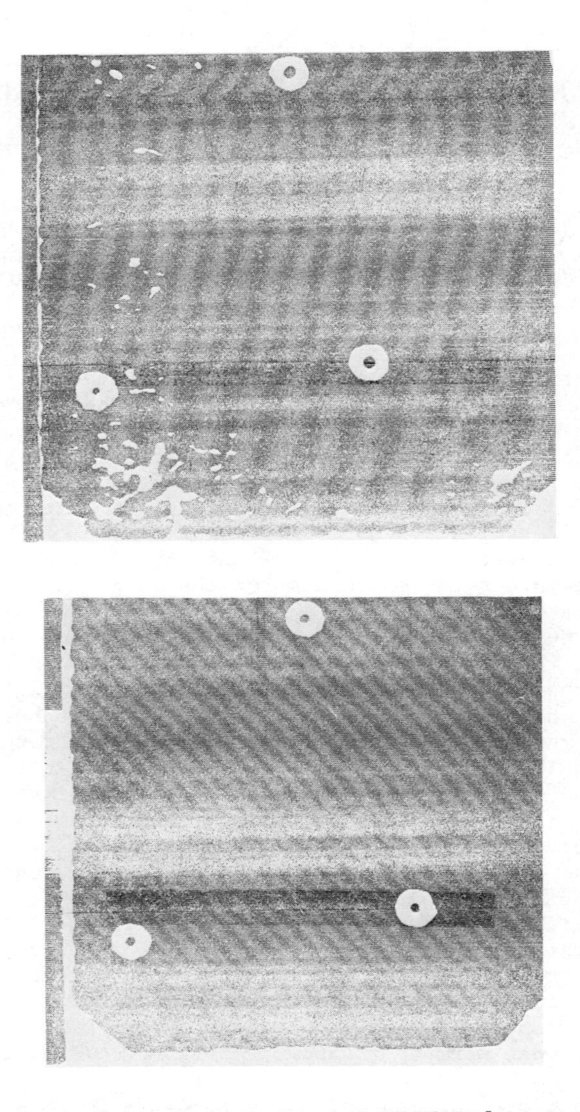

Figure 1. Top: ultrasonic C-scan results for F-178/T300 laminates, (showing voids); bottom: ultrasonic C-scan results for F-178/T300 laminates, (void free)

Experimental

Material. The polyimide/graphite prepreg (F-178/T300) was obtained directly from the manufacturer, Hexcel. The resin is described as a bismaleimide type of polyimide which cures through a free-radical addition mechanism. The prepreg had been included for testing in the NRL V/STOL program because it appeared to have a number of desired features: high operational temperature, 500 hrs at 246°C (475°F); it contains no solvents; no volatile products are formed during cure; and it can be processed like state-of-the-art epoxies. Recent work at this Laboratory has shown that this resin system has a higher fracture toughness than high temperature epoxy systems. The glass transition temperature, determined by torsional pendulum measurements, was 360°C for this resin compared to 260°C for an epoxy (2).

Aging and Fabrication. Fresh prepreg material was cut, packed in kits and stored at room temperature in three environmental chambers controlled at 16%, 50% and 95% relative humidity, respectively. The kits were essentially small aluminum shelves with 17 open slots. Each kit therefore contained seventeen pieces of prepreg, 15 cm x 15 cm in size with one piece in each slot. Aging extended over a period of twenty-eight weeks. At the end of every four-week period, a kit was removed from storage and a 16-ply laminate prepared on the same day. The extra sheet of aged prepreg from each kit was used for proton nmr analysis.

Unidirectional 16-ply laminates were fabricated in a Wabash Press using conventional vacuum bag techniques. The sample lay-up included one glass bleeder ply per every four plies of prepreg with top bleeding only. A porous felt material was used around the sample to form a dam to prevent overbleeding from the edge of the sample. Initially the processing in the press was based on a manufacturer recommended cure cycle originally designed for auto-clave operation. The resulting laminates, however, had high void contents as evident from the result of an ultrasonic C-scan study of the laminates (see Fig. 1a). Variations from that cure cycle were made to adjust for the press operation. Processing parameters such as dwell time, temperature and time of pressure application were varied until high quality, reproducible laminates could be fabricated. These laminates had a void content of less than 0.5%, and a typical C-scan result is shown in Fig. 1b. This modified cure cycle, which is given graphically in Fig. 2, essentially included a full vacuum cycle at 121°C (250°F) and a pressure cycle at 177°C (350°F) and a pressure of 6.9×10^5 Pa (100 psi). Post-cure requirement was 10 hours at 249°C (480°F). This cure cycle was then kept the same for processing all aged samples in order to detect any change of processability in the material due to aging.

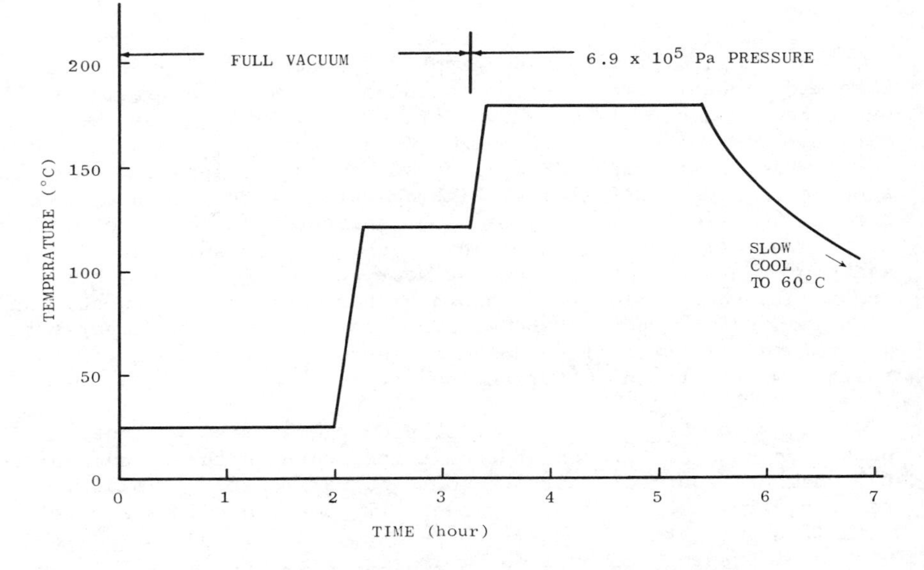

Figure 2. Cure schedule for fabricating F-178/T300 laminates in press

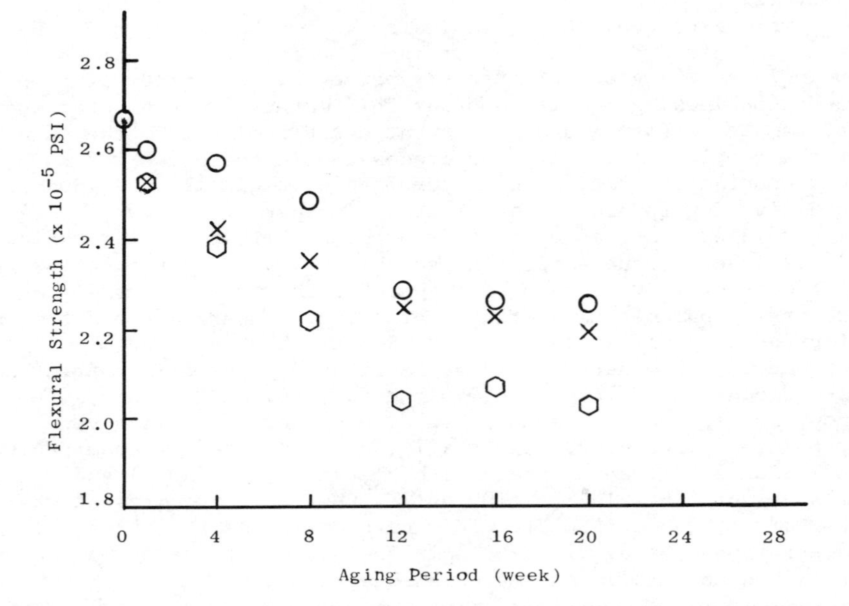

Figure 3. Laminate flexural strength as a function of aging period; F-178/T300, room-temperature aging; R.H.: (○) 16%, (×) 50%, (⬡) 95%

Mechanical Testing. Samples were cut from the laminate and tests were carried out to determine if any change occurred in their mechanical properties relative to composites prepared from unaged prepreg. The short-beam shear test was chosen to determine the interlaminar shear strength, a resin-dominated property. The test was carried out according to the ASTM standard D2344. The dimensions of the shear specimens were 0.5" x 0.16" x 0.28", and 20 specimens were tested for each laminate sample. On the other hand, the flexural test was selected to measure the laminate flexural strength and modulus, which are fiber-dominated properties. The flexural specimens were 2" x 1" x 0.08", satisfying the requirement in ASTM standard D790. Six specimens were tested in this case for each laminate sample. Both mechanical tests were carried out on an INSTRON testing machine at a crosshead speed of 0.05 in/min.

Proton NMR. Weighed 2 cm x 2 cm samples of aged prepreg were placed in tared vials; 1 gram of acetone-d_6 was added to each vial. The samples were extracted for ~2 hours at room temperature while being stirred on a magnetic stirrer. The resulting solutions were filtered directly into dry nmr sample tubes. The residues were dried and weighed. The spectra were obtained on a JEOL FX60Q spectrometer system with the following parameters: frequency, 59.75 MHz; 90° pulse (8μsec); 600 Hz spectral width; 500 Hz, 4-pole Butterworth filter; quadrature phase detection; 8 sec pulse repetition rate; 100 pulses; 8192 data points. The internal deuterium lock was provided by the solvent. The spectra were run at 25°C.

Results

Mechanical Properties. Figure 3 shows the flexural strength of the graphite/polyimide composite plotted as a function of prepreg aging period in weeks. It can be seen that laminate flexural property steadily decreases as aging continues. The effect is more pronounced at higher humidity levels. After twenty weeks of aging at 95% relative humidity, the flexural strength is reduced by nearly 25% from that of the original control sample. The same trend is followed by the flexural modulus data.

Figure 4 presents the interlaminar shear strength data as a function of prepreg aging period. In this case, the results show a rapid increase in shear strength initially as aging proceeds. Then it becomes constant at a level approximately 55% higher than that of the initial control. The increase seems to be independent of the humidity level at which prepreg samples were aged.

At first glance these results seemed surprising because any resin degradation taking place due to moisture absorption in the prepreg would lead to poorer resin-dominated property such as interlaminar shear strength. The fact that interlaminar shear strength was improved and flexural properties deteriorated may be

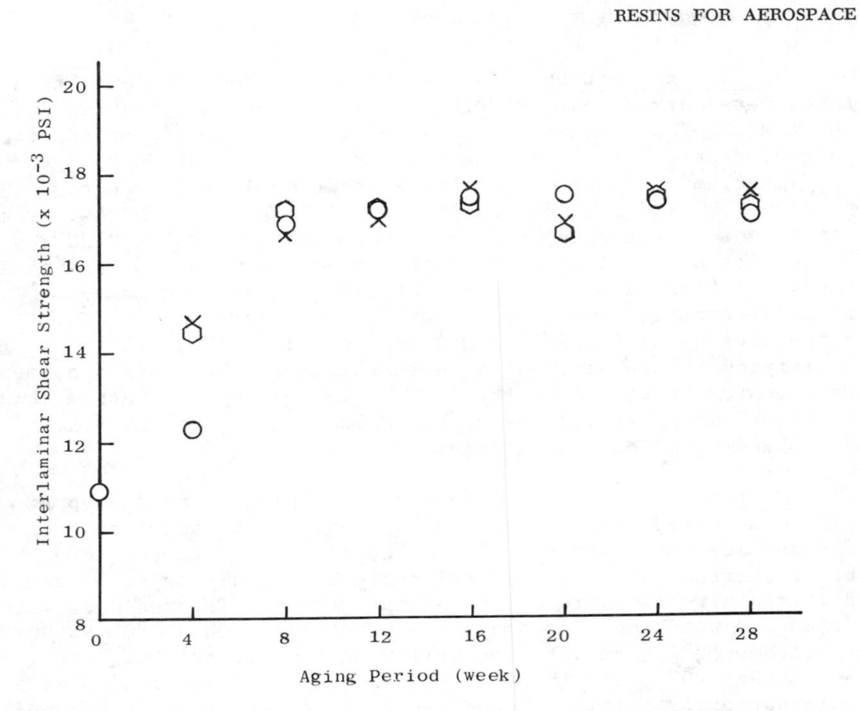

Figure 4. Interlaminar shear strength of aged samples; F-178/T300, room-temperature aging; R.H.: (○) 16%, (×) 50%, (⬡) 95%

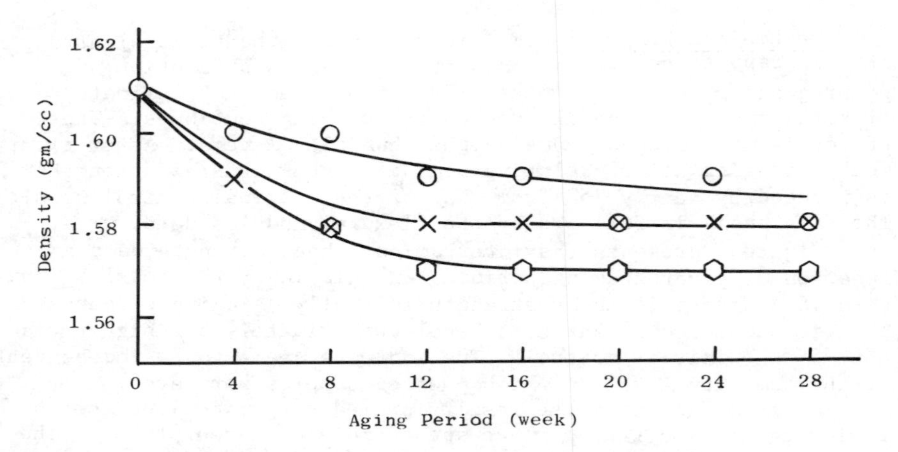

Figure 5. Laminate density as a function of aging period; F-178/T300, room-temperature aging; R.H.: (○) 16%, (×) 50%, (⬡) 95%

resolved by considering some physical changes to the prepregs which occurred during aging.

First of all, during the aging period the prepreg material clearly showed a steady loss of tackiness. As aging continued, the prepreg gradually became dry and brittle, making it difficult to keep all plies properly aligned and adhered during lay-up. Secondly, the bleeding during curing was greatly reduced for aged samples. Wetting of the bleeder plies became less and less, particularly around the corners of the square plies, as aging continued. These changes seem to suggest that the amount of resin flow was greatly reduced due to aging.

These observations suggest that exposure to high humidity caused little or no chemical degradation in the resin system. But room temperature storage produced a prepreg material that was excessively B-staged. The aged prepregs essentially had a higher degree of crosslinking than a fresh sample at the beginning of the fabrication process. This change in degree of crosslinking was not detected by proton NMR, but the material was essentially one with higher molecular weights having greater melt viscosities. This, therefore, lowered the flow when the sample was subjected to the processing cycle originally designed for the control sample. The result was a composite laminate having higher resin content, but poorer flexural properties. Laminate density measurements and fiber/matrix volume determination confirmed this. The results, shown in Figs. 5 and 6, indicated that the resin content in the laminates increased with increased age of the prepreg, resulting in decreased laminate density.

Proton NMR. The two major components of the F-178 resin system were identified through proton and carbon-13 nmr. In the proton spectrum, Figure 7, the lines at 4.2, 7.1 and 7.4 ppm arise from the methylene, olefinic and aromatic protons of the bismaleimide of methylene dianiline, I. All of the lines in the 4.4 to 6.3 ppm region are due to the allyl protons of triallyl isocyanurate, II. There are a number of low intensity lines which we have not attempted to identify.

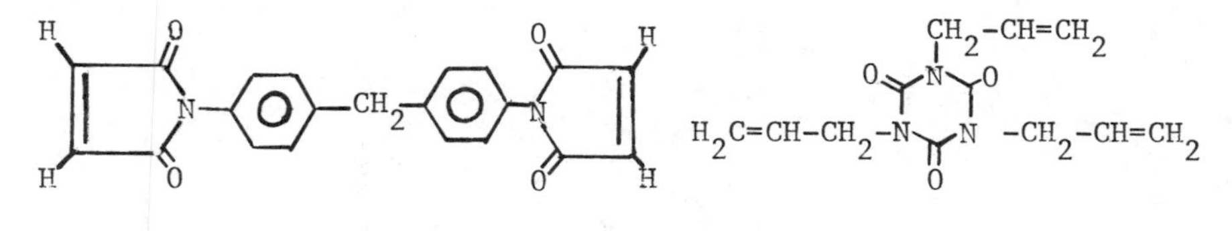

I

II

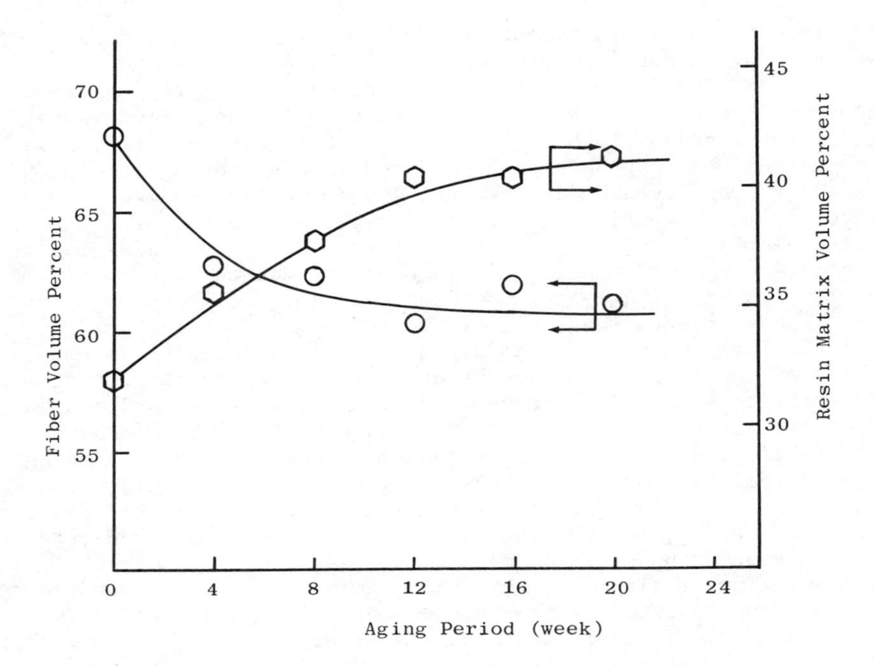

Figure 6. Volumetric fractions of fiber and matrix vs. aging period in weeks; F-178/T300, room-temperature aging at 95% R.H.

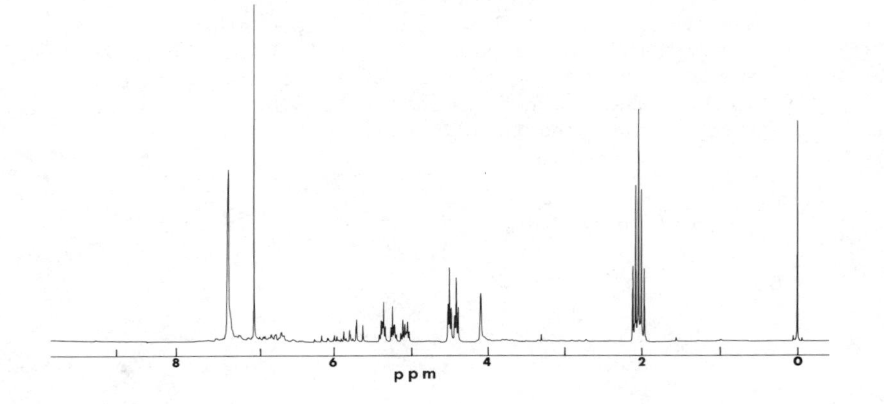

Figure 7. 60-MHz proton NMR spectrum of material extracted from unaged F-178/T300 prepreg. The line at 0 ppm is that of the reference, tetramethylsilane. The multiplet at 2 ppm is attributable to the residual protons in the solvent, acetone —d_6.

The aging in the F-178 resin system is a continuation of the curing reactions initiated during the manufacture of the prepreg. The system cures through free radical polymerization of the olefinic bonds. Some typical reactions are shown below. Reactions (1) and (2) are two of the chain propagating reactions for this system where R· and R'· are any free radicals.

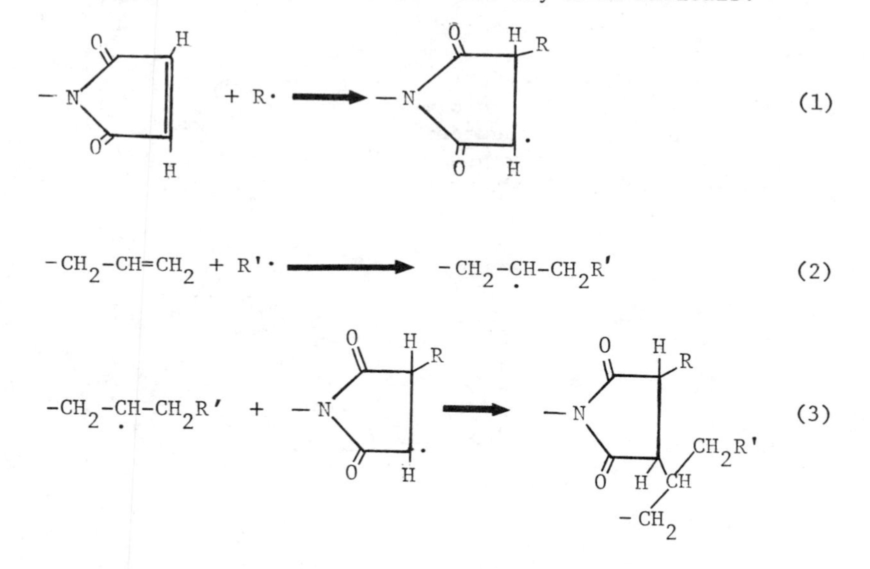

The new free radicals formed in (1) and (2) may continue to similarly attack other double bonds or they may combine in a chain terminating reaction as shown in (3).

Thus, as the cure proceeds the double bonds are consumed and the system becomes more "aliphatic". Either carbon-13 or proton nmr could be expected to monitor the changes occurring in the system. We decided to use proton nmr because it is faster, requires less sample and is easier to quantify. We looked for new proton lines growing in the 2 to 4 ppm region similar to those observed by Crivello from polyaspartimides which he had prepared from bis-maleimides (3). We did not detect any notable difference in this region of the spectra of the aged samples compared to that of the control sample.

We also measured the ratios of the intensity of the olefinic peak of I (7 ppm) to those of the aromatic peak of I (7.4 ppm) and the CH_2-N peak (4.6 ppm) of II. Changes in the 7 ppm/7.4 ppm ratio would indicate molecular change in the extracted maleimide. Changes in the 7.0 ppm/4.6 ppm ratio would indicate changes in the composition of the extracted resin. In each case, the value of the ratio showed no significant variations from the one measured on the control sample.

The failure of proton nmr to reveal any aging effects in this study is probably because the initial products of the reactions are not soluble in acetone or that they do not reach a detectable concentration before reacting further. The extraction procedure was designed to have the least purturbation on the system. The contact time had to be short, for we had previously found that acetone solutions of the F-178 resin degrade rapidly, even overnight. Extraction with hot acetone was not attempted because of possible side reactions. Acetone-d_6 was chosen over dimethyl sulfoxide-d_6 (DMSO) and dimethyl formamide-d_7 (DMF) because its residual proton absorption occurs at 2 ppm, whereas that of DMSO occurs at 2.5 ppm and DMF has two absorptions, 2.7 and 2.9 ppm, which would start obscuring the region of interest.

References

1. Reference to a specific product does not imply approval of or recommendation by the U. S. Department of Defense.

2. W. D. Bascom, J. L. Bitner and R. L. Cottington, Preprint, ACS Organic Coatings and Plastics, 38, 477 (1978).

3. J. V. Crivello, Polym. Prepr., American Chemical Society, Div. of Polym. Chem., 14(1), 294 (1973).

RECEIVED January 22, 1980.

Characterization of Aging Effects of LARC-160

PHILIP R. YOUNG and GEORGE F. SYKES

National Aeronautics and Space Administration, Langley Research Center, Hampton, VA 23665

The quality control of high-performance polymer matrix resins is an area vitally important to the aerospace community. Simple, reliable, and acceptable techniques must be available to assure that resins and prepregs are of consistent quality if the timely introduction of these advanced materials into aircraft and spacecraft structures is to be successful. The present study is an initial characterization of LARC-160 polyimide precursor resin, including resin aging effects, and was made in an effort to establish meaningful quality control procedures for this new and potentially useful aerospace material. The specific objectives of this study were to develop characterization techniques for the resin and to correlate resin changes due to aging with processing and composite properties.

Freshly prepared LARC-160 resin and resin aged at room temperature up to 45 days were monitored by high pressure liquid chromatography and several different thermal analyses. Graphite reinforced (Hercules HTS-1)(1) composites made from fresh and aged resin were then fabricated and tested to determine if changes observed by these techniques were significant.

Experimental

Materials. LARC-160 was prepared as solventless resin from benzophenone tetracarboxylic acid dianhydride (BTDA), nadic anhydride, Jeffamine AP-22, and ethanol by the method of St. Clair and Jewell (2). Using this method, an initial batch of resin was prepared, stored at room temperature, and characterized as a function of time. After 45 days, a second batch of resin (fresh material) was formulated. Hercules HTS-1 graphite prepreg was then prepared according to the procedure outlined in reference 3 by diluting the two batches of resin to 50 percent solids in tetrahydrofuran. Eight and 16-ply (1.4 mm and 3.1 mm thickness) unidirectional composites from each prepreg were then molded using the cure cycle given in reference 2. The completed 7.6 x 12.7-cm panels were ultrasonically examined and test specimens were cut from the areas that exhibited clear C-scans. The molded composites were not postcured.

Instrumentation. A Waters Associates Model ALC202/R401 Liquid Chromatograph equipped with a Model 6000A Solvent Delivery System, U6K Injector, and Model 440 Absorbance Detector was used to make the chromatographic separations. Analyses were performed at 254 nm UV on a Waters μBondapak CN column (3.9 mm ID x 30 cm) using glass distilled 1-propanol (Burdick and Jackson Laboratories, Muskegon, MI) as both the sample solvent and the mobile phase. A Finnigan Model 3300 Quadrapole Mass Spectrometer equipped with a Model 6000 Data System was used to obtain mass spectra of compounds prepped on the liquid chromatograph. Spectra were obtained by heating the sample at 3K/min in the ion source using a programmable temperature solid inlet probe.

A parallel plate plastometer attached to the DuPont Model 943 Thermomechanical Analyzer/Model 990 console and interfaced with a Wang 2200 Programmable Calculator was used to make viscosity determinations on thermally staged resins (4). A DuPont Model 990 Thermal Analyzer in combination with a standard differential scanning calorimetric (DSC) cell and a Model 943 Thermomechanical Analyzer was used for the DSC and glass transition measurements, respectively. A description of the torsional braid analyzer and experimental procedure is given in reference 5. Isothermal weight loss was determined on short beam shear specimens in a forced air oven at 589K.

Mechanical Testing. All mechanical tests were performed at room temperature on a Model TT6 Instron Testing Machine. Flexural strengths and moduli were determined on 1.3 x 6.4-cm 8-ply (1.4 mm thick) composites according to ASTM D790. Short-beam-shear strengths, using a 4:1 span-to-thickness ratio, were obtained on 0.6 x 1.6-cm specimens machined from the 16-ply (3.1 mm thick) composite.

Results and Discussion

An outline of this study is given in Figure 1. This scheme shows the type of LARC-160 resin sample analyzed, the treatment given, if any, and the particular characterization technique employed. High pressure liquid chromatography (HPLC) and mass spectrometry (MS) played the major role in the characterization of the neat resin. After an acceptable separation was developed and the major peaks identified, HPLC and torsional braid analysis (TBA) were used to characterize the initial batch of resin as it aged. Resin samples were also imidized to simulate B-staging and studied by differential scanning calorimetry (DSC), thermal gravimetric analysis (TGA), and parallel plate plastometry. Finally, after the initial batch of resin had aged for 45 days at room temperature, a fresh batch of resin was prepared. HTS-1 graphite composites fabricated from these two resins were then characterized by thermomechanical analysis (TMA), and mechanical property and isothermal weight loss measurements. The various areas outlined in Figure 1 will be discussed in greater detail.

Characterization of Fresh and Aged Resin. HPLC was selected to play a major roll in the characterization of fresh and aged resin because of its ability to rapidly separate these materials into individual components. However, the chromatographic fingerprint obtained depended largely on the efficiency of the column used and an appreciation for column performance in quality control work was soon established. For example, Figure 2 shows chromatograms for the same LARC-160 resin sample run on three different µBondapak CN columns. Each column had a different efficiency as determined by injecting a naphthalene standard under analytical conditions and calculating the number N of theoretical plates. Shown in the figure are chromatograms of the resin on a new column, N=4600, a used column, N=650, and an unacceptable column, N=100. The three chromatograms are different and might be interpreted as resulting from different resins.

Figure 3 shows how the column used in the present study behaved. This column was dedicated to running only LARC-160 samples and N was determined each day an analysis was made. The trend shown by the decrease in the number of theoretical plates with the number of samples analyzed is striking. Figures 2 and 3 demonstrate that column performance must be carefully monitored when using HPLC for quality control purposes. Otherwise, different investigators using columns of differing chromatographic histories may be unable to duplicate the findings of others or obtain the separation required for adequate quality control. Column plate count is specified for all separations made during this study. Experience does not yet permit a lower limit to be established on N.

Figure 4 gives a chromatogram of room-temperature aged (103 days) LARC-160 resin and the identification of most peaks. The identification was made by collecting fractions at the detector, evaporating the mobile phase, and transferring the residue to the mass spectrometer where spectra were obtained. Nadic ester-acid was not detected in the free state because the compound does not exhibit a UV chromaphore at 254 nm. The first two peaks, corresponding respectively to the unresolved BTDA esters (I) and the AP-22 amine mixture (II), were present when the resin was formulated. AP-22 is approximately 80 percent 4,4'-diaminodiphenyl methane. Three other compounds (III, IV, and V) formed during aging. In actual practice, these three compounds could also be formed during the prepregging and molding

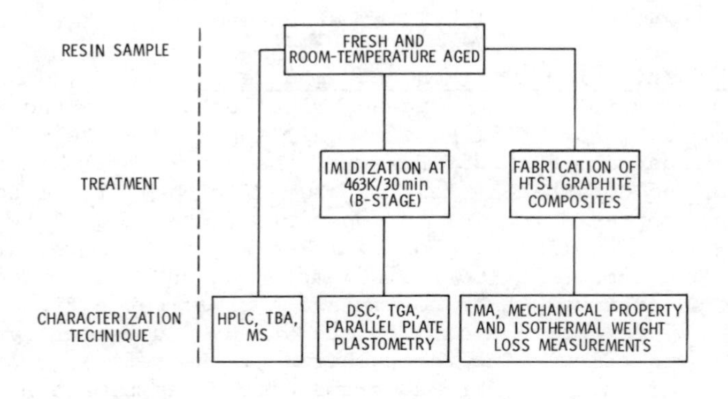

Figure 1. Scheme for LARC-160 resin characterization

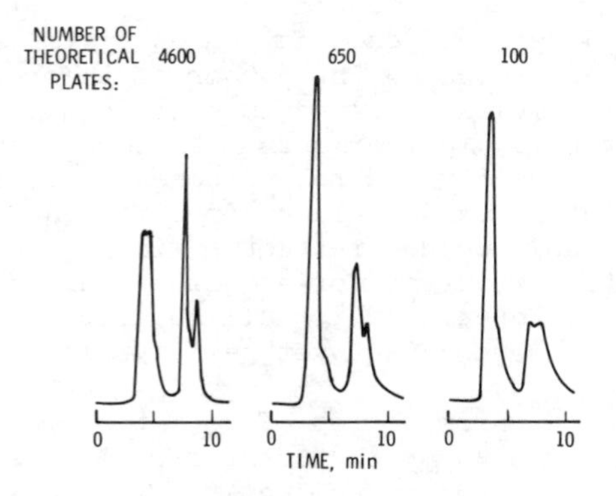

Figure 2. HPLC of the same LARC-160 resin sample run on three different columns. Conditions: column, μBondapak CN; solvent, 1-propanol; flow rate, 0.5 mL/min; size, 1-μL(1-μg/μL); room temperature; detector wavelength, 254 nm.

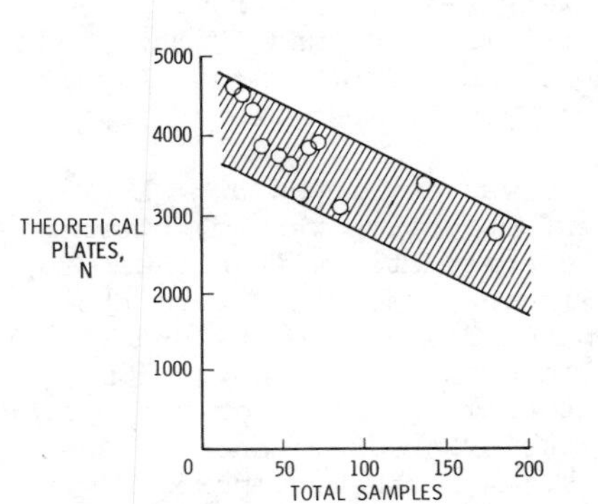

Figure 3. Decrease in the number of theoretical plates, N, *as a function of the number of LARC-160 resin samples analyzed. Conditions: standard, naphthalene; column, μBondapak CN; solvent, 1-propanol; flow rate, 0.5 mL/min; size, 1-μL (2μg/μL); room temperature; detector wavelength, 254 nm.*

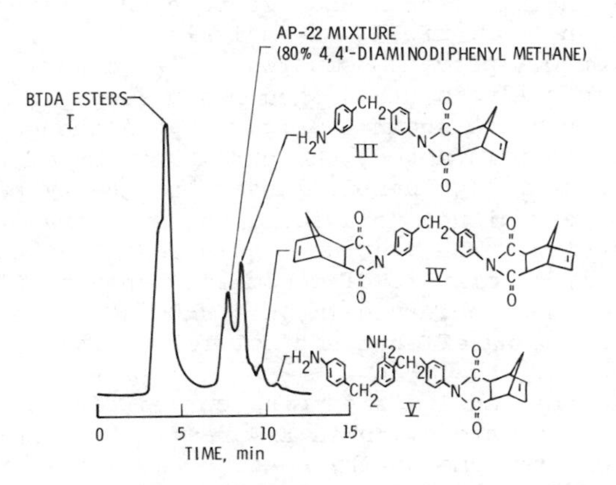

Figure 4. HPLC of LARC-160 resin aged at room temperature for 103 days. Conditions: column, μBondapak CN; solvent, 1-propanol; flow rate, 0.5 mL/min; sample size, 1-μL (1-μg/μL); room temperature; detector wavelength, 254 nm; N = *2750.*

operations. However, the bis-nadimide IV, which resulted from the reaction and subsequent imidization of two molecules of nadic ester with one molecule of 4,4'-diaminodiphenyl methane, is an undesirable product regardless of how or when it is formed since its formation upsets the designed LARC-160 stoichiometry. Fresh resin is formulated to yield an average imidized molecular weight of 1600 (3). These 1600-molecular-weight-segments crosslink during cure. The cure of resin containing compound IV may yield more than the expected amount of lower molecular weight, highly crosslinked segments. More linear polyimide must then form to maintain stoichiometry.

Once an acceptable HPLC analysis was developed and the major peaks were identified, the next objective was to determine if the observed resin aging could be correlated with either processing or composite properties. Figure 5 shows the chromatographic fingerprint of freshly prepared resin (time 0) and fingerprints of the resin after aging at intervals to 34 days. Only two peaks are present in the chromatogram of the freshly prepared resin. These two peaks are due to the BTDA esters and the AP-22 amine mixture identified in Figure 4. A further comparison with Figure 4 will show that the amine peak decreased with aging while peaks for compounds III, IV, and V increased.

To prove that the differences observed in Figure 5 are caused by resin aging effects and not to a change in column performance as shown in Figure 3, an analysis of the initial time 0 HPLC specimen, a 0.1 percent solution in 1-propanol, was repeated each day that an analysis of an aged resin sample was conducted. These chromatograms are shown in Figure 6. Although a deterioration in plate count is noted, the chromatograms of this dilute solution are essentially identical over the 34-day period. Apparently, in very dilute solutions, the starting materials have difficulty reacting. The viscous liquid resin, as a solventless (100 percent solids) mixture of monomers, behaves differently. Thus, the differences observed in this HPLC study were concluded to be true resin aging effects and not related to changes in column performance.

Torsional braid analysis damping curves for the series of aged LARC-160 resins are given in Figure 7. All TBA samples were prepared on the same day by coating the glass braids with freshly formulated resin diluted in tetrahydrofuran. The coated braids were then aged in a desiccator at room temperature until tested. Differences below 400K probably resulted from the evaporation of residual solvent. However, between 450 and 515K, where imidization occurs, the observed difference in damping probably indicates that aging alters the ability of the material to absorb mechanical energy. The decreasing magnitude with age could mean that the aged resin is already partially imidized. This would agree with the HPLC results. Differences between damping properties for fresh and aged resin may be small above

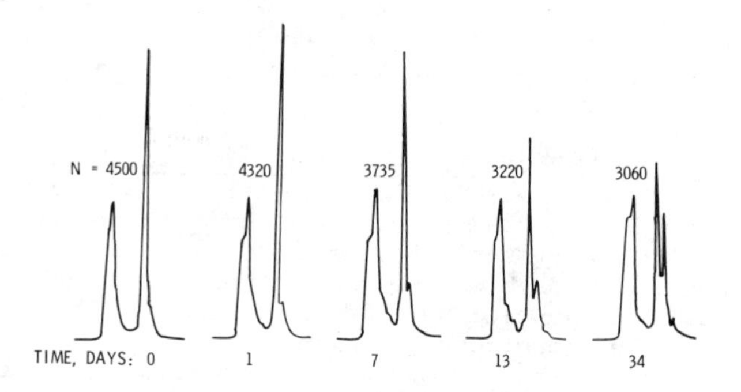

Figure 5. Effect of room temperature aging on HPLC of LARC-160 resin. Conditions: column, μBondapak CN; solvent, 1-propanol; flow rate, 0.5 mL/min; sample size, ~ 2.5μL (1-μg/μL); room temperature; detector wavelength, 254 nm; N *as noted.*

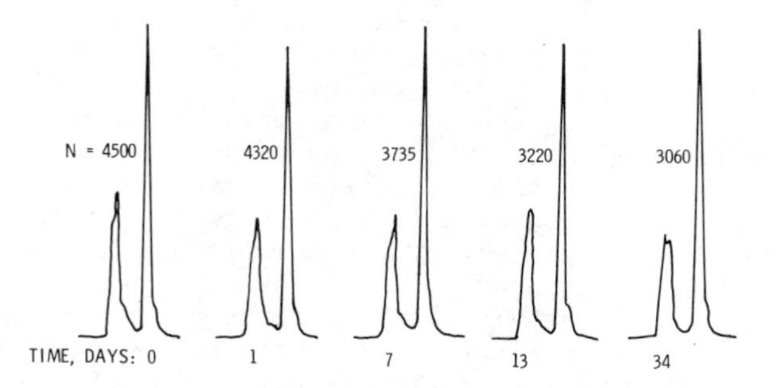

Figure 6. Repetitive analysis of dilute (0.1%) time 0 LARC-160 resin solution over a 34-day period. Conditions: same as for Figure 5.

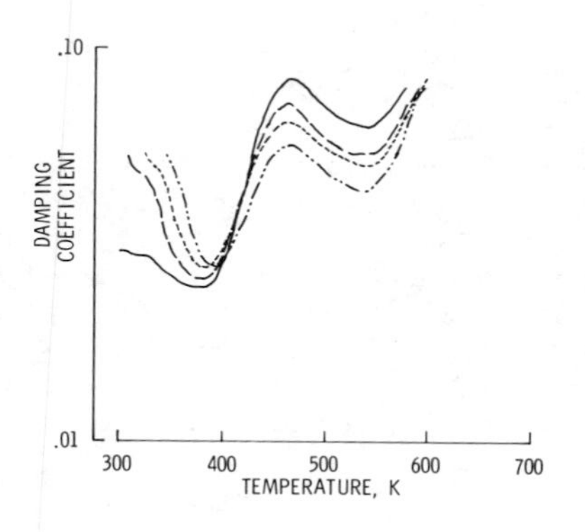

Figure 7. TBA damping curves for room-temperature-aged LARC-160 resin. Conditions: temperature rise rate, 3 K/min; atmosphere, nitrogen; (——) fresh, (– – –) 1-day aged, (– – – –) 6-day aged, (— – – – –) 21-day aged.

600K since the curves tend to coincide in that temperature range. These TBA results suggest that aging may influence the processing properties of the resin.

Characterization of Imidized Fresh and Aged Resin. A sample of fresh resin was imidized at 463K for 30 minutes to simulate B-staging, cooled to room temperature, ground up, and stored in a desiccator until tested. This process was repeated periodically for 36 days as the resin aged at room temperature. After all imidized samples had been prepared, they were characterized using the techniques given in Figure 1.

Figure 8 shows the DSC curves of both fresh resin and resin that had aged for 36 days before being imidized. DSC did not prove to be an effective technique for studying aging. Calorimetric calculations could not be made because a stable baseline was not established after the crosslinking exotherm due to partial sample decomposition. However, the difference in the onset of the curing exotherm for these two curves is probably significant. Even after imidization, the aged resin appears to be more advanced.

Programmed temperature TGA also was not very informative. Only slight differences in thermal stability were observed to 700K. However, the fresh resin appeared to be slightly more thermally stable than the aged resin.

In contrast to DSC and TGA data, melt viscosity measurements showed significant differences in the flow characteristics of thermally staged fresh and aged resin. Figure 9 indicates at least an order-of-magnitude difference between the viscosity/temperature curves over most of the temperature range tested. The fact that aged resin had a lower viscosity when treated thermally indicated that its ability to be processed could be dramatically affected. Indeed, greater flow was observed visually during the molding of aged-resin/graphite composites than was observed during molding of fresh-resin/graphite composites. This indicates that processing conditions for aged resin may have to be adjusted to allow for this greater flow. Changes in resin chemistry fingerprinted by HPLC apparently play a role in altering resin viscosity behavior.

Characterization of HTS-1 Graphite Composites. After the initial batch of resin had aged for 45 days, a fresh batch was formulated. These two resins were immediately used to prepare HTS-1 graphite prepreg. The prepregs were laid up at the same time to form 8- and 16-ply composites and B-staged in the same vacuum bag. The prepregs were then molded on the same day under identical conditions. This approach was adopted in an effort to minimize processing variables and as an alternative to performing all processing on the same batch of resin. A repeat of the composition of fresh resin was considered to be easier to achieve than would be a repeat of processing conditions over a 45-day period.

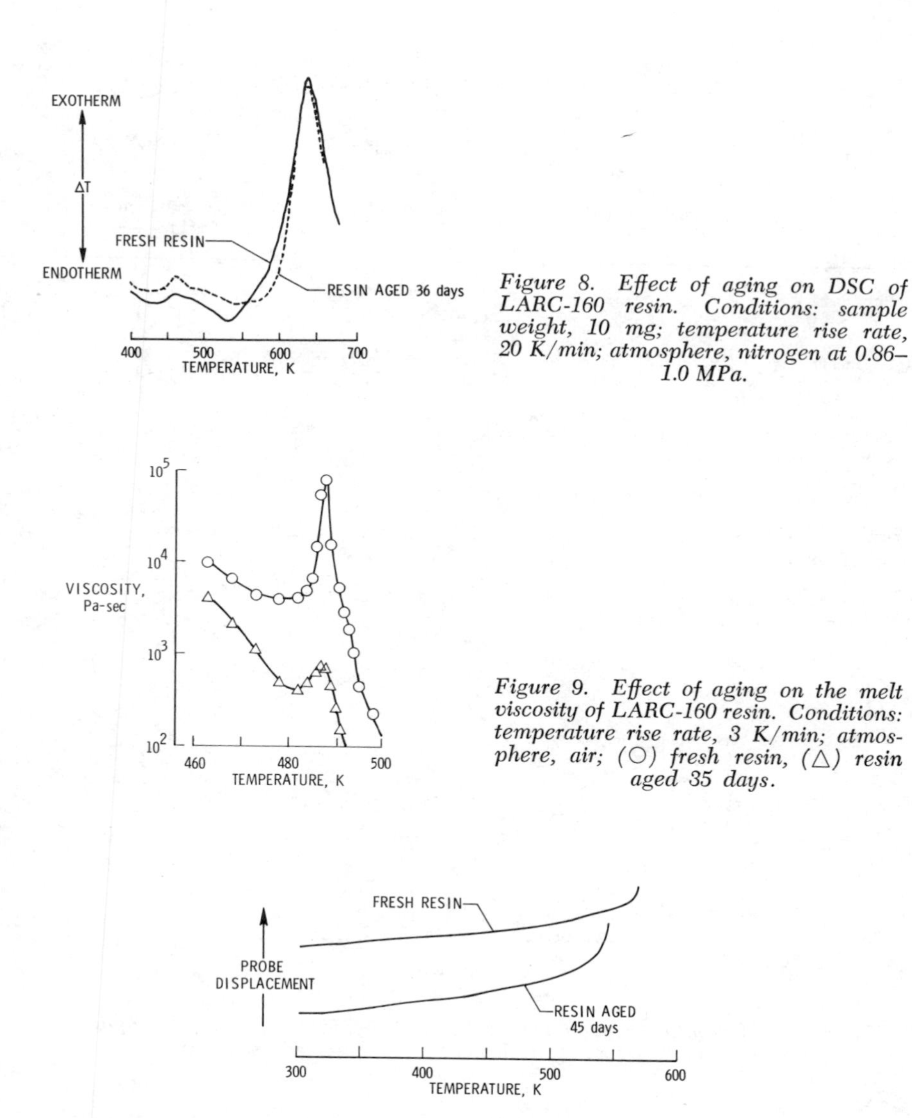

Figure 8. Effect of aging on DSC of LARC-160 resin. Conditions: sample weight, 10 mg; temperature rise rate, 20 K/min; atmosphere, nitrogen at 0.86–1.0 MPa.

Figure 9. Effect of aging on the melt viscosity of LARC-160 resin. Conditions: temperature rise rate, 3 K/min; atmosphere, air; (○) fresh resin, (△) resin aged 35 days.

Figure 10. Effect of resin aging on TMA of HTS-1/LARC-160 composites. Conditions: mode, thermal expansion; temperature rise rate, 5 K/min; atmosphere, air.

TABLE I. ROOM TEMPERATURE MECHANICAL PROPERTIES[a] OF HTS-1 GRAPHITE/LARC-160 COMPOSITES MADE FROM FRESH AND AGED[b] RESIN

Composite	Flexural Strength[c] MPa[ksi]	Flexural Modulus[c] MPa[ksi]	Short beam shear[d] MPa[ksi]
Fresh Resin	2000 (± 170) [288 (± 25)]	148,000 (± 6000) [21,400 (± 800)]	74 (± 14) [10.6 ± 2.0]
Aged Resin	2100 (± 130) [305 (± 19)]	151,000 (± 7000) [21,700 (± 1000)]	82 (± 8) [11.9 (± 1.1)]

[a] all values are the average of ten tests; numbers in parentheses are one standard deviation.

[b] resin aged 45 days at room temperature.

[c] flexural measurements made on 8-ply (0.14 x 1.3 x 6.4-cm) unidirectional specimens.

[d] short beam shear measurements made on 16-ply (0.31 x 0.6 x 1.6-cm) unidirectional specimens.

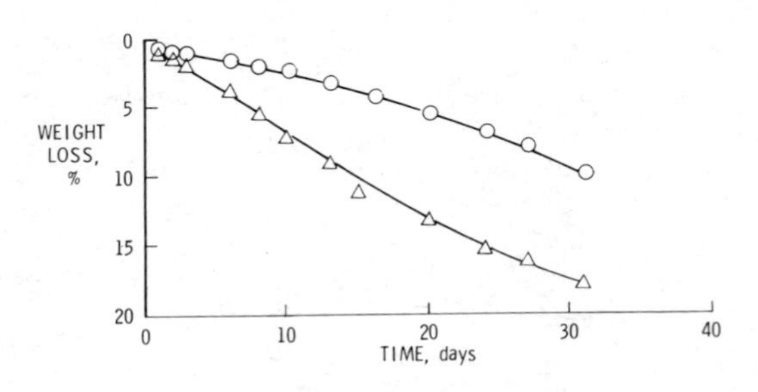

Figure 11. Isothermal weight loss at 589 K for HTS-1/LARC-160 composites made from (○) fresh and (△) 45-day-aged resin

Figure 10 shows the TMA of 16-ply short beam shear specimens fabricated from freshly prepared resin and from the initial batch of resin that had aged for 45 days at room temperature. The change in slope after 550K corresponds to the glass transition temperature, T_g. The composite made from fresh resin has a significantly higher T_g. This suggests that aging of LARC-160 matrix resin might affect the T_g of composites.

Table I gives the flexural and short-beam-shear strengths and the flexural modulus for graphite composites fabricated from fresh and aged resin. The room temperature mechanical properties of these sets of composites are probably identical within experimental error. Therefore, changes in resin chemistry with time, as fingerprinted by HPLC, do not seem to be reflected in the room temperature mechanical properties of composites made from these resins, at least not those mechanical properties measured in this study. It is possible that specific environmental exposure, such as moisture or elevated temperature, would have shown differences due to resin aging.

Short beam shear specimens fabricated from fresh and aged resin were isothermally aged at 589K and their weight loss was measured periodically. These data are presented in Figure 11 where each point represents the average weight loss of three specimens. The greater weight loss of the composite made with aged resin is significant. TGA data on the imidized resins may have suggested this behavior. The weight loss data indicates that if mechanical property measurements were repeated after periodic isothermal aging at 589K, greater differences than those seen in Table I would be observed for composites made from fresh and aged resin on HTS-1.

Concluding Remarks

High pressure liquid chromatography is a valuable tool for fingerprinting changes in LARC-160 resin due to aging. The major aging products, identified by mass spectrometry, resulted from the reaction of nadic ester-acid with Jeffamine AP-22. Torsional braid analysis and parallel plate plastometry were also useful techniques and suggested that the processing of composites might have to be altered to allow for resin changes due to aging. Differential scanning calorimetry and thermal gravimetric analysis were not effective techniques for studying LARC-160 resin aging.

The fabrication and testing of graphite composites made with fresh and aged resin showed that resin aging resulted in greater flow during processing and poorer composite isothermal stability. However, aging was not reflected in the room temperature flexural or short-beam-shear strengths of composites made with fresh and aged resin. Finally, column plate count was identified as a parameter which must be monitored if acceptable chromatographic procedures for the quality control of LARC-160 are to be established.

Literature Cited

1. Use of trade names or names of manufacturers in this report does not constitute an official endorsement of such products or manufacturers, either expressed or implied, by the National Aeronautics and Space Administration.
2. St. Clair, T. L., Jewell, R. A., 8th Nat. SAMPE Tech. Conf. (Seattle, October 1976) Preprints, vol. 8, p. 82.
3. St. Clair, T. L., Jewell, R. A., 23rd Nat. SAMPE Symp. (Anaheim, May 1978) Preprints, vol. 23, p. 520.
4. Price, H. L., Burks, H. D., Dalal, S. K., Soc. Plastics Ind., 33rd Ann. Tech. Conf. Proc. (Washington, D.C., February 1978) Paper 16-A.
5. Dalal, S. K., Carl, G. L., Inge, A. T., Johnston, N. J., Am. Chem. Soc., Div. Polymer Chem., Preprints, vol. 15, 576 (1974).

RECEIVED January 21, 1980.

INDEX

N

O

P

T

U

V